ANNALS OF MATHEMATICS STUDIES

Number 21

ANNALS OF MATHEMATICS STUDIES

Edited by Marston Morse and Emil Artin

7. Finite Dimensional Vector Spaces, *by* PAUL R. HALMOS
11. Introduction to Nonlinear Mechanics, *by* N. KRYLOFF *and* N. BOGOLIUBOFF
14. Lectures on Differential Equations, *by* SOLOMON LEFSCHETZ
15. Topological Methods in the Theory of Functions of a Complex Variable, *by* MARSTON MORSE
16. Transcendental Numbers, *by* CARL LUDWIG SIEGEL
17. Problème Général de la Stabilité du Mouvement, *by* M. A. LIAPOUNOFF
18. A Unified Theory of Special Functions, *by* C. A. TRUESDELL
19. Fourier Transforms, *by* S. BOCHNER *and* K. CHANDRASEKHARAN
20. Contributions to the Theory of Nonlinear Oscillations, *edited by* S. LEFSCHETZ
21. Functional Operators, Vol. I, *by* JOHN VON NEUMANN
22. Functional Operators, Vol. II, *by* JOHN VON NEUMANN
23. Existence Theorems in Partial Differential Equations, *by* DOROTHY BERNSTEIN
24. Contributions to the Theory of Games, *edited by* A. W. TUCKER
25. Contributions to Fourier Analysis, *by* S. BOCHNER, A. P. CALDERON, M. MORSE, W. TRANSUE, *and* A. ZYGMUND

FUNCTIONAL OPERATORS

BY JOHN VON NEUMANN

Volume I: Measures and Integrals

PRINCETON
PRINCETON UNIVERSITY PRESS
1950

LONDON: GEOFFREY CUMBERLEGE, OXFORD UNIVERSITY PRESS

PRINTED IN THE UNITED STATES OF AMERICA

FOREWORD

The lectures on "Operator Theory", of which the present volume constitutes the first part, were given in the academic years 1933-34 and 1934-35, at the Institute for Advanced Study. The notes were prepared in these years by Dr. Robert S. Martin and Dr. Charles C. Torrance, respectively. They were multigraphed and distributed by the Institute for Advanced Study shortly thereafter, but the original edition has been completely exhausted for several years. The interest in these lecture notes appears to have been continuing, and therefore a new edition is now being brought out. The present volume comprises Chapters I - XI, dealing with preliminaries, namely, with the theory of Measures and Integrals. The second volume, on Operator Theory proper, will be published subsequently. The present edition is identical with the original one, except that typographical errors have been corrected and some notations and references have been elaborated. I would like to express my warmest thanks to Dr. H. H. Goldstine, for his advice on this edition, and also for having most obligingly undertaken the exacting task of proof-reading the typescript.

JOHN VON NEUMANN

The Institute for Advanced Study
Princeton, New Jersey
November 1949.

FUNCTIONAL OPERATORS

CHAPTER I.

POINT SET THEORY

The points P of the space under consideration are ordered sets of n real numbers $(x_1,\ldots,x_n)$; x_i is called the ith coordinate of the point.

Definition 1.1: The distance between two points $(x_1,\ldots,x_n)$ and $(y_1,\ldots,y_n)$ is $\sqrt{\sum_{1}^{n}(x-y)^2}$.

Definition 1.2: An open interval I is determined by two sets of numbers X_ν and Y_ν ($\nu = 1,\ldots,n$; $X_\nu < Y_\nu$ for each ν) and consists of all points $(x_1,\ldots,x_n)$ satisfying the condition $X_\nu < x_\nu < Y_\nu$ ($\nu = 1,\ldots,n$).

Definition 1.3: A closed interval I is the set of all points satisfying the condition $X_\nu \leqq x_\nu \leqq Y_\nu$ ($\nu = 1,\ldots,n$; $X_\nu < Y_\nu$).

Definition 1.4: The edges of an interval are $s_\nu = Y_\nu - X_\nu$ and the volume of an interval is $\prod_{1}^{n} s_\nu$. If all s_ν are equal, the interval is a cube.

By Definition 1.2 and 1.3, the volume of every interval is positive.

Definition 1.5: If P is a point of space and r a positive number, the open sphere with radius r and center P is the set of points Q such that the distance $\overline{P,Q}$ is less than r; the closed sphere is the set of points for which the distance $\overline{P,Q}$ is less than or equal to r.

Definition 1.6: A sequence of points $P^{(i)}$: $(x_1^{(i)},\ldots,x_n^{(i)})$ is said to have the limit P: $(x_1,\ldots,x_n)$ if, for each ($\nu = 1,\ldots,n$), $\lim_{i\to\infty} x_\nu^{(i)} = x_\nu$.

By Definitions 1.2 and 1.6 it follows that almost all $P^{(i)}$ (that is, all but a finite number of $P^{(i)}$) are in any open interval containing P.

Let P be a point and let M and N be point sets. The notation $P \in M$ indicates that P is an element of M , $M \supset N$ indicates that M includes N (that is, each point of N is a point of M), and $N \subset M$ indicates that N is a part of M ($M \supset N$ and $N \subset M$ being equivalent).

Definition 1.7: P *is called a condensation point of a set* M *if every open interval containing* P *contains infinitely many points of* M. (P *itself need not belong to* M.)

It is easily shown that a sequence of points can have at most one limit point while a set may have more than one condensation point.

Definition 1.8: A point set M *is called closed if each condensation point of* M *is an element of* M.

Definition 1.9: A point set M is called open if, for each $P \in M$, *there exists an open interval* I *containing* P *such that* $I \subset M$.

Since closed and open intervals have been defined separately from closed and open point sets, it is necessary to show the compatibility of the two pairs of definitions.

THEOREM 1.1: A closed interval is a closed point set and an open interval is an open point set.

Proof: The second part of the theorem is obvious after consideration of Definitions 1.2 and 1.9, so that only the first part requires formal proof. Let $P: (\xi_1, \dots, \xi_n)$ be any condensation point of the closed interval I : $X_\nu \leqq x_\nu \leqq Y_\nu$ ($\nu = 1, \dots, n$). Then, by Definition 1.7, the open interval $\xi_\nu - \epsilon < \eta_\nu < \xi_\nu + \epsilon$ contains infinitely many points $(\bar{\eta}_1, \dots, \bar{\eta}_n)$ of I, where ϵ is any positive number. Hence $X_\nu - \epsilon \leqq \bar{\eta}_\nu - \epsilon < \xi_\nu < \eta_\nu + \epsilon \leqq Y_\nu + \epsilon$. Since ϵ is arbitrary, $X_\nu \leqq \xi_\nu \leqq Y_\nu$, so that $P \in I$.

Definition 1.10: The complement $-M$ *of a point set* M *is the set of all the points of the space which are not elements of* M.

THEOREM 1.2: *If M is an open set, then -M is closed, and conversely, if M is closed, then -M is open.*

Proof: Suppose M is open and let P be a condensation point of -M. It must be shown that $P \in -M$. Suppose $P \in M$. By Definition 1.9 there would exist an open interval containing P and containing no point of -M. But by Definition 1.7 every open interval containing P contains infinitely many points of -M. Hence $P \in -M$. Conversely, suppose that M is closed. If P is any point of -M, it must be shown that there exists an open interval containing P and contained in -M. Suppose that every open interval containing P contained infinitely many points of M. Then P would be a condensation point of M and, by Definition 1.8, $P \in M$. Therefore there exists an open interval containing P and only a finite number of points of M. Hence there exists an open interval (sufficiently small) containing P but no point of M, and -M is open.

A set of point sets M is said to *cover* a point set N if each point of N is contained in at least one of the sets M.

THEOREM 1.3: *If a finite set of intervals* $J^{(i)}$: $X_\nu^{(i)} < x_\nu < Y_\nu^{(i)}$ $(i = 1,\ldots,k;\ \nu = 1,\ldots,n)$ *covers an interval* I: $X_\nu < x_\nu < Y_\nu$, *then the sum of the volumes of the intervals* $J^{(i)}$ *is not less than the volume of* I. (In this theorem I may equally well be open or closed, as may also the intervals $J^{(i)}$. The proof is essentially the same in all cases.)

Proof: Let N_I be the number of integral points in I and $N_{J^{(i)}}$ the number of integral points in $J^{(i)}$. (A point P is integral if all its coordinates are integers.) Since I is covered by the set of intervals $J^{(i)}$,

$$N_I \leqq \sum_{i=1}^{k} N_{J^{(i)}} . \tag{1}$$

Let $\overline{X}_\nu$ be the smallest integer $> X_\nu$ and $\overline{Y}_\nu$ the largest integer $< Y_\nu$. The number of integers in the interval $\overline{X}_\nu \leqq x_\nu \leqq \overline{Y}_\nu$ is $(\overline{Y}_\nu - \overline{X}_\nu + 1)$. Hence the number of integral points in I is

(2) $$N_I = \prod_{\nu=1}^{n} (\overline{Y}_\nu - \overline{X}_\nu + 1).$$

But $X_\nu < \overline{X}_\nu \leqq X_\nu + 1$, $Y_\nu - 1 \leqq \overline{Y}_\nu < Y_\nu$, $Y_\nu - X_\nu - 1 \leqq \overline{Y}_\nu - \overline{X}_\nu + 1 < Y_\nu - X_\nu + 1$. Hence, by (2)

(3) $$\prod_{\nu=1}^{n} (Y_\nu - X_\nu - 1) \leqq N_I < \prod_{\nu=1}^{n} (Y_\nu - X_\nu + 1).$$

Similarly,

(4) $$\prod_{\nu=1}^{n} (Y_\nu^{(i)} - X_\nu^{(i)} - 1) \leqq N_{J^{(i)}} < \prod_{\nu=1}^{n} (Y_\nu^{(i)} - X_\nu^{(i)} + 1).$$

By (1), (3), and (4),

(5) $$\prod_{\nu=1}^{n} (Y_\nu - X_\nu - 1) < \sum_{i=1}^{k} [\prod_{\nu=1}^{n} (Y_\nu^{(i)} - X_\nu^{(i)} + 1)].$$

Let the space be subjected to the coordinate transformation $x_\nu = \frac{1}{\epsilon} x'_\nu$. Since this is merely a change of scale none of the above relations are altered and (5) becomes

(6) $$\prod_{\nu=1}^{n} (\frac{1}{\epsilon} Y_\nu - \frac{1}{\epsilon} X_\nu - 1) < \sum_{i=1}^{k} [\prod_{\nu=1}^{n} (\frac{1}{\epsilon} Y_\nu^{(i)} - \frac{1}{\epsilon} X_\nu^{(i)} + 1)].$$

The theorem follows when each factor of (6) is multiplied by ϵ and ϵ then allowed to approach zero.

THEOREM 1.4: <u>If a closed interval</u> I : $X_\nu \leqq x_\nu \leqq Y_\nu$ <u>is covered by an infinite set</u> S <u>of open intervals</u> J, <u>then there exists a finite subset of</u> S <u>which covers</u> I. (This is known as the Heine-Borel covering theorem.)

Proof: Suppose the theorem false. Divide I into 2^n closed subintervals $I^{(1)}, I^{(2)}, \ldots, I^{(2^n)}$ by breaking up each interval $X_\nu \leqq x_\nu \leqq Y_\nu$ into the two intervals $X_\nu \leqq x_\nu \leqq \frac{X_\nu + Y_\nu}{2}$ and $\frac{X_\nu + Y_\nu}{2} \leqq x_\nu \leqq Y_\nu$, $\nu = 1,\ldots,n$. At least one of those subintervals, say $I^{(\rho)}$, is not covered by any finite subset of S. Divide $I^{(\rho)}$ into 2^n subintervals $I^{(\rho,j)}$ in a manner analogous to that described for I. At least one $I^{(\rho,j)}$, say $I^{(\rho,\sigma)}$ is not covered by any finite subset of S. By repetition of this process there is obtained an infinite sequence of intervals

$$(1) \qquad I,\ I^{(\rho)},\ I^{(\rho,\sigma)},\ I^{(\rho,\sigma,\tau)},\ \ldots ,$$

no one of which is covered by any finite subset of S and each of which contains all of its successors. It is a well-known theorem that the intervals (1) have a point P* in common. But P* is in some interval J* of S. Since J* is open, almost all of the intervals (1) are contained in J*. This is contrary to the first assertion made of the intervals (1), and the theorem follows.

THEOREM 1.5: *If a closed interval I is covered by a set S (finite or infinite) of open intervals J, then the sum of the volumes of the intervals J is not less than the volume of I.*

Proof: By Theorem 1.4 a finite subset of S covers I. By Theorem 1.3 the sum of the volumes of the intervals of this finite subset is not less than the volume of I. The theorem follows from the fact that the volume of any interval is positive.

Definition 1.11: *A point set M is called bounded if there exists a finite interval containing it.*

THEOREM 1.6: *If a closed and bounded point set M is covered by an infinite set S of open point sets N, then there exists a finite subset of S which covers M.*

Proof: By Definition 1.9 each open set N can be covered by a set $\overline{S}$ of open intervals each of which is contained in N.

Suppose first that M is a closed interval. The set of all the open intervals in all the sets $\overline{S}$ covers M. By Theorem 1.4 a finite number of those open intervals covers M. But a finite number of the sets N cover this finite set of open intervals and the theorem is proved in this case.

If M is any closed, bounded set there exists a closed interval $\overline{I}$ containing it. Adjoin the open set -M to the set S and denote the resulting set of open sets by T. T covers all space and hence it covers $\overline{I}$. By Theorem 1.4 a finite subset T' of T covers I. T' therefore covers M. If -M is in T' it may be omitted from T' without impairing the covering of M by T'. This completes the proof.

THEOREM 1.7: *If S is any sequence of real numbers x_i, $a < x_i < b$, then there exist real numbers in this interval which do not appear in S.*

Proof: Suppose first that a = 0 and b = 1. Let the numbers x_i be written in decimal form, i.e., $x_i = 0.p_1^i p_2^i p_3^i \ldots$. If x_i be rational let it be agreed to write it as $0.p_1^i \ldots p_k^i 000\ldots$ and not as $0.p_1^i \ldots p_{k-1}^i 999\ldots$. Any number $= 0.q_1q_2q_3\ldots$, where $q_i \neq p_i^i$, $q_i \neq 9$, is in the interval $0 < x < 1$ and is distinct from each x_i. The fact that there is more than one such number is obvious. The theorem follows by considering the transformation $x' = \frac{x-a}{b-a}$ which carries the interval $a < x < b$ over into the interval $0 < x' < 1$.

Corollary: *No sequence of points exhausts the points of space.*

Proof: It is necessary merely to apply Theorem 1.7 to the first coordinate of the points of space.

THEOREM 1.8: *The set of rational points is everywhere dense in space, that is, each point of space is a condensation point of the rational points, or again, every interval of space contains infinitely many rational points.*

Proof: The theorem follows immediately from the well-known theorem that between any two real numbers there exist infinitely many rational numbers.

THEOREM 1.9: A sequence of points may be everywhere dense in space.

Proof: The proof will be conducted by showing that all rational points in space may be ordered into a sequence. Theorem 1.8 then applies. If $P:(\frac{N_1}{M_1}, \frac{N_2}{M_2}, \dots, \frac{N_n}{M_n})$ is any rational point, let its coordinates be supposed written in lowest terms with positive denominators. Let the degree of P be $d_p = |N_1| + |M_1| + |N_2| + |M_2| + \dots + |N_n| + |M_n|$. The number of rational points of given degree d is finite since each numerator and each denominator is numerically not greater than d. The desired sequence is constructed by ordering all rational points of degree one, all of degree two, and so on, and writing these ordered sets in a sequence.

THEOREM 1.10: If any point set M is covered by a set S of open sets $\emptyset$ (the letter O is written $\emptyset$ to distinguish it from zero), then there exists a sequence Σ of open cubes I which covers M, where each I is contained in at least one of the sets $\emptyset$.

Proof: Let P be any rational point of space. If P is an element of any set $\emptyset$, then, by Definition 1.9, it is possible to associate with P an open cube I_P with center P and contained in $\emptyset$. If P is contained in no set $\emptyset$, then no cube is associated with P. If Q is an irrational point of a set $\emptyset$, then there exists an open interval J containing Q and contained in $\emptyset$. Hence not all the cubes I_P need be so small but that at least one of them contains Q. The set of cubes I_P covers the sets $\emptyset$ and therefore M. Since the points P may be ordered in a sequence, the cubes I_P may also.

Closed point sets will be called C-sets, open sets will be called $\emptyset$-sets, and the elements of a set S of point sets will be called S-sets.

Definition 1.12: The sum M + N + ... of two or more point sets is the set of all those points contained in at least one of the sets M, N,...; the product M·N· ... is the set of all those points common to all the sets M, N,...

THEOREM 1.11: The product of any set of closed sets is closed and the sum of any set of open sets is open. Likewise the sum of any finite set of closed sets is closed and the product of any finite set of open sets is open.

Proof: The theorem follows immediately from Definitions 1.8, 1.9, and 1.12.

Let S be a set of point sets M and let $\sum$ be any sequence (finite or infinite) of sets M selected from S. Let σ be the sum and π the product of the sets in $\sum$. Suppose that $\sum$ is formed in every possible way from S. The set of all sums $\sum$ will be denoted by S_σ and the set of all products $\prod$ will be denoted by S_π.

If C_σ and C_π are formed from the set C of all closed sets, the elements of C_π are closed (by Theorem 1.11), but those of C_σ may not be closed. Similarly if $\emptyset_\sigma$ and $\emptyset_\pi$ are formed from the set $\emptyset$ of all open sets, the elements of $\emptyset_\sigma$ are open, but those of $\emptyset_\pi$ may not be open. From C_σ it is possible to obtain $C_{\sigma\sigma}$ and $C_{\sigma\pi}$. But every element of $C_{\sigma\sigma}$ is an element of C_σ since an element of $C_{\sigma\sigma}$ is the sum of a sequence $\sum$ of sums of sequences of C sets , and the C sets in $\sum$ may be reordered into an ordinary sequence. Similarly $\emptyset_{\sigma\pi}$ and $\emptyset_{\pi\pi}$ arise from $\emptyset_\pi$, but, by the same reasoning every element of $\emptyset_{\pi\pi}$ is an element of $\emptyset_\pi$. From $C_{\sigma\pi}$ arise $C_{\sigma\pi\sigma}$ and $C_{\sigma\pi\pi}$, but every element of $C_{\sigma\pi\pi}$ is an element of $C_{\sigma\pi}$. This process may be continued indefinitely.

Definition 1.13: All point sets arising from sets of closed or open sets by repeatedly forming sums and products of sequences of sets are called Borel sets.

Definition 1.14: The distance $\overline{P,M}$ between a point P and a point set M is the greatest lower bound of all distances $\overline{P,Q}$, where Q runs through the set M. The distance $\overline{M,N}$ between two point sets M and N is the greatest lower bound of all distances $\overline{P,Q}$, where P runs through M and Q runs through N.

The distance between any two point sets is obviously non-negative.

THEOREM 1.12: *If N is a closed point set and P a point, there exists at least one point Q ∈ N such that* $\overline{P,Q} = \overline{P,N}$.

Proof: Consider first the case where N is bounded. Suppose there were no such point Q. Let R be any point of N and let I_R be the set of all points R' of space for which $\overline{P,R'} > \overline{P,R}$. Each I_R is open. The set of all I_R, where R ∈ N, covers N since there is no point Q nearest P. By Theorem 1.6 a finite set of I_R, say $I_{R_1}, I_{R_2}, \ldots, I_{R_k}$, covers N. Hence *every* point S ∈ N is such that $\overline{P,S}$ is greater than either $\overline{P,R_1}$ or $\overline{P,R_2}$ or ... or $\overline{P,R_k}$. This is manifestly absurd since R_1, R_2, ..., R_k are themselves points S of N. Therefore the point Q of the theorem exists.

If N is not bounded, let R be some point of N and let N' be the set of all points R' ∈ N for which $\overline{P,R'} \leq \overline{P,R}$. N' is closed and bounded so that there exists a point Q ∈ N' for which $\overline{P,Q} = \overline{P,N'}$. Since any point R" in N but not in N' is such that $\overline{P,R''} > \overline{P,R} > \overline{P,Q}$, it follows that $\overline{P,Q} = \overline{P,N}$.

THEOREM 1.13: *If M and N are any two closed point sets, there exists a point P ∈ M and a point Q ∈ N such that* $\overline{P,Q} = \overline{M,N}$.

Proof: If $\overline{M,Q}$ is used instead of $\overline{P,Q}$, the method of Theorem 1.12 shows the existence of a Q ∈ N such that $\overline{M,Q} = \overline{M,N}$. Then, by Theorem 1.12, there is a P ∈ M such that $\overline{P,Q} = \overline{M,Q}$. Hence $\overline{P,Q} = \overline{M,N}$.

CHAPTER II.

OUTER MEASURE

Definition 2.1: If M is any point set, if S is a sequence (finite or infinite) of open intervals $I_1, I_2, \ldots$ whose sum covers M, if $\sum$ is the sum of the volumes of all the intervals of S, and if S is formed in every possible way to cover M, then the greatest lower bound of the $\sum$'s so arising is called the outer measure, $\mu^*(M)$, of M.

THEOREM 2.1: For any set M, $0 \leqq \mu^*(M) \leqq \infty$.

Proof: Outer measure is non-negative since the volume of any interval is non-negative.

The outer measure of a single point is zero. The outer measure of the entire space is infinite.

THEOREM 2.2: If $M \subset N$, $\mu^*(M) \leqq \mu^*(N)$.

Proof: Any sequence S of open intervals which covers N also covers M, so that $\mu^*(M)$ is not greater than $\mu^*(N)$. But there may exist a sequence S which covers M but not N.

THEOREM 2.3: If I is any interval, $\mu^*(I)$ is the volume of I.

Proof: If I is closed, then, by Theorem 1.5, $\mu^*(I)$ is not less than the volume of I. But I can be covered by a single open interval J whose volume is greater than that of I by an arbitrarily small amount. Hence $\mu^*(I)$ is the volume of I. If I is the open interval $X_\nu < x_\nu < Y_\nu$ and if S is a sequence of open intervals covering I, then S can be made to cover the closed interval $\overline{I} : X_\nu \leqq x_\nu \leqq Y_\nu$ by adjoining to it a finite set T of open intervals whose total volume is arbitrarily small. Hence $\mu^*(I) = \mu^*(\overline{I})$. Since the volumes of I and $\bar{I}$ are equal, the theorem holds whether I is open or closed.

THEOREM 2.4: *If M is bounded, $\mu^*(M) < \infty$, and if M contains an interval $\mu^*(M) > 0$.*

Proof: The theorem follows immediately from Definition 1.11, Theorems 2.2 and 2.3, and the fact that the volume of any interval is positive.

THEOREM 2.5: *The intervals used in Definition 2.1 may be restricted, without loss of generality, to arbitrarily small rational open cubes.*

Proof: An open interval I may be covered by a finite set S of open intervals J with edges δ (δ positive and rational) so that the total volume of the intervals J exceeds the volume of I by an arbitrarily small amount ϵ. (The intervals J must obviously overlap, and all that is necessary is to make the "thickness" of the overlapping parts sufficiently small.) If $I_1, I_2, \ldots$ is a sequence of open intervals covering a given point set M and if the volume of I_i is V_i, then I_i may be replaced by a finite set S_i of open intervals having a total volume $V_i + \frac{\epsilon}{2^i}$. The total volume of all the intervals in all the sets S_i is $\sum_{i=1}^{\infty} V_i + \epsilon$. Since ϵ is arbitrarily small, the greatest lower bound in Definition 2.1 is not altered by this restriction on the intervals used.

THEOREM 2.6: *If $M_1, M_2, \ldots$ is a finite or infinite sequence of point sets, then* $\mu^*(M_1 + M_2 + \ldots) \leq \mu^*(M_1) + \mu^*(M_2) + \ldots$.

Proof: M_i can be covered by a sequence S_i of open intervals of total volume $\mu^*(M_i) + \frac{\epsilon}{2^i}$. The intervals in the sequences S_i can be ordered into a single sequence S which covers the sum of the sets M_i. The total volume of all the intervals in S is $\sum_{i=1}^{\infty} \mu^*(M_i) + \epsilon$. Since ϵ can be made arbitrarily small, $\mu^*(M_1 + M_2 + \ldots)$ is not greater than $\sum_{i=1}^{\infty} \mu^*(M_i)$. But the sequences S may not comprise all the sequences which cover $M_1 + M_2 + \ldots$, so that

$\mu^*(M_1 + M_2 + \dots)$ may be less than $\sum_{i=1}^{\infty} \mu^*(M_i)$.

THEOREM 2.7: *If* M *and* N *have a positive distance* D *(and therefore no common point),* $\mu^*(M + N) = \mu^*(M) + \mu^*(N)$.

Proof: By Theorem 2.6 it is sufficient to show that $\mu^*(M + N) \geqq \mu^*(M) + \mu^*(N)$. Let M + N be covered by a sequence S of open intervals I_i so small that the distance between any two points in I_i is less than D. By Theorem 2.5 it is possible to make the total volume of the intervals in S less than $\mu^*(M + N) + \epsilon$. It is also possible to omit from S those intervals which have no point in M + N. S consists of a covering T of M and a covering U of N, where T contains no point of N and U contains no point of M. $\mu^*(M)$ is not greater than the total volume of the intervals in T and $\mu^*(N)$ is not greater than the total volume of the intervals in U. Hence $\mu^*(M) + \mu^*(N) \leqq \mu^*(M + N) + \epsilon$. The theorem follows from the fact that ϵ can be made arbitrarily small.

Theorems 2,1, 2.2, 2.6, and 2.7 state the following properties of outer measure:

I. For any set M, $0 \leqq \mu^*(M) \leqq \infty$.

II. If $M \subset N$, $\mu^*(M) \leqq \mu^*(N)$.

III. For any sequence $M_1, M_2, \dots, \mu^*(M_1 + M_2 + \dots) \leqq \mu^*(M_1) + \mu^*(M_2) + \dots$.

IV. If M and N have a positive distance, $\mu^*(M + N) = \mu^*(M) + \mu^*(N)$.

CHAPTER III

MEASURE

If M and N are any two point sets, then $N = MN + (N - MN)$. By Theorem 2.6, $\mu^*(N) \leqq \mu^*(MN) + \mu^*(N - MN)$. Since MN and N - MN have no point in common, it would be natural to expect the equality to hold, rather than the inequality, when M and N have more or less of the nature of intervals. This suggests

Definition 3.1: *If M is a point set and if the relation*

$$(1) \qquad \mu^*(N) = \mu^*(MN) + \mu^*(N - MN)$$

holds for every point set N, then M is said to be measurable and its measure, $\mu(M)$ is taken to be $\mu^*(M)$.

It may be well to point out that, in proving relation (1) for a given set to show its measurability, it is always sufficient, by Theorem 2.6, to prove the inequality $\mu^*(N) \geqq \mu^*(MN) + \mu^*(N - MN)$. Furthermore, since this inequality necessarily holds whenever $\mu^*(N)$ is infinite, it is sufficient to consider only those sets N having a finite outer measure.

THEOREM 3.1: *If M and N are two point sets for which $MN = 0$ and if one of them, say M, is measurable, then $\mu^*(M + N) = \mu^*(M) + \mu^*(N)$.*

Proof: This follows immediately when the set N of Definition 3.1 is taken to be the set M + N of the theorem.

Let $\emptyset$ be an open set and let P be any point of space. By Theorem 1.2, $-\emptyset$ is closed. By Theorem 1.12, if P is such that $\overline{P,-\emptyset} = 0$, then $P \in -\emptyset$. Obviously, if $\overline{P,-\emptyset} > 0$, then $P \in \emptyset$. Let M_K be the set of points P for which $\overline{P,-\emptyset} \geqq \frac{1}{K}$. Then $M_1 \subset M_2 \subset \dots \subset \emptyset$ and any point of $\emptyset$ is in some M_K. The sets

M_K and $-M_{K+1}$ have a positive distance, for, if P is any point of M_K and Q any point of $-M_{K+1}$, then $\overline{Q,-\emptyset} < \frac{1}{K+1}$. Hence there exists a point $R \in -\emptyset$ such that $\overline{Q,R} < \frac{1}{K+1}$. But $\overline{P,R} \geqq \frac{1}{K}$ and $\overline{P,Q} + \overline{Q,R} \geqq \overline{P,R}$. Therefore $\overline{P,Q} > \frac{1}{K(K+1)}$ and $\overline{M_K,-M_{K+1}} \geqq \frac{1}{K(K+1)} > 0$.

Let N be any point set with finite outer measure. From the preceding paragraph it follows that $M_1N \subset M_2N \subset \dots \subset \emptyset N$ and any point of $\emptyset N$ is in some M_KN. By Theorem 2.2, $\mu^*(M_1N) \leqq \mu^*(M_2N) \leqq \dots \leqq \mu^*(\emptyset N)$. The sequence $\mu^*(M_KN)$ is increasing and bounded, and so has a limit $\mu \leqq \mu^*(\emptyset N)$.

<u>Lemma</u>: $\mu = \mu^*(\emptyset N)$, <u>where the notation is as in the preceding paragraph</u>.

Proof: It is sufficient to show that $\mu \geqq \mu^*(\emptyset N)$. Form the set

$$S_K = (M_KN - M_{K-1}N) + (M_{K-2}N - M_{K-3}N) + \dots + \begin{cases} (M_2N - M_1N), & K \text{ even.} \\ M_1N, & K \text{ odd.} \end{cases}$$

Then $S_{K+2} - S_K = M_{K+2}N - M_{K+1}N$ is contained in $(-M_{K+1})$. S_K is contained in M_K. Since $\overline{M_K,-M_{K+1}} > 0$, the sets $(S_{K+2} - S_K)$ and S_K have a positive distance. But $S_{K+2} = (S_{K+2} - S_K) + S_K$. By Theorem 2.7, $\mu^*(S_{K+2}) = \mu^*(S_{K+2} - S_K) + \mu^*(S_K)$, or

$$(1) \qquad \mu^*(S_{K+2} - S_K) = \mu^*(S_{K+2}) - \mu^*(S_K).$$

By (1),

$$(2) \qquad \mu^*(S_{K+2} - S_K) + \mu^*(S_K - S_{K-2}) + \dots + \begin{Bmatrix} \mu^*(S_2), & K \text{ even} \\ \mu^*(S_3 - S_1), & K \text{ odd} \end{Bmatrix} =$$

$$= \mu^*(S_{K+2}) + \begin{Bmatrix} 0 \\ -\mu^*(S_1) \end{Bmatrix} \leqq \mu^*(S_{K+2}) \leqq \mu^*(N) < \infty.$$

Hence the two series with positive terms

$$\mu^*(S_3 - S_1) + \mu^*(S_5 - S_3) + \cdots ,$$

$$\mu^*(S_2) + \mu^*(S_4 - S_2) + \cdots ,$$

together with their sum

(3) $$\mu^*(M_2N - M_1N) + \mu^*(M_3N - M_2N) + \cdots ,$$

are convergent, since, by (2), the partial sums of each of the series are bounded. By taking $\underline{K}$ sufficiently large the remainder R_{K-1} of (3) can be made less than ϵ, where R_i is the series (3) with the first $\underline{i}$ terms omitted. But

$$\emptyset N - M_KN = (M_{K+1}N - M_KN) + (M_{K+2}N - M_{K+1}N) + \cdots .$$

By Theorem 2.6,

(4) $$\mu^*(\emptyset N - M_KN) \leqq \mu^*(M_{K+1}N - M_KN) + \mu^*(M_{K+2}N - M_{K+1}N) + \cdots = R_{K-1} < \epsilon .$$

Since $\emptyset N = M_K(\emptyset N) + [\emptyset N - M_K(\emptyset N)]$, it follows by Theorem 2.6 that

$$\mu^*(\emptyset N) \leqq \mu^*[M_K(\emptyset N)] + \mu^*[\emptyset N - M_K(\emptyset N)] =$$

$$= \mu^*(M_KN) + \mu^*(\emptyset N - M_KN) < \qquad \text{(since } M_K\emptyset = M_K\text{)}$$

$$< \mu^*(M_KN) + \epsilon \leqq \qquad \text{(by (4))}$$

$$\leqq \mu + \epsilon \qquad \text{(since } \mu^*(M_KN) \text{ is increasing with } \underline{K}\text{).}$$

Since ϵ is arbitrarily small, $\mu \geqq \mu^*(\emptyset N)$.

<u>THEOREM 3.2</u>: *Any open set $\emptyset$ is measurable.*

Proof: It is sufficient to prove that if N is any set of finite outer measure, $\mu^*(N) \geqq \mu^*(\emptyset N) + \mu^*(N - \emptyset N)$. In the notation of the preceding lemma,

$\overline{M_K, -\emptyset} \geqq \frac{1}{K}$. Since $(N - \emptyset N) \subset -\emptyset$ and $M_K N \subset M_K$, $\overline{M_K N, (N - \emptyset N)} > 0$. By Theorem 2.7,

$$(1) \qquad \mu^*[M_K N + (N - \emptyset N)] = \mu^*(M_K N) + \mu^*(N - \emptyset N).$$

Since $N \supset [M_K N + (N - \emptyset N)]$, it follows by (1) that

$$\mu^*(N) \geqq \mu^*(M_K N) + \mu^*(N - \emptyset N).$$

The theorem follows from the preceding lemma when *K* becomes infinite.

THEOREM 3.3: *If M is measurable, then -M is also.*

Proof: M is measurable if $\mu^*(N) = \mu^*(MN) + \mu^*[(-M)N]$ for every N. But this relation remains unchanged if M is replaced by -M, since $-(-M) = M$.

THEOREM 3.4: *Any closed set is measurable.*

Proof: This follows directly from Theorems 3.2, 1.2, and 3.3.

THEOREM 3.5: *If M_1 and M_2 are measurable, then $M_1 + M_2$ is also.*

Proof: Since M_1 is measurable,

$$(1) \qquad \mu^*(N) = \mu^*(M_1 N) + \mu^*[(-M_1)N],$$

where N is any set. Since M_2 is measurable,

$$(2) \qquad \mu^*[(-M_1)N] = \mu^*[M_2(-M_1)N] + \mu^*[(-M_2)(-M_1)N].$$

But

$$(3) \qquad (-M_1)(-M_2) = -(M_1 + M_2).$$

By (1), (2), and (3),

$$(4) \qquad \mu^*(N) = \mu^*(M_1 N) + \mu^*[M_2(-M_1)N] + \mu^*[\{-(M_1 + M_2)\} N].$$

Again, since M_1 is measurable,

(5) $$\mu^*[(M_1 + M_2)N] = \mu^*[M_1(M_1 + M_2)N] + \mu^*[(-M_1)(M_1 + M_2)N] =$$
$$= \mu^*(M_1N) + \mu^*[(-M_1)M_2N].$$

The theorem follows by substitution of (5) in (4).

THEOREM 3.6: *If* M_1 *and* M_2 *are measurable, then* M_1M_2 *is also.*

Proof: By Theorems 3.3 and 3.5, $-M_1$, $-M_2$, and $-M_1 + (-M_2)$ are measurable. But $-M_1 + (-M_2)$ is $-(M_1M_2)$. Hence M_1M_2 is measurable.

Theorems 3.5 and 3.6 may be extended immediately to any finite number of measurable sets.

THEOREM 3.7: *If* M_1 *and* M_2 *are measurablo, then* $M_1 - M_1M_2$ *is also.*

Proof: $M_1 - M_1M_2$ is simply $M_1(-M_2)$.

Lemma 1: *If* $M_1 \subset M_2 \subset \dots$ *is an increasing sequence of measurable point sets with sum* M, *then* $\lim_{K \to \infty} \mu^*(M_KN) = \mu^*(MN)$, *where* N *is any set.*

Proof: If $\mu^*(M_KN)$ is infinite for any k, then it is infinite for all larger k and the lemma is trivial. Hence it is necessary to prove only the case where $\mu^*(M_KN)$ is finite for all K. Since $M_1N \subset M_2N \subset \dots \subset MN$ and since MN is the sum of all the sets M_KN, it is possible to write

$$MN = M_1N + (M_2N - M_1N) + (M_3N - M_2N) + \dots .$$

By Theorem 2.6,

(1) $$\mu^*(MN) \leqq \mu^*(M_1N) + \mu^*(M_2N - M_1N) + \dots .$$

Since M_{K-1} is measurable,

$$\mu^*(M_KN) = \mu^*(M_{K-1}M_KN) + \mu^*(M_KN - M_{K-1}M_KN) =$$
$$= \mu^*(M_{K-1}N) + \mu^*(M_KN - M_{K-1}N),$$

so that

(2) $\mu^*(M_K N) - \mu^*(M_{K-1}N) = \mu^*(M_K N - M_{K-1}N)$, $\mu^*(M_{K-1}N)$ being finite.

By (2) the sum of K terms of (1) is $\mu^*(M_K N) \leqq \mu^*(MN)$. The sum S of the entire series (1) is the limit of the sum of K terms, i.e., $S = \lim_{K\to\infty} \mu^*(M_K N) \leqq \mu^*(MN)$. But, by (1), $S \geqq \mu^*(MN)$. Hence $S = \lim_{K\to\infty} \mu^*(M_K N) = \mu^*(MN)$.

Corollary: If in Lemma 1 the set N be taken as M, then $\lim_{K\to\infty} \mu^*(M_K) = \mu^*(M)$.

Lemma 2: *If* $M_1 \supset M_2 \supset \ldots$ *is a decreasing sequence of measurable point sets with product M, then* $\lim_{K\to\infty} \mu^*(M_K N) = \mu^*(MN)$, *where N is any set with finite outer measure.*

Proof: It is evident that $-M_1 \subset -M_2 \subset \ldots$. By Lemma 1,

(1) $$\lim_{K\to\infty} \mu^*[(-M_K)N] = \mu^*[(-M)N].$$

Since M_K is measurable,

$$\mu^*(N) = \mu^*(M_K N) + \mu^*[(-M_K)N],$$

so that

$$\mu^*(N) = \lim_{K\to\infty} \mu^*(M_K N) + \lim_{K\to\infty} \mu^*[(-M_K)N] =$$

(2) $$= \lim_{K\to\infty} \mu^*(M_K N) + \mu^*[(-M)N] \geqq \qquad \text{by (1)}$$

$$\geqq \mu^*(MN) + \mu^*[(-M)N], \qquad \text{since } M_K N \supset MN.$$

Hence M is measurable and

(3) $$\mu^*(N) = \mu^*(MN) + \mu^*[(-M)N].$$

The lemma follows by comparison of (2) and (3), and the fact that $\mu^*(N)$ is finite.

Corollary: If in Lemma 2 some $\mu^*(M_K)$ is finite, then N can be taken as this M_K and $\lim_{K\to\infty} \mu^*(M_K) = \mu(M)$.

The following theorem was proved incidentally in the proof of Lemma 2:

THEOREM 3.8: If $M_1 \supset M_2 \supset \dots$ is a decreasing sequence of measurable point sets with product M, then M is measurable.

THEOREM 3.9: If $M_1 \subset M_2 \subset \dots$ is an increasing sequence of measurable point sets with sum M, then M is measurable.

Proof: Since M_K is measurable, $\mu^*(N) = \mu^*(M_K N) + \mu^*[(-M_K)N]$. The theorem follows immediately from the preceding lemmas when K becomes infinite.

THEOREM 3.10: If M_1, N_2, ... is any finite or infinite sequence of measurable sets, then the sum of those sets is measurable.

Proof: Form the partial sums $S_K = M_1 + M_2 + \dots + M_K$. Then $S_1 \subset S_2 \subset S_3 \subset \dots$, the sum of the S_K is the sum of the M_K, and Theorem 3.9 applies to the S_K.

THEOREM 3.11: If M_1, M_2, ... is any finite or infinite sequence of measurable sets, then the product of these sets is measurable.

Proof: This follows from Theorem 3.8 when applied to the partial products of the M_K.

Theorems 3.2, 3.4, 3.10 and 3.11 show that any Borel set is measurable.

THEOREM 3.12: If $\mu^*(M) = 0$, then M is measurable with measure zero.

Proof: For any set N, $\mu^*(MN) \leq \mu^*(M) = 0$ and $\mu^*[(-M)N] \leq \mu^*(N)$, hence $\mu^*(MN) + \mu^*[(-M)N] \leq \mu^*(N)$, and this is all that requires proof.

THEOREM 3.13: If M or N is measurable and $\mu^*(MN)$ is finite, then $\mu^*(M + N) = \mu^*(M) + \mu^*(N) - \mu^*(MN)$.

Proof: Suppose M measurable. Then $\mu^*(N) = \mu^*(MN) + \mu^*(N-MN)$. Since $\mu^*(MN)$ is finite this can be written

$$(1) \qquad \mu^*(N - MN) = \mu^*(N) - \mu^*(MN).$$

Again since M is measurable,

$$(2) \qquad \mu^*(M + N) = \mu^*[(M + N) M] + \mu^*[(M + N) - (M + N)M] =$$

$$= \mu^*(M) + \mu^*(N - MN).$$

Substitution of (1) in (2) gives the result stated.

THEOREM 3.14: *If M_1, M_2,... is a finite or infinite sequence of measurable point sets such that no two of them have a common point, then $\mu^*(M_1 + M_2 + \dots) = \mu^*(M_1) + \mu^*(M_2) + \dots$.*

Proof: This is a generalization of Theorem 3.1 and obviously holds if the sequence is finite. If the sequence is infinite it is necessary merely to note that both $\mu^*(M_1 + M_2 + \dots)$ and $\mu^*(M_1) + \mu^*(M_2) + \dots$ are representable as $\lim_{K \to \infty} \mu^*(M_1 + M_2 + \dots + M_K)$.

THEOREM 3.15: *The outer measure $\mu^*(M)$ of any set M may be defined as the greatest lower bound of the measures $\mu(N)$ of those elements N of a certain set S of measurable sets which contain M. Herein S may be any set of measurable sets, provided only that it includes all open sets; in particular, S may be the set of Borel sets.*

Proof: Since S contains the set of open sets containing M, it contains the sums of sequences of open intervals covering M. Hence the outer measure is not increased. Neither is it decreased, for the measure of any one of the sets N is its outer measure, and its outer measure may be approximated to any desired accuracy by the sum of the volumes of a sequence of open intervals which cover N.

THEOREM 3.16: *Any measurable set M can be represented as a Borel set of type $\emptyset_\pi$ minus a set of measure zero.*

Proof: Consider the case where M is a bounded measurable set. There exists an open set $\phi_i \supset M$ such that $\mu(\phi_i) < \mu^*(M) + \frac{1}{i} = \mu(M) + \frac{1}{i}$. Let $\overline{M} = \prod_{i=1}^{\infty} \phi_i$. $\overline{M}$ is a Borel set of type ϕ_π and is measurable. $\overline{M} \supset M$. Hence $\mu(\overline{M}) \geqq \mu(M)$. But $\overline{M} \subset \phi_i$ and $\mu(\overline{M}) \leqq \mu(\phi_i) < \mu(M) + \frac{1}{i}$. Therefore $\mu(\overline{M}) = \mu(M)$. But $\overline{M} = M + (\overline{M} - M)$. By Theorem 3.1, $\mu(\overline{M}) = \mu(M) + \mu(\overline{M} - M)$. Hence $\mu(\overline{M} - M) = 0$. The theorem follows from the fact that $M = \overline{M} - (\overline{M} - M)$, providing that M is bounded. The case where M is not bounded will be taken up immediately after the proof of

THEOREM 3.17: <u>Any measurable set M can be represented as the sum of a Borel set of type C_σ and a set of measure zero.</u>

Proof: Let M be a bounded measurable set and let I be a finite closed interval containing M. I-M is measurable and, by Theorem 3.16 (proved for bounded sets), $I-M = \overline{\overline{M}}-Z$, where $\overline{\overline{M}}$ is a Borel set of type ϕ_π and Z a set of measure zero. Hence $M = I - (\overline{\overline{M}} - Z) = (I)(-\overline{\overline{M}}) + IZ$. Since I is closed and $-\overline{\overline{M}}$ is a Borel set of type C_σ, $(I)(-\overline{\overline{M}})$ is also a Borel set of type C_σ. Since IZ is a set of measure zero, the theorem is proved for the case where M is bounded. If M is not bounded, let I_N be the closed interval $-N \leqq x_\nu \leqq N$. $M = MI_1 + M(I_2 - I_1) + M(I_3 - I_2) + \dots$. Each of these summands is bounded and measurable so that, by the case just proved, $M(I_K - I_{K-1}) = B_K + Z_K$, where B_K is a Borel set of type C_σ and Z is a set of measure zero. Hence $M = (B_1 + B_2 + \dots) + (Z_1 + Z_2 + \dots)$. But $(B_1 + B_2 + \dots)$ is a Borel set of type C_σ and $(Z_1 + Z_2 + \dots)$ is a set of measure zero. Thus the theorem holds for any measurable set M.

It remains to complete the proof of Theorem 3.16. Suppose M of that theorem is not bounded. By Theorem 3.17, it is possible to write $-M = C_\sigma + Z$. Then $M = (-C_\sigma)(-Z) = (-C_\sigma) - (-C_\sigma) Z$. But $-C_\sigma$ is a Borel set of type ϕ_π and $(-C_\sigma)Z$ is a set of measure zero.

CHAPTER IV.

INNER MEASURE

Definition 4.1: The inner measure, $\mu_*(M)$, of a point set M is the least upper bound of the measures of all measurable sets N contained in M.

THEOREM 4.1: For any set M, $\mu_*(M) \leq \mu^*(M)$, and if M is measurable, $\mu_*(M) = \mu^*(M) = \mu(M)$.

Proof: The first part of the theorem is obvious and the second part follows from the fact that in Definition 4.1 one of the sets N may be M itself.

THEOREM 4.2: $\mu_*(M)$ is the least upper bound of the measures of all the closed sets contained in M.

Proof: There exists a measurable set N contained in M such that $\mu(N) > \mu_*(M) - \varepsilon$. It is shown in the proof of Theorem 3.17 that there is a closed set N' contained in N such that $\mu(N') > \mu(N) - \varepsilon$. Hence the least upper bound of $\mu(N')$ is not less than $\mu_*(M)$. The theorem follows from Definition 4.1.

In Theorem 4.2 the set of all closed sets may be replaced by any set S of measurable sets, providing S includes the set of all closed sets. In particular, S might be the set of Borel sets.

THEOREM 4.3: If M is any point set for which $\mu^*(M)$ is finite and if $\mu_*(M) = \mu^*(M)$, then M is measurable.

Proof: There exist an open set $\emptyset_K \supset M$ and a closed set $C_K \subset M$ such that $\mu(\emptyset_K) < \mu^*(M) + \frac{1}{K}$ and $\mu(C_K) > \mu_*(M) - \frac{1}{K}$. If $\emptyset_\pi = \prod_{K=1}^{\infty} \emptyset_K$ and if $C_\sigma = \sum_{K=1}^{\infty} C_K$, then $\emptyset_\pi \supset M \supset C_\sigma$. Hence $\mu(\emptyset_\pi) \geq \mu^*(M)$ and $\mu_*(M) \geq \mu(C_\sigma)$. Since $\emptyset_\pi \subset \emptyset_K$ and $C_\sigma \supset C_K$, $\mu(\emptyset_\pi) < \mu^*(M) + \frac{1}{K}$ and $\mu(C_\sigma) > \mu_*(M) - \frac{1}{K}$. Since k may become infinite, $\mu(\emptyset_\pi) \leq \mu^*(M)$ and $\mu(C_\sigma) \geq \mu_*(M)$. But $\mu_*(M) = \mu^*(M)$, so that $\mu(\emptyset_\pi) = \mu(C_\sigma)$. All these numbers are finite because $\mu^*(M)$ is finite. Since $\emptyset_\pi = C_\sigma + (\emptyset_\pi - C_\sigma)$, then, by Theorem 3,1, $\mu(\emptyset_\pi) = \mu(C_\sigma) + \mu(\emptyset_\pi - C_\sigma)$, so that $\mu(\emptyset_\pi - C_\sigma) = 0$. Again, $(M - C_\sigma) \subset (\emptyset_\pi - C_\sigma)$, $\mu^*(M - C_\sigma) \leqq \mu(\emptyset_\pi - C_\sigma) = 0$. By Theorem 3.12, $(M - C_\sigma)$ is measurable with measure zero. Since $M = C_\sigma + (M - C_\sigma)$, M is the sum of two measurable sets and hence is measurable.

<u>THEOREM 4.4:</u> <u>If</u> $MN = 0$, $\mu_*(M) + \mu_*(N) \leqq \mu_*(M + N)$.

Proof: If P and Q are any two measurable sets contained in M and N respectively, then, by Theorem 3.1, $\mu(P) + \mu(Q) = \mu(P + Q) \leqq \mu_*(M + N)$. The theorem follows from the fact that $\mu_*(M)$ and $\mu_*(N)$ are the least upper bounds of $\mu(P)$ and $\mu(Q)$.

<u>THEOREM 4.5:</u> <u>If</u> $M_1, M_2, \ldots$ <u>is any sequence of sets such that no two have a common point, then</u> $\mu_*(M_1 + M_2 + \ldots) \geqq \mu_*(M_1) + \mu_*(M_2) + \ldots$.

The proof is omitted.

Properties I, II, and IV of outer measure obviously hold for inner measure and Property III is the analogue of Theorem 4.5. (Note, however, that these properties could not be used as a starting point for the theory of measure in which one began with μ_* instead of μ^*. The $\leqq$ of μ_* is less appropriate for such a purpose than the $\geqq$ of μ^*.

<u>THEOREM 4.6:</u> If $MN = 0$, then (1) $\mu_*(M + N) \leqq \mu_*(M) + \mu^*(N) \leqq \mu^*(M + N)$, and (2) $\mu_*(M + N) \leqq \mu^*(M) + \mu_*(N) \leqq \mu^*(M + N)$.

Proof: It is sufficient to prove part (1). Let P be any measurable set contained in M. By Theorem 3.1, $\mu(P) + \mu^*(N) = \mu^*(P + N) \leqq \mu^*(M + N)$. Since $\mu_*(M)$ is the least upper bound of $\mu(P)$, $\mu_*(M) + \mu^*(N) \leqq \mu^*(M + N)$.

Again, let P be any measurable set contained in M + N, let B be any measurable set containing N, and let $A = P - PB$. Since $AB = 0$, $A \subset M$. By Theorem 3.7, A is measurable. Hence $\mu(A) \leqq \mu_*(M)$. Since $P \subset (A + B)$, then, by Theorem 3.1, $\mu(P) \leqq \mu(A + B) = \mu(A) + \mu(B) \leqq \mu_*(M) + \mu(B)$. Since $\mu_*(M + N)$

is the least upper bound of $\mu(P)$ and $\mu^*(N)$ the greatest lower bound of $\mu(B)$, $\mu_*(M + N) \leqq \mu_*(M) + \mu^*(N)$.

<u>THEOREM 4.7: If M is a set of finite outer measure and if N is any set containing M and having a finite measure, then</u> $\mu_*(M) = \mu(N) - \mu^*(N - M)$.

Proof: If the set $(M + N)$ of Theorem 4.6 is measurable with finite measure, then $\mu_*(M) + \mu^*(N) = \mu(M + N)$, where $MN = 0$. Since $\mu^*(N)$ is finite, $\mu_*(M) = \mu(M + N) - \mu^*(N)$. $M + N$ may be regarded as the set N of the theorem.

<u>Definition 4.2: If M is any point set for which</u> $\mu^*(M)$ <u>is finite, then</u> $\mu^*(M) - \mu_*(M)$ <u>is a measure of the non-measurability of M and is called the non-measure,</u> $\nu(M)$, <u>of M</u>.

It follows from Theorem 4.1 that $\nu(M) \geqq 0$ and from Theorem 4.3 that, if $\nu(M) = 0$, M is measurable.

<u>THEOREM 4.8: If</u> $MN = 0$, <u>then</u> (1) $\nu(M) + \nu(N) \geqq \nu(M + N)$, (2) $\nu(M + N) + \nu(M) \geqq \nu(N)$, <u>and</u> (3) $\nu(M + N) + \nu(N) \geqq \nu(M)$.

Proof: Part (1) follows immediately from Theorems 2.6 and 4.4. By Theorem 4.6, $\mu_*(M) + \mu^*(N) \leqq \mu^*(M + N)$ and $\mu^*(M) + \mu_*(N) \geqq \mu_*(M + N)$. Part (2) follows from these relations and part (3) is analogous to part (2).

The relations in Theorem 4.8 are analogous to the triangular law of distances.

CHAPTER V.

INVARIANCE OF MEASURE UNDER TRANSFORMATIONS

Only one-to-one transformations with an inverse are considered here.

Definition 5.1: *A transformation is called measure-preserving if it leaves outer and inner measure invariant and preserves measurability.*

It is obvious that measure itself is also invariant under a measure preserving transformation.

THEOREM 5.1: *If a tranformation T leaves outer measure invariant, then T is measure preserving.*

Proof: If M is a measurable set and N an arbitrary set, then $\mu*(N) = \mu*(MN) + \mu*(N - MN)$. If M' is the transform of M under T and N' that of N, then $\mu*(N') = \mu*(M'N') + \mu*(N' - M'N')$ since outer measure is invariant under T. But N' may be regarded as arbitrary. Hence M' is measurable and $\mu(M) = \mu(M')$. It follows from Definition 4.1 that inner measure is also invariant under T.

It is obvious that if T is measure preserving, the inverse of T is also. Likewise the product of two measure preserving transformations is measure preserving.

THEOREM 5.2: *If T is a transformation such that* $\mu(I) \geqq \mu*(I')$ *and* $\mu(J') \geqq \mu*(J)$, *where I is any interval with transform I' and J' any interval with inverse-transform J, then T is measure preserving.*

Proof: Suppose M to be any point set. There exists a sequence of open intervals $I_1, I_2, \dots$ covering M such that $\mu(I_1) + \mu(I_2) + \mu(I_3) + \dots \leqq \mu*(M) + \epsilon$. M' is covered by $I_1' + I_2' + I_3' + \dots$ and $\mu*(M') \leqq \mu*(I_1' + I_2' + \dots) \leqq \mu*(I_1') + \mu*(I_2') + \dots \leqq \mu(I_1) + \mu(I_2) + \mu(I_3) + \dots \leqq \mu*(M) + \epsilon$. It may be shown in a similar way that $\mu^*(M) \leq \mu^*(M') + \epsilon$. Hence $\mu^*(M) = \mu^*(M')$. The theorem follows from Theorem 5.1.

It is obviously possible to restrict the intervals I and J' of Theorem 5.2 to rational cubes without loss of generality.

The remainder of this chapter is devoted to showing that the general linear unimodular transformation is measure preserving. This will be effected by resolving such a transformation into a product of measure preserving transformations of the types occurring in the following

Lemma: The transformations

I $\qquad x'_\nu = x_\nu + b_\nu \qquad (\nu = 1, \ldots, n),$

II $\qquad x'_i = x_j, \quad x'_j = x_i, \quad x'_\nu = x_\nu \qquad (\nu \neq i, j),$

III $\qquad x'_i = cx_i, \quad x'_j = \frac{1}{c} x_j, \quad x'_\nu = x_\nu \qquad (\nu \neq i, j);$

IV $\qquad x'_i = x_i + cx_j, \quad x'_\nu = x_\nu \qquad (\nu \neq i)$

are measure preserving.

Proof: The proofs of parts I - III are trivial. With regard to part IV, let I be the interval $X_\nu < x_\nu < X_\nu + s_\nu$ $(\nu = 1, \ldots, n)$. Under the transformation IV this interval goes over into the set I' of points satisfying the conditions

$$X_i + cx'_j < x'_i < X_i + cx'_j + s_i,$$

$$X_\nu < x'_\nu < X_\nu + s_\nu \qquad (\nu \neq i).$$

This point set is covered by the intervals S_K:

$$X_i + c[X_j + \frac{K-1}{N} s_j] \leqq x_i' \leqq X_i + c[X_j + \frac{K}{N} s_j] + s_i \quad (K = 1,\dots,N),$$

$$X_j + \frac{K-1}{N} s_j \leqq x_j' \leqq X_j + \frac{K}{N} s_j,$$

$$X_\nu \leqq x_\nu' \leqq X_\nu + s_\nu \qquad (\nu \neq i, j).$$

The volume of S_K is $(s_i + \frac{cs_j}{N}) \frac{s_j}{N} \prod s_\nu$ $(\nu \neq i, j)$. The total volume of all the S_K is $(s_i + \frac{cs_j}{N}) s_j \prod s_\nu$. Hence $\mu^*(I') \leqq \mu(I)$ since N may be arbitrarily large. Part IV of the lemma follows from Theorem 5.2, since the same argument holds for the inverse transformation which also has the form IV with $-c$ instead of c.

THEOREM 5.3: *If* T *is the transformation* $x_\nu' = b_\nu + \sum_{\mu=1}^{n} a_{\nu\mu} x_\mu$ *with determinant* $D = \pm 1$, *where*

$$D = \begin{vmatrix} a_{11} & \cdots & a_{1n} \\ & \vdots & \\ a_{n1} & \cdots & a_{nn} \end{vmatrix},$$

then T *is measure preserving.*

Proof: T may be resolved into the product T"T', where T' is the transformation $x_\nu'' = \sum_{\mu=1}^{n} a_{\nu\mu} x_\mu$ and T" is the transformation $x_\nu' = x_\nu'' + b_\nu$. T" is measure preserving according to the Lemma. Since $D \neq 0$, some $a_{1j} \neq 0$. Let T_1 be the transformation II with $i = 1$. Let T_2 be the transformation III with $c = a_{1j}$ and $i = 1$. Then T' may be resolved into the product $T_3 T_2 T_1$, where T_3 is a homogeneous transformation with determinant of the form

$$D_3 = \begin{vmatrix} 1 & b_{12} & \cdots & b_{1n} \\ b_{21} & b_{22} & \cdots & b_{2n} \\ & \vdots & & \\ b_{n1} & b_{n2} & \cdots & b_{nn} \end{vmatrix}$$

and where the b's are certain functions of the a's. $D_3 = \pm 1$ since the determinants of T', T_1, and T_2 are each equal to ± 1. By the Lemma, T_1 and T_2 are measure preserving. It remains to show that T_3 is also.

T_3 may be resolved into the product $T_b T_a$ where T_a is the transformation

$$x_1'' = x_1 + b_{12}x_2 + \dots + b_{1n}x_n,$$

$$x_\nu'' = x_\nu \qquad (\nu \neq 1)$$

and T_b is the transformation

$$x_1' = x_1'',$$

$$x' = b_{\nu 1}x_1'' + c_{\nu 2}x_2'' + \dots + c_{\nu n}x_n'' \qquad (\nu \neq 1).$$

But T_b may be resolved into $T_d T_c$, where T_c is the transformation

$$x_1''' = x_1'',$$

$$x_\nu''' = c_{\nu 2}x_2'' + \dots + c_{\nu n}x_n'' \qquad (\nu \neq 1)$$

and T_d is the transformation

$$x_1' = x_1''',$$

$$x_\nu' = b_{\nu 1}x_1''' + x_\nu'' \qquad (\nu \neq 1).$$

T_a may be resolved into the series of transformations

$$x_1' = x_1 + b_{12}x_2, \qquad x_1'' = x_1' + b_{13}x_3',$$
$$x_\nu' = x_\nu, \qquad x_\nu'' = x_\nu', \qquad \text{etc.} \qquad (\nu \neq 1).$$

T_d may be resolved into the series of transformations

$$x_1^{(1)} = x_1''' , \qquad x_1^{(2)} = x_1^{(1)} ,$$

$$x_2^{(1)} = b_{21}x_1''' + x_2''' , \qquad x_2^{(2)} = x_2^{(1)} ,$$

$$x_\nu^{(1)} = x_\nu''' , \quad (\nu>2) \qquad x_3^{(2)} = b_{31}x_1^{(1)} + x_3^{(1)} , \qquad \text{etc.}$$

$$x_\nu^{(2)} = x_\nu^{(1)} , \quad (\nu > 3)$$

By the Lemma, each of the transformations in these two series is measure preserving. The determinants of T_a and T_d are each equal to unity. Hence the minor

$$M_c = \begin{vmatrix} c_{22} & \cdots & c_{2n} \\ & \vdots & \\ c_{n2} & \cdots & c_{nn} \end{vmatrix}$$

of T_c has one of the values ± 1. If the n-1 dimensional transformation T_{n-1} with determinant M_c is measure preserving, then T_c is also. Thus the problem of showing that T' is measure preserving has been reduced to a similar one of one less dimension. This reduction may be repeated (n-1) times and the theorem follows immediately.

Corollary: *If in Theorem 5.3 the determinant D of T has the absolute value K, then the measure of any measurable set is multiplied by K under T. If K = 0, then T is singular and any set is transformed into a measurable set.*

The proof of this corollary is apparent.

CHAPTER VI.

COVERING THEOREMS

THEOREM 6.1: If $N_1, N_2, \ldots$ is an infinite sequence S of measurable sets such that $\mu(N_K) \geqq \epsilon > 0$ for every K and such that $\mu^*(\sum_{K=1}^{\infty} N_K) < \infty$, if P is a point contained in any infinite subset of S, and if $\overline{N}$ is the set of all points P, then $\overline{N}$ is measurable and $\mu(\overline{N}) \geqq \epsilon$. (This is known as the Arzela-Young theorem.)

Proof: Let $S_K = N_K + N_{K+1} + \ldots$. Then $\overline{N} = \prod_{K=1}^{\infty} S_K$. By Theorem 3.10 and 3.11, S_K and $\overline{N}$ are measurable. But $S_1 \supset S_2 \supset \ldots$ and $\mu(S_K) \geqq \mu(N_K) \geqq \epsilon$. By the corollary of Lemma 2 of Chapter III, $\mu(\overline{N}) = \lim_{K\to\infty} \mu(S_K) \geqq \epsilon$.

THEOREM 6.2: If M is a measurable set, if $\emptyset$ is an open set containing M, if $C_K(P)$ is an infinite sequence of cubes $X_\nu - \frac{s_K}{2} < x_\nu < X_\nu + \frac{s_K}{2}$ $(\nu = 1, \ldots, n)$ with center P: $(X_1, \ldots, X_n)$ such that s_K is a decreasing sequence with limit zero, if such a sequence of cubes is associated with each point $P \in M$, and if with each cube $C_K(P)$, $P \in M$, there is associated a closed set $N_K(P) \subset C_K(P)$ such that $\frac{\mu[N_K(P)]}{\mu[C_K(P)]} \geqq \epsilon > 0$ for every k and P, then there exists a sequence S: $N_{K_1}(P_1), N_{K_2}(P_2), \ldots$ of the sets $N_K(P)$ such that (1) no two of the sets of S have a common point, (2) each set of S is contained in $\emptyset$, and (3) $\sum_{i=1}^{\infty} N_{K_i}(P_i)$ covers M except for a set of measure zero. (This is known as the Vitali covering theorem.)

Proof: It may be assumed that $\mu(M) > 0$, for otherwise the theorem is trivial. Suppose that M is bounded. Let $\emptyset'$ be an open set such that $M \subset \emptyset' \subset \emptyset$ and

$$(1) \qquad \mu(\emptyset' - M) \leqq \frac{\epsilon}{4 \cdot 3^n} \mu(M).$$

(Such a set $\emptyset'$ obviously exists.) For each $P \in M$ there is a smallest subscript $K_{(P)}$ such that $C_{K_{(P)}}(P) \subset \emptyset'$. $\sum C_{K_{(P)}}(P)$ covers M. By Theorem 1.10, a sequence T of the cubes $C_{K_{(P)}}(P)$ covers M. Suppose T to be selected so that no two cubes in it have the same center. Let T be ordered so that the edges of its elements form a monotonically decreasing sequence and then let the cubes in T be renumbered $C_{K^{(\nu)}}(P^{(\nu)})$, where $C_{K^{(\nu)}}(P^{(\nu)})$ is written instead of $C_{K_{(P^{(\nu)})}}(P^{(\nu)})$. Let $\sum_{\nu=1}^{\rho} C_{K^{(\nu)}}(P^{(\nu)})$ be denoted by $\sum_{\rho}$. $M - M\sum_{\rho}$ is a decreasing sequence, $\prod_{\rho=1}^{\infty} (M - M\sum_{\rho}) = 0$, and $\lim_{K \to \infty} u(M - M\sum_{\rho}) = 0$. Hence there exists a value ρ_1' of ρ sufficiently large so that $\mu(M - M\sum_{\rho_1'}) < \delta$, where δ is an arbitrary positive number. Then

$$(2) \qquad \mu(M) - \delta < \mu(M\textstyle\sum_{\rho_1'}) \leqq \mu[C_{K^{(1)}}(P^{(1)})] + \dots + \mu[C_{K^{(\rho_1')}}(P^{(\rho_1')})].$$

Some of the cubes $C_{K^{(1)}}(P^{(1)}), \dots, C_{K^{(\rho_1')}}(P^{(\rho_1')})$ are to be discarded by a weeding-out process. The cubes retained, called the survivors, are determined as follows: $C_{K^{(1)}}(P^{(1)})$ is a survivor and $C_{K^{(\nu)}}(P^{(\nu)})$ is a survivor if and only if it has no point in common with any preceding survivor. Let the survivors be renumbered $C_{K_1}(P_1), \dots, C_{K_{\rho_1}}(P_{\rho_1})$, where the lowered indices are to be distinguished from the raised indices in the preceding enumeration.

(a) No two survivors have a common point. (b) Each survivor is contained in $\emptyset'$. (c) By (2), the total volume of the survivors is greater than $\frac{\mu(M) - \delta}{3^n}$, for if the edges of the survivors be tripled (with a multiplication of volume by 3^n) without displacement of the centers, the expanded survivors cover the original set of cubes $C_{K^{(\nu)}}(P^{(\nu)})$, $\nu = 1,\ldots,\rho_1'$. If δ is taken to be $\frac{1}{2}\mu(M)$, then the total volume of the survivors is greater than $\frac{1}{2\cdot 3^n}\mu(M)$. Let $N_{K_1}(P_1), \ldots, N_{K_{\rho_1}}(P_{\rho_1})$ be the sets $N_{K(P)}(P)$ associated with the survivors. These sets are the first ρ_1 sets of the sequence S of the theorem. They obviously satisfy conditions (1) and (2). Although they do not satisfy condition (3), yet their total measure is greater than $\frac{\epsilon}{2\cdot 3^n}\mu(M)$.

Let $\Sigma_N = N_{K_1}(P_1) + \ldots + N_{K_{\rho_1}}(P_{\rho_1})$. Then

$$\mu(M-M\Sigma_N) = \mu(M) - \mu(M\Sigma_N). \tag{3}$$

But $M\Sigma_N = \Sigma_N - (\emptyset' - M)\Sigma_N$ and

$$\mu(M\Sigma_N) = \mu(\Sigma_N) - \mu[(\emptyset' - M)\Sigma_N] \geqq \mu(\Sigma_N) - \mu(\emptyset' - M). \tag{4}$$

By (1), (3), and (4),

$$\begin{aligned}\mu(M-M\Sigma_N) &\leqq \mu(M) - \mu(\Sigma_N) + \mu(\emptyset' - M)\\ &< \mu(M) - \frac{\epsilon}{2\cdot 3^n}\mu(M) + \frac{\epsilon}{4\cdot 3^n}\mu(M)\\ &= \left(1 - \frac{\epsilon}{4\cdot 3^n}\right)\mu(M)\\ &= \overline{\theta}\,\mu(M).\end{aligned}$$

Let θ be any number in the interval $0 < \theta < 1$ such that there exists a finite sequence S': $N_{K_1}(P_1), \ldots, N_{K_\rho}(P_\rho)$ of the sets $N_K(P)$ such that (1) no two of the sets of S' have a common point, (2) each set of S' is contained in $\emptyset$, and (3) $\mu(M-M\Sigma_N) \leqq \theta\,\mu(M)$, where $\Sigma_N = N_{K_1}(P_1) + \ldots + N_{K_\rho}(P_\rho)$. Such a number θ exists, for θ may be $\overline{\theta}$. $M-M\Sigma_N$ is measurable and contains no point of Σ_N.

$\emptyset - \Sigma_N$ is open and contains no point of Σ_N. By an argument similar to that above, there exists a set T: $N_{K_{\rho+1}}(P_{\rho+1})$, ..., $N_{K_\sigma}(P_\sigma)$ of the sets $N_K(P)$ such that (1) no set of T has a point in Σ_N and no two sets of T have a common point, (2) each set of T is contained in $\emptyset - \Sigma_N$, and hence in $\emptyset$, and (3)

$$\mu\{(M-M\Sigma_N) - (M - M\Sigma_N)[N_{K_{\rho+1}}(P_{\rho+1}) + \dots N_{K_\sigma}(P_\sigma)]\} =$$

$$= \mu[M-M(\Sigma_N + N_{K_{\rho+1}}(P_{\rho+1}) + \dots + N_{K_\sigma}(P_\sigma))] \leqq$$

$$\leqq \overline{\theta}\ \mu(M-M\Sigma_N) \leqq \overline{\theta}\theta\ \mu(M).$$

This process is inductive and, after r repetitions,

$$\mu[M-M(\sum_{\nu=1}^{\rho_r} N_{K_\nu}(P_\nu))] \leqq \overline{\theta}^r\ \mu(M).$$

(When passing from r-1 to r, ρ_{r-1} and ρ_r play the roles of ρ and σ respectively in the above discussion.) But $\lim_{K\to\infty} \overline{\theta}^r\ \mu(M) = 0$ and the set of all the sets $N_{K_\nu}(P_\nu)$ obtained by this process is the sequence S of the theorem.

If M is not bounded, let R_N be the set of all points of space such that $N-1 < |x_1| + \dots + |x_n| < N$. The set of all the sets R_N covers the whole of space except for a set of points which is obviously of measure zero. The above argument holds for each of the pairs of sets MR_N and $\emptyset R_N$. The theorem follows from the fact that no two of the sets R_N have a common point and that the set of the sets R_N is countable.

<u>THEOREM 6.3</u>: <u>If</u> M <u>is</u> <u>a</u> <u>measurable point set such that for each</u> $P \in M$ <u>and for any</u> $\delta > 0$ <u>there exists a cube</u> $C_a(P)$ <u>with center</u> P <u>and edge</u> $a < \delta$ <u>such that</u> $\dfrac{\mu[MC_a(P)]}{\mu[C_a(P)]} < 1 - \epsilon$, <u>then</u> $\mu(M) = 0$.

Proof: First suppose that M is bounded. There exists an open set $\emptyset$

such that $\mu(\emptyset) < \mu(M) + \eta$. Let $C'_a(P)$ be a closed cube contained in $C_a(P)$ such that $\dfrac{\mu[C'_a(P)]}{\mu[C_a(P)]} > K > 0$, where K is a fixed constant. The assumptions of Theorem 6.2 hold, and so there exists a sequence $C'_{a_\nu}(P_\nu)$ of the cubes $C'_a(P)$ which covers M except for a set of measure zero. No two of the cubes $C'_{a_\nu}(P_\nu)$ have a common point and each of them is contained in $\emptyset$. Hence $\mu(M) =$ $= \mu(\sum MC'_{a_\nu}(P_\nu)) = \sum\mu[MC'_{a_\nu}(P_\nu)] \leqq \sum\mu[MC_{a_\nu}(P_\nu)] < (1-\varepsilon)\sum\mu[C_{a_\nu}(P_\nu)] =$ $= (1-\varepsilon)\,\mu[\sum C_{a_\nu}(P_\nu)] \leqq (1-\varepsilon)\,\mu(\emptyset) \leqq (1-\varepsilon)(\mu(M) + \eta)$. Since η may be arbitrarily small, $\varepsilon\,\mu(M) \leqq 0$. Hence $\mu(M) = 0$. If M is unbounded it may be broken up into parts as was done in the proof of Theorem 6.2.

<u>THEOREM 6.4:</u> <u>If M is any measurable set and if N is the set of those points P ∈ M for which there exists an $\varepsilon > 0$, such that, for every $\delta > 0$, there exists a cube $C_a(P)$ with center P, with side a less than δ, and such that</u> $\dfrac{\mu[MC_a(P)]}{\mu[C_a(P)]} < 1-\varepsilon$, <u>then N has measure zero.</u>

Proof: Let A, B, and C be any three measurable sets such that ABC = 0. Then $\mu(MA) \leqq \mu[M(A+B)] \leqq \mu(MA) + \mu(MB)$. Hence $0 \leqq \mu[M(A+B)] - \mu(MA) \leqq \mu(MB) \leqq$ $\leqq \mu(B)$. Similarly $0 \leqq \mu[M(A+C)] - \mu(MA) \leqq \mu(MC) \leqq \mu(C)$. Hence $-\mu(B+C) \leqq -\mu(C)$ $\leqq \mu[M(A+B)] - \mu[M(A+C)] \leqq \mu(B) \leqq \mu(B+C)$. But A+B may be regarded as an arbitrary set D and A+C as an arbitrary set E. Then $|\mu(MD) - \mu(ME)| \leqq \mu(D+E-DE)$. Let Q and Q' be two points of space, and let $C_a(Q)$ and $C_{a'}(Q')$ be two cubes with sides <u>a</u> and <u>a'</u> and centers Q and Q'. Then by the preceding relation it follows that $\mu[MC_a(Q)] - \mu[MC_{a'}(Q')] \leqq \mu[C_a(Q) + C_{a'}(Q') - C_a(Q)C_{a'}(Q')]$. It is obvious that the term on the right side of the inequality can be made arbitrarily small by taking Q' sufficiently near Q or by making <u>a'</u> sufficiently close to <u>a</u>. Hence $\mu[MC_a(Q)]$ is a continuous function of <u>a</u> and Q. Hence $f(a,Q) = \dfrac{\mu[MC_a(Q)]}{\mu[C_a(Q)]}$ is continuous except at a = 0. Let the set of points Q such that $f(a,Q) < 1 - \varepsilon$ for some $a < \delta$, be $N_{\delta\varepsilon}$. For a fixed value of a the set of points Q is open. Since $N_{\delta\varepsilon}$ is the sum of such sets for all $a < \delta$, $N_{\delta\varepsilon}$ is open. Let $\phi_\varepsilon = \prod_{K=1}^{\infty} N_{\frac{1}{K},\varepsilon}$. Owing to the fact that $\delta \leqq \delta'$ implies $N_{\delta\varepsilon} \subset N_{\delta'\varepsilon}$, it follows that $\phi_\varepsilon = \prod_\delta N_{\delta\varepsilon}$. Since ϕ_ε is a Borel set of type ϕ_π it is measurable. If $M\phi_\varepsilon = M_\varepsilon$ and if P is any point of M_ε, then there exists a sequence of cubes $C_{a_1}(P)$, $C_{a_2}(P)$, ... with sides a_ν decreasing to zero such that $\frac{\mu[MC_{a_\nu}(P)]}{\mu[C_{a_\nu}(P)]} < 1-\varepsilon$. Hence $\frac{\mu[M_\varepsilon C_{a_\nu}(P)]}{\mu[C_{a_\nu}(P)]} < 1-\varepsilon$. By Theorem 6.3 it follows that $\mu(M_\varepsilon) = 0$. Since $\varepsilon \leqq \varepsilon'$ implies $N_{\delta\varepsilon} \supset N_{\delta\varepsilon'}$, and therefore $\phi_\varepsilon \supset \phi_{\varepsilon'}$, $\phi = \sum_{K=1}^{\infty} \phi_{\frac{1}{K}} = \sum_\varepsilon \phi_\varepsilon$. But $M\phi = N$ and $\mu(N) = 0$.

<u>Definition 6.1: If M is any measurable point set, if P is a point of M, and if $\lim_{a\to 0} f(a,P)$ exists, where $f(a,P) = \frac{\mu[MC_a(P)]}{\mu[C_a(P)]}$, then this limit is called the density of M at P.</u>

<u>THEOREM 6.5: Any measurable set M has density at every point of the space, except possibly for a set of measure zero, and the density is, again with the possible exception of a set of measure zero, 1 in the points of M and 0 in the points of -M.</u>

Proof: It follows from Theorem 6.4 that the condition $\liminf_{a\to 0} f(a,P) < 1$ where $P \in M$, holds over only a set of measure zero. Let M_1 be the subset of M such that if $P \in M_1$, $\liminf_{a\to 0} f(a,P) \geqq 1$. According to the preceding remark, $\mu(M-M_1) = 0$. Since it is always the case that $f(a,P) \leqq 1$ for $P \in M_1$, $1 \geqq \limsup_{a\to 0} f(a,P) \geqq \liminf_{a\to 0} f(a,P) \geqq 1$. Hence $\lim_{a\to 0} f(a,P)$ exists and is unity for $P \in M_1$. It follows in a similar manner that, if M_2 arises from -M as M_1 does from M, then $\mu[(-M)-(M_2)] = 0$ and the density exists and is zero over M_2.

CHAPTER VII.

NON-MEASURABLE SETS

Definition 7.1. *If M is a set of real numbers, then M is said to have the period a if (x+a) $\in$ M when and only when x $\in$ M.*

THEOREM 7.1: *If P is the set of periods of a set M of real numbers, then (1) a $\in$ P implies -a $\in$ P, (2) a $\in$ P and b $\in$ P imply (a+b) $\in$ P, and (3) a $\in$ P implies na $\in$ P, where n is any integer.*

Proof: Parts (1) and (2) follow immediately from Definition 7.1, and part (3) follows from parts (1) and (2), where in part (2) the period b is taken to be a, 2a, ... successively.

THEOREM 7.2: *If P is the set of periods of a set M of real numbers, then either (1) P is empty (except for the trivial period zero), or (2) P contains a smallest positive period a and all other periods in P are integral multiples of a, or (3) the periods in P are everywhere dense over the set of real numbers.*

Proof: It follows from part (1) of Theorem 7.1 that P contains a positive period if it contains any period other than zero. Suppose P contains a smallest positive period *a*. Let *b* be any other period in P. There exists an integer *n* such that $n \leq \frac{b}{a} < n+1$. Then $0 \leq b-na < a$. By Theorem 7.1, b-na is a period in P. Since *a* is the smallest positive period in P, b-na = 0 and b = na.

Suppose now that P contains no smallest positive period and is not empty. Let $\bar{a}$ be the greatest lower bound of the positive periods in P. Since $\bar{a}$ is not a positive period in P, there exist a period *b* in P such that $\bar{a} < b < \bar{a}+\varepsilon$ and a period *c* in P such that $\bar{a} < c < b$. Hence $0 < b-c < \varepsilon$ and b-c is a period in P. Since ε may be arbitrarily small, P contains arbitrarily small positive periods. If α is any number, there exists a period a in P and an integer n such that $na \leq \alpha < na+\varepsilon$. Hence the periods in P are everywhere dense over the set of real numbers.

Definition 7.2: *If M is any set of real numbers and if I_{ab} is the interval $a \leq x < b$, then $f(a,b)$ is $\mu^*(I_{ab}M)$.*

It is obvious that f is a non-negative function defined for every pair of real numbers and that $f(a,b) \leq b-a$, where $a < b$.

THEOREM 7.3: *If p is a period of the set M, then* $f(a+p, b+p) = f(a,b)$.

THEOREM 7.4: *If* $a \leq a'$ *and* $b \geq b'$, *then* $f(a,b) \geq f(a',b')$.

The proofs of these two theorems are apparent.

THEOREM 7.5: *If M has an everywhere dense set of periods and if* $b-a = b'-a'$, *then* $f(a,b) = f(a',b')$, *so that* $f(a,b)$ *is really a function* $g(b-a)$ *of the interval length alone.*

Proof: M has a period p such that $a \leq a'+p < a+\varepsilon$ and $b \leq b'+p < b+\varepsilon$. By Theorems 7.3 and 7.4, $f(a',b') = f(a'+p, b'+p) \leq f(a, b+\varepsilon) \leq f(a,b) + \varepsilon$. Similarly, $f(a,b) \leq f(a',b')+\varepsilon$.

In the sequel it is always assumed that the function g arises from a set M having an everywhere dense set of periods, as otherwise g is meaningless.

THEOREM 7.6: *For any two positive numbers u and v,* $g(u) + g(v) = g(u+v)$.

Proof: This is merely a restatement of the obvious fact that $f(a,b) + f(b,c) = f(a,c)$, where $u = b-a$ and $v = c-b$.

THEOREM 7.7: $g(ku) = kg(u)$, *where k is any positive rational number.*

Proof: The theorem follows immediately from Theorem 7.6 if k is a positive integer. If k is not integral, let $k = \frac{m}{n}$. Then $g(\frac{m}{n} u) = \frac{1}{n} ng(\frac{m}{n} u) = \frac{1}{n} g(mu) = \frac{m}{n} g(u)$.

THEOREM 7.8: $g(u) = cu$, where c is a constant and $0 \leq c \leq 1$.

Proof: By Theorem 7.6, $0 \leq g(u+v) - g(u) = g(v)$, that is $0 \leq g(x) - g(y) \leq x-y$, where $y = u$ and $x = u+v$. Hence $g(u)$ is a continuous function of u. Hence $g(x) = xg(1) = cx$, and, since $0 \leq f(a,b) \leq b-a$, $0 \leq c \leq 1$.

THEOREM 7.9: If $g(u) = cu$ arises from the set M, then $c = 1$ unless M is a set of measure zero, in which case $c = 0$.

Proof: Let I be any interval such that $m = \mu^*(IM) < \infty$. IM can be covered by a sequence of open intervals J_i of total length less than $m+\varepsilon$. Since $IM \subset J_1M + J_2M + \ldots$, $m = \mu^*(IM) \leq \mu^*(J_1M) + \mu^*(J_2M) + \ldots \leq c(\ell_1 + \ell_2 + \ldots) \leq c(m+\varepsilon)$, where ℓ_i is the length of J_i. Hence $m \leq cm$ and $c = 1$ unless $m = 0$. If there exists an interval I such that $m \neq 0$, then $c = 1$. But if $m = \mu^*(IM) = 0$ for every interval I, then $\mu(M) = 0$.

THEOREM 7.10: If M has an everywhere dense set of periods, then either (1) M is a set of measure zero, or (2) $-M$ is a set of measure zero, or (3) $\mu^*(M) = \mu^*(-M) = \infty$ and $\mu_*(M) = \mu_*(-M) = 0$.

Proof: If M is not a set of measure zero then $c = 1$. Let N be any measurable set contained in $-M$. By Theorem 3.1, $\mu^*[I(M+N)] = \mu^*(IM) + \mu^*(IN)$, where I is any interval. Hence $0 = \mu^*(I) - \mu^*(IM) \geq \mu^*(IN)$. Hence by Definition 4.1, $\mu_*(-M) = 0$. Therefore, either $\mu(M) = 0$ or $\mu_*(-M) = 0$. Similarly, either $\mu(-M) = 0$ or $\mu_*(M) = 0$, since M and $-M$ have the same periods. If $\mu^*(M) \neq 0$ and $\mu^*(-M) \neq 0$, then $\mu_*(-M) = \mu_*(M) = 0$ and $\mu^*(M) = \mu^*(-M) = \infty$.

THEOREM 7.11: There exists a non-measurable set.

Proof: Let the space be the set of real numbers. x is said to be congruent to y, $x \sim y$, if $x-y$ is rational. The following properties are apparent: (1) $x \sim x$, (2) if $x \sim y$, then $y \sim x$, and (3) if $x \sim y$ and $y \sim z$, then $x \sim z$. Let C_x be the class of all real numbers congruent to x. It follows from the properties of congruence that every real number belongs to one and only one congruence class. C_0 is the class of all rational numbers. $C_x C_{-x} = 0$ if x is irrational. Consider all the pairs $[C_x, C_{-x}]$ of congruence classes obtained by allowing x to run through the set of irrational numbers. Regard the pair $[C_{-x}, C_x]$ as being the same as the pair $[C_x, C_{-x}]$. Then the pair $[C_y, C_{-y}]$ is the same as the pair $[C_x, C_{-x}]$ if y is congruent to either x or $-x$. Assume the axiom of selection to be valid. Then it is possible to select a C_x from each of the distinct pairs $[C_x, C_{-x}]$. Let the C_x's so selected be denoted by $C_{\overline{x}}$, let the sum of all these congruence classes $C_{\overline{x}}$ be denoted by $\overline{C}$, and let the sum of all the congruence classes $C_{-\overline{x}}$ be denoted by $\overline{\overline{C}}$. Then $(\overline{C})(\overline{\overline{C}}) = 0$. $C_0 + \overline{C} + \overline{\overline{C}}$ is the set of all real numbers, and $-\overline{C} = \overline{\overline{C}} + C_0$. Each of the sets C_0, $\overline{C}$, and $\overline{\overline{C}}$ has an everywhere dense set of periods (all rational numbers). $\mu(C_0) = 0$. If $\overline{C}$ is measurable, then $\overline{\overline{C}}$ is also and their measures are equal since $y = -x$ transforms $\overline{C}$ into $\overline{\overline{C}}$. Therefore neither $\overline{C}$ nor $\overline{\overline{C}}$ can have measure zero. By part (3) of Theorem 7.10, $\mu^*(\overline{C}) = \mu^*(\overline{\overline{C}}) = \infty$ and $\mu_*(\overline{C}) = \mu_*(\overline{\overline{C}}) = 0$. Hence $\overline{C}$ and $\overline{\overline{C}}$ are not measurable.

CHAPTER VIII.

LEBESGUE INTEGRAL.

Definition 8.1: Let $f(P)$ be a function of P, that is, let a real number $f(P)$ be associated with each point P of the space R_n. If in some way there is associated with $f(P)$ a number $\int f(P)\,dv_P$ which has the following properties:

1. $\int c\, f(P)dv_P = c\int f(P)\,dv_P$, where c is a non-negative real number,
2. $\int [f(P) + g(P)]\,dv_P = \int f(P)\,dv_P + \int g(P)\,dv_P$,
3. $\int 1_I(P)\,dv_P = \mu(I)$, where I is any finite interval and

$$1_I(P) = \begin{cases} 1 & \text{for } P \varepsilon I, \\ 0 & \text{for } P \varepsilon -I, \end{cases}$$

4. $\int f(P)\,dv_P \geqq 0$ if $f(P) \geqq 0$ for all P,
5. $\int f[T(P)]\,dv_P = \int f(P)\,dv_P$, where $T(P)$ is the transform of P by any inhomogeneous linear transformation with determinant ± 1,

then $\int f(P)\,dv_P$ is called an integral of $f(P)$.

It has been shown by Banach and Tarski that $\int f(P)\,dv_P$ can be defined in such a manner as to be finite and not always zero for the set of bounded functions in R_1. $\int f(P)\,dv_P$ can be similarly defined for the set of bounded functions in R_2 if only orthogonal transformations are allowed. But an integral can not be so defined for the set of bounded functions in R_n, $n > 2$. Hence it must be assumed that $f(P)$ belongs to some restricted set of functions in order that $\int f(P)\,dv_P$ may be meaningful.

It can be shown that postulate 3 is dependent upon postulates 2, 4, and 5, when written in the form $\int 1_I(P)\, dv_P = k\, \mu(I)$, where k is some non--negative constant. Hence postulate 3 really asserts merely that $k = 1$.

An integral as defined above arises from a function defined over the entire space. If M is a set of points over which f(P) is defined, then it is readily possible to define an integral of f(P) over M by introducing the notation $f_M(P) = \begin{cases} f(P) \text{ for } P \in M, \\ 0 \quad \text{ for } P \in -M, \end{cases}$ and defining $\int_M f(P)\, dv_P$ to be $\int f_M(P)\, dv_P$. If M is the one-dimensional interval $a \leqq x < b$, then postulate 5 assumes the form $\int_{a-c}^{b-c} f(x+c)\, dx = \int_a^b f(x)\, dx$. It can be shown that the property

$$\int_a^b f(x)\, dx + \int_b^c f(x)\, dx = \int_a^c f(x)\, dx$$

is dependent upon the above postulates.

It is now necessary to introduce the concept of an ordinate set in order to be able to give a definition of an integral. Let $f(P) \geqq 0$ be defined over R_n. Let R_{n+1} be the space of points with coordinates $(x_1, \ldots, x_n, x_{n+1})$. In R_{n+1} let B be the point set $0 < x_{n+1} < f(P)$, $-\infty < x_i < \infty$, $i = 1, \ldots, n$ and let C be the set $0 \leqq x_{n+1} \leqq f(P)$, $-\infty < x_i < \infty$, $i = 1, \ldots, n$. If I is any finite interval of R_n then the sets B_I and C_I may be defined analogously. Any set S of points such that $C \supset S \supset B$ (or any set S_I such that $C_I \supset S_I \supset B_I$) will be called an ordinate set of f(P); C is the outer ordinate set of f(P) and B is the inner ordinate set.

THEOREM 8.1: *If any ordinate set S arising from $f(P) \geqq 0$ is measurable, then every ordinate set is measurable and all their measures are equal.*

Proof: Consider first the case of a measurable ordinate set S_I corresponding to a finite interval I in R_n and suppose f(P) to be bounded. Let T_1 be the transformation $x'_{n+1} = x_{n+1} - \varepsilon$, $x'_i = x_i$ and let T_2 be the transformation $x'_{n+1} = x_{n+1} + \varepsilon$, $x'_i = x_i$. Let I_ε be the interval $-\varepsilon < x_{n+1} < \varepsilon$, $x_i \in I$. (In this discussion $i=1, \ldots, n$.) It is obvious that (1) $T_1(S_I) \subset B_I + I_\varepsilon$ and (2) $T_2(S_I) + I_\varepsilon \supset C_I$. By (1), $\mu(S_I) = \mu[T_1(S_I)] \leqq \mu_*(B_I + I_\varepsilon) \leqq \mu_*(B_I) + \mu(I_\varepsilon) = \mu_*(B_I) + 2\varepsilon\mu(I)$. Hence $\mu_*(B_I) \geqq \mu(S_I)$. Similarly by (2), $\mu^*(C_I) \leqq \mu^*[T_2(S_I) + I_\varepsilon] \leqq \mu(S_I) + 2\varepsilon\mu(I)$. Hence $\mu^*(C_I) \leqq \mu(S_I)$ and $\mu^*(C_I) \leqq \mu_*(B_I)$. But it is obvious that $\mu^*(C_I) \geqq \mu_*(B_I)$. Hence $\mu^*(C_I) = \mu_*(B_I)$. Therefore the outer and inner measures of any ordinate set over I are equal. Since the entire space may be divided up into finite intervals, the theorem is immediate in case $f(P)$ is bounded. If $f(P)$ is not bounded, let $f_N(P)$ = minimum of $f(P)$ and N, where N is a positive integer. $f_N(P)$ is bounded. Let S_N be the outer ordinate set arising from the function $g(P) \equiv N$. Since S is a measurable ordinate set of $f(P)$, $S_N S$ is a measurable ordinate set of $f_N(P)$. If S' is any ordinate set of $f(P)$ and if C is the outer ordinate set of $f(P)$, then $S_N S'$ and $S_N C$ are measurable ordinate sets of $f_N(P)$, $\mu(S_N S) = \mu(S_N S') = \mu(S_N C)$, and $\mu(S_N C - S_N S') = 0$. $\sum_{N=1}^{\infty} S_N S' = S'$ and $\sum_{N=1}^{\infty} S_N C = C$. Hence $\mu(C - S') = 0$ and the proof is complete.

Definition 8.2: If $f(P) \geqq 0$ has a measurable ordinate set with measure μ, then (1) $f(P)$ is called measurable, (2) $f(P)$ is called summable if $\mu < \infty$, and (3) $\int f(P)\, dv_P = \mu$.

Lemma: If M is a set of points $(x_1, \ldots, x_n)$ in R_n and if M_1 is the set of points $(x_1, \ldots, x_n, x_{n+1})$ in R_{n+1} where $(x_1, \ldots, x_n) \in M$ and $0 \leqq x_{n+1} \leqq 1$, then (1) $\mu(M) = \mu(M_1)$ if M is an interval, (2) $\mu^*(M) \geqq \mu^*(M_1)$, (3) $\mu_*(M) \leqq \mu_*(M_1)$, (4) if M is measurable, then M_1 is also and $\mu(M) = \mu(M_1)$, (5) $\mu^*(M) \leqq \mu^*(M_1)$, (6) $\mu^*(M) = \mu^*(M_1)$, (7) $\mu_*(M) = \mu_*(M_1)$, and (8) if M_1 is measurable, then M is also and $\mu(M_1) = \mu(M)$.

Proof: Part (1) is obvious.

Part (2). There exists a sequence of open intervals $I_1, I_2, \ldots$

covering M such that $\sum \mu(I_i) \leqq \mu^*(M) + \varepsilon$. Let I_i' be the interval containing all the points $(x_1, \ldots, x_n, x_{n+1})$ where $(x_1, \ldots, x_n) \in I_i$ and $0 \leqq x_{n+1} \leqq 1$. The sequence of intervals $I_1', I_2', \ldots$ covers M_1 and, by Part (1), $\mu^*(M_1) \leqq \sum \mu(I_i') = \sum \mu(I_i) \leqq \mu^*(M) + \varepsilon$. Hence $\mu^*(M) \geqq \mu^*(M_1)$.

Part (3). Suppose M bounded and let C be a cube containing M. By Theorem 4.7, $\mu_*(M) = \mu(C) - \mu^*(C-M)$. If C_1 is the set of points $(x_1, \ldots, x_n, x_{n+1})$ where $(x_1, \ldots, x_n) \in C$ and $0 \leqq x_{n+1} \leqq 1$, then $C_1 \supset M_1$ and $\mu_*(M_1) = \mu(C_1) - \mu^*(C_1 - M_1)$. By Part (1), $\mu(C) = \mu(C_1)$, and by Part (2), $\mu^*(C_1 - M_1) \leqq \mu^*(C-M)$. Hence $\mu_*(M) \leqq \mu_*(M_1)$. If M is not bounded it is possible to divide up the whole of space into a sequence of finite non-overlapping intervals and the condition $\mu_*(M) \leqq \mu_*(M_1)$ holds in each interval.

Part (4). This follows immediately from Theorem 4.3 if M is bounded. The proof of the case where M is not bounded is apparent.

Part (5). Let $\emptyset'$ be an open set containing M_1 such that $\mu(\emptyset') \leqq \mu^*(M_1) + \varepsilon$. $-\emptyset'$ is closed. Let P' be any point $(x_1, \ldots, x_n, x_{n+1})$ of $-\emptyset'$ such that $0 \leqq x_{n+1} \leqq 1$ and let P be the point $(x_1, \ldots, x_n)$ whose coordinates are the first $\underline{n}$ coordinates of P'. Let $-\emptyset$ be the set of all points P arising in this way from all points P' of $-\emptyset'$. The set of points P' is closed (since it is the intersection of two closed sets) and $-\emptyset$ is closed since the coordinate x_{n+1} is bounded. Hence $\emptyset$ is open and contains M. Let $\emptyset_1$ be the set arising from $\emptyset$ in the same way that M_1 arises from M. Then $M_1 \subset \emptyset_1 \subset \emptyset'$ and $\mu^*(M) \leqq \mu(\emptyset) = \mu(\emptyset_1) \leqq \mu(\emptyset') \leqq \mu^*(M_1) + \varepsilon$. Hence $\mu^*(M) \leqq \mu^*(M_1)$.

Part (6) follows from Parts (2) and (5).

Part (7) follows from Part (6) and the relations occurring in the proof of Part (3).

Part (8) follows from Parts (6) and (7).

It is obvious that the preceding Lemma holds if either or both of the equality signs in the condition $0 \leqq x_{n+1} \leqq 1$ are omitted.

Theorem 8.2: *If $f(P)$ is measurable and non-negative, then the sets of points P in R_n such that $f(P) > a$, $\geqq a$, $< a$, $\leqq a$, where a is any real number, are measurable.*

Proof: Let S_a be the set of all points of R_{n+1} such that $x_{n+1} > a \geqq 0$. Let C be the outer ordinate set of $f(P)$. Then S_aC is the measurable set of points of R_{n+1} such that $a < x_{n+1} \leqq f(P)$. If the transformation $x_1' = x_1, \ldots, x_n' = x_n$, $x_{n+1}' = k(x_{n+1} - a)$ be performed in R_{n+1}, then S_aC becomes the measurable set of points of R_{n+1} such that $0 < x_{n+1} \leqq k[f(P)-a]$. The sum $\sum$ of all these point sets corresponding to the values 1, 2, ... of k is also measurable. Let A_a be the set of all points P of R_n such that $f(P) > a$. Then $\sum$ is the set of all points $(x_1, \ldots, x_n, x_{n+1})$ such that $(x_1, \ldots, x_n) \in A_a$. The subset of $\sum$ such that $0 \leqq x_{n+1} \leqq 1$, being the intersection of two measurable sets, is measurable. Hence, by the preceding Lemma, A_a is measurable. If $a < 0$, A_a is the whole of space and therefore measurable. Thus A_a is measurable for every a. If D_a is the set of points P of R_n such that $f(P) \leqq a$, then D_a is seen to be measurable by considering $\prod_{k=1}^{\infty} A_{(a-\frac{1}{k})}$. The remaining parts of the theorem follow from Theorem 3.3.

Corollary: *If $f(P)$ is measurable and non-negative, and if M is a Borel set of real numbers, then the set of points P of R_n such that $f(P) \in M$ is measurable.*

Proof: The corollary follows immediately from the preceding theorem, the definition of a Borel set, and Theorems 3.10 and 3.11.

THEOREM 8.3: *If M is a set of real numbers, if $f(P)$ is non-negative, if $S_P[f(P) \in M]$ denotes the set of points P of R_n such that $f(P) \in M$, and if there exists a set of numbers a everywhere dense over the set of real numbers such that for each of them the set $S_P[f(P) > a]$ is measurable, then $f(P)$ is* measurable.

Proof: The sets $S_P[f(P) > a]$ and $S_P[f(P) \leqq a]$ are measurable for all real a. For let $a_1, a_2, \dots$ be a sequence of the numbers a of the theorem approaching a given number a as a limit from above. Then all the sets $S_P[f(P) > a]$ are measurable, as is also their sum which is simply $S_P[f(P) > a]$; the complementary set, $S_P[f(P) \leqq a]$ is likewise measurable. Let a and b be two numbers such that $a < b$. By Theorems 3.3 and 3.6 the set $S_P[a < f(P) \leqq b]$ is measurable since it is the common part of $S_P[f(P) > a]$ and $S_P[f(P) \leqq b]$. Let A_{ab} be the set of points $(x_1, \dots, x_n, x_{n+1})$ such that $(x_1, \dots, x_n) \varepsilon S_P[a < f(P) \leqq b]$ and $0 \leqq x_{n+1} \leqq a$; likewise let B_{ab} be the set of points $(x_1, \dots, x_n, x_{n+1})$ such that $(x_1, \dots, x_n) \varepsilon S_P[a < f(P) \leqq b]$ and $0 \leqq x_{n+1} \leqq b$. By the above Lemma it follows that A_{ab} and B_{ab} are measurable. In fact, $\mu(A_{ab}) = a\mu S_P[a < f(P) \leqq b]$ and $\mu(B_{ab}) = b\mu S_P[a < f(P) \leqq b]$. Let $\emptyset$ be the outer ordinate set of $f(P)$ and let $0 = \alpha_0 < \alpha_1 < \dots$ be a sequence of real numbers approaching ∞. It is obvious that $\sum_{n=1}^{\infty} A_{\alpha_{n-1}\alpha_n} \subset \emptyset \subset \sum_{n=1}^{\infty} B_{\alpha_{n-1}\alpha_n}$. Let C' be the interval $-N < x_i < N$, $i = 1, \dots, n$, $-\infty < x_{n+1} < \infty$. Then $\sum_{n=1}^{\infty} C'A_{\alpha_{n-1}\alpha_n} \subset C'\emptyset \subset \sum_{n=1}^{\infty} C'B_{\alpha_{n-1}\alpha_n}$. Since the sets $C'A_{\alpha_{n-1}\alpha_n}$ are non-intersecting, as are also the sets $C'B_{\alpha_{n-1}\alpha_n}$, $\mu(\sum_{n=1}^{\infty} C'A_{\alpha_{n-1}\alpha_n}) =$

$$= \sum_{n=1}^{\infty} \alpha_{n-1}\mu\left\{CS_P[\alpha_{n-1} < f(P) \leqq \alpha_n]\right\} \leqq \mu(C'\emptyset) \leqq \mu(\sum_{n=1}^{\infty} C'B_{\alpha_{n-1}\alpha_n}) =$$

$$= \sum_{n=1}^{\infty} \alpha_n\mu\left\{CS_P[\alpha_{n-1} < f(P) \leqq \alpha_n]\right\},$$ where C is the cube $-N < x_i < N$,

$i = 1, \dots, n$. If $f(P)$ is bounded, then, for some value $\bar{n}$ of n, $f(P) \leqq \alpha_{\bar{n}}$, and all the preceding sums are finite. If $\alpha_n - \alpha_{n-1} \leqq \delta$ for all n, then

$$\mu(\sum_{n=1}^{\infty} C'B_{\alpha_{n-1}\alpha_n}) - \mu(\sum_{n=1}^{\infty} C'A_{\alpha_{n-1}\alpha_n}) = \sum_{n=1}^{\infty} (\alpha_n - \alpha_{n-1})\, \mu\left\{CS_P[\alpha_{n-1} < f(P) \leqq \alpha_n]\right\} \leqq$$

$\leqq \delta\, \mu(C)$. Hence $\lim\limits_{\delta \to 0} \sum\limits_{n=1}^{\infty} (\alpha_n - \alpha_{n-1})\, \mu \left\{ CS_P[\alpha_{n-1} < f(P) \leqq \alpha_n] \right\} = 0$. But

$\lim\limits_{\delta \to 0} \mu(\sum\limits_{n=1}^{\infty} C'A_{\alpha_{n-1}\alpha_n})$ is a lower bound of $\mu_*(C'\phi)$ and $\lim\limits_{\delta \to 0} \mu(\sum\limits_{n=1}^{\infty} C'B_{\alpha_{n-1}\alpha_n})$

is an upper bound of $\mu^*(C'\phi)$. Hence $C'\phi$ is measurable. By summing $C'\phi$ over all intervals C' it is seen that $f(P)$ is measurable. The theorem still holds if $f(P)$ is not bounded; this may be proved by considering $f_N(P)$=minimum of $f(P)$ and N, and allowing N to become infinite.

<u>Corollary: Theorem 8.3 holds if the condition that the sets</u> $S_P[f(P) < a]$ <u>be measurable be replaced by any one of the conditions that the sets</u> $S_P[f(P) \geqq a]$, $S_P[f(P) < a]$, <u>or</u> $S_P[f(P) \leqq a]$ <u>be measurable.</u>

Proof: The third of these conditions is immediately equivalent to that in Theorem 8.3. The second is equivalent to the first, and the first leads to the condition of Theorem 8.3 by taking a sequence of values $a_1, a_2, \ldots$ approaching any given number <u>a</u> from above and summing the resulting set of sets $S_P[f(P) \geqq a_i]$.

It is now desired to set up an explicit formula for $\int f(P)\, dv_P$. Suppose $f(P) = 0$ except over a measurable set M of finite measure. The proof of Theorem 8.3 applies verbatim from the point where the ordinate set ϕ is introduced, except that the sets C and C' are no longer needed. It is evident that $\sum\limits_{n=1}^{\infty} \alpha_{n-1} \mu \left\{ S_P[\alpha_{n-1} < f(P) \leqq \alpha_n] \right\} \leqq \mu(\phi) = \int f(P)\, dv_P \leqq$

$\leqq \sum\limits_{n=1}^{\infty} \alpha_n \mu \left\{ S_P[\alpha_{n-1} < f(P) \leqq \alpha_n] \right\}$. Instead of considering merely one sequence α_n, let $0 = \alpha_0^{\rho} < \alpha_1^{\rho} < \ldots$ be an infinite set of sequences such that $\lim\limits_{n \to \infty} \alpha_n^{\rho} = \infty$ and that $\max(\alpha_n^{\rho} - \alpha_{n-1}^{\rho}) = \delta_{\rho} \to 0$ as $\rho \to \infty$. Let ξ_n^{ρ} be any number such that $\alpha_{n-1}^{\rho} < \xi_n^{\rho} \leqq \alpha_n^{\rho}$. Then the following theorem is obvious:

<u>THEOREM 8.4</u>: *If* $S^\rho = \sum_{n=1}^{\infty} \xi_n^\rho\, \mu\left\{S_P[\alpha_{n-1}^\rho < f(P) \leq \alpha_n^\rho]\right\}$, *then* $\lim_{\rho\to\infty} S^\rho = \int f(P)\, dv_P$.

<u>THEOREM 8.5</u>: *If* $f_1(P), f_2(P), \ldots$ *is a monotonically increasing sequence of measurable non-negative functions with a limit function* $f(P)$, *then* $f(P)$ *is measurable and* $\lim_{n\to\infty} \int f_n(P)\, dv_P = \int f(P)\, dv_P$.

Proof: Let B_n be the inner ordinate set arising from $f_n(P)$. Then $B_1, B_2, \ldots$ is an increasing sequence of measurable point sets. By Theorem 3.9, $\sum_{n=1}^{\infty} B_n$ is measurable and $\lim_{n\to\infty} \mu(B_n) = \mu(\sum_{n=1}^{\infty} B)$. But $\sum_{n=1}^{\infty} B_n$ is the inner ordinate set of $f(P)$. The theorem follows from the fact that $\mu(B_n) = \int f_n(P) dv_P$ and $\mu(\sum_{n=1}^{\infty} B_n) = \int f(P)\, dv_P$.

<u>THEOREM 8.6</u>: *If* $f_1(P), f_2(P), \ldots$ *is a monotonically decreasing sequence of measurable non-negative functions with a limit function* $f(P)$, *then* $f(P)$ *is measurable and* $\lim_{n\to\infty} \int f_n(P)\, dv_P = \int f(P)\, dv_P$ *provided that at least one of the functions* $f_n(P)$ *is summable.*

The proof is analogous to that of the preceding theorem.

<u>THEOREM 8.7</u>: *If* $f_1(P), f_2(P), \ldots$ *is any sequence of measurable non-negative functions with a limit function* $f(P)$ *and if there exists a summable function* $g(P)$ *such that* $f_n(P) \leq g(P)$ *for all* n, *then* $f(P)$ *is measurable and* $\lim_{n\to\infty} \int f_n(P)\, dv_P = \int f(P)\, dv_P$.

Proof: Let

$$F_{m,k}(P) = \text{minimum of } f_m(P), f_{m+1}(P), \ldots, f_{m+k}(P),$$

$$F_m(P) = \text{greatest lower bound of } f_m(P), f_{m+1}(P), \ldots$$

It follows that $F_{m,1}(P) \geqq F_{m,2}(P) \geqq \dots$ and that $F_{m,k}(P)$ approaches $F_m(P)$ from above as $k \to \infty$. Similarly, $F_1(P) \leqq F_2(P) \leqq \dots$ and $F_m(P)$ approaches $f(P)$ from below as $m \to \infty$. All these functions are obviously measurable. For any n relations $f_n(P) \geqq F_{m,k}(P)$ and $\int f_n(P)\, dv_P \geqq \int F_{m,k}(P)\, dv_P$ hold for all $m \leqq n$ and all $k \geqq n-m$. Hence, by Theorem 8.6, if $n \geqq m$, $\int f_n(P)\, dv_P \geqq$

$\geqq \lim_{k\to\infty} \int F_{m,k}(P)\, dv_P = \int \lim_{k\to\infty} F_{m,k}(P)\, dv_P = \int F_m(P)\, dv_P$. Hence

$\liminf_{n\to\infty} \int f_n(P)\, dv_P \geqq \int F_m(P)\, dv_P$ for each m. Hence, by Theorem 8.5,

$\liminf_{n\to\infty} \int f_n(P)\, dv_P \geqq \lim_{m\to\infty} \int F_m(P)\, dv_P = \int \lim_{m\to\infty} F_m(P)\, dv_P = \int f(P)\, dv_P$.

It follows by a similar argument that $\liminf_{n\to\infty} \int (g(P) - f_n(P))\, dv_P \geqq$

$\geqq \int (g(P) - f(P))\, dv_P$. But the left side of this inequality is

$\liminf_{n\to\infty} [\int g(P)\, dv_P - \int f_n(P)\, dv_P] = \int g(P)\, dv_P - \limsup_{n\to\infty} \int f_n(P)\, dv_P$.

Hence $\limsup_{n\to\infty} \int f_n(P)\, dv_P \leqq \int f(P)\, dv_P$. Therefore $\limsup_{n\to\infty} \int f_n(P)\, dv_P \leqq$

$\leqq \liminf_{n\to\infty} \int f_n(P)\, dv_P$. Hence $\lim_{n\to\infty} \int f_n(P)\, dv_P$ exists and equals $\int f(P)\, dv_P$.

Definition 8.3: _If $f(P)$ is a non-negative function which assumes only the values 0, $v_1, \dots, v_n$, if S_{v_K} is the set of points P at which $f(P) = v_K$, and if $\mu(S_{v_K})$ exists and is finite for each v_K, then $f(P)$ is called a finitely valued function._

THEOREM 8.8: _If $f(P)$ is a finitely valued function, then in the notation of Definition 8.3, $\int f(P)\, dv_P = \sum_{K=1}^{n} v_K\, \mu(S_{v_K})$. The sum of two finitely valued functions is finitely valued and the integral of their sum exists and equals the sum of their integrals._

The proof of this theorem is apparent.

THEOREM 8.9: *Every real non-negative measurable function* $f(P)$ *is the limit of a sequence* $f_1(P)$, $f_2(P)$, ... *of finitely valued functions. This sequence can be chosen in such a manner that* $0 \leqq f_1(P) \leqq f_2(P) \leqq \dots \leqq f(P)$ *everywhere.*

Proof: Let K_N be the closed cube in R_n with edge N and center at the origin, where N is a positive integer. If P is a point such that $f(P) < 2^N$, then there exists a positive integer $\nu = \nu(P)$ such that $\frac{\nu-1}{2^N} \leqq f(P) < \frac{\nu}{2^N}$. Let

$$f_N(P) = \begin{cases} \frac{\nu-1}{2^N} & \text{if } P \in K_N \text{ and } f(P) < 2^N, \\ 0 & \text{for all other } P. \end{cases}$$

$f_N(P)$ is finitely valued and is measurable, $f_N(P)$ is monotonically increasing in N, and $\lim_{N \to \infty} f_N(P) = f(P)$.

THEOREM 8.10: *If* $f(P)$ *and* $g(P)$ *are real non-negative measurable functions, then* $f(P) + g(P)$ *is also measurable, and* $\int [f(P) + g(P)]\, dv_P = \int f(P)\, dv_P + \int g(P)\, dv_P$.

Proof: This is obviously true for finitely valued functions, and it follows from Theorem 8.9 that this is true for all functions.

THEOREM 8.11: *If* $f_1(P) - f_2(P) = g_1(P) - g_2(P)$, *then* $\int f_1(P)\, dv_P - \int f_2(P)\, dv_P = \int g_1(P)\, dv_P - \int g_2(P)\, dv_P$, *providing that* $f_1(P)$, $f_2(P)$, $g_1(P)$, *and* $g_2(P)$ *are non-negative.*

Proof: The theorem follows immediately from Theorem 8.10 upon transposing terms.

Theorem 8.10 states that Postulate 2 in Definition 8.1 is satisfied by the integral of Definition 8.2. It is apparent that the other postulates are also satisfied by this integral when $f(P)$ is real and $\geqq 0$.

If two such functions are identical except over a set of measure zero, then it is obvious that their integrals are equal. Hence it is possible to generalize Theorem 8.7 as follows:

THEOREM 8.12: *If $f_1(P)$, $f_2(P)$, ... is a sequence of measurable non-negative functions which approach a limit function $f(P)$ except over a set of measure zero and if there exists a summable function $g(P)$ such that each $f_n(P) \leqq g(P)$ except for a set of measure zero, then $f(P)$ is measurable and*

$$\lim_{n\to\infty} \int f_n(P)\, dv_P = \int f(P)\, dv_P.$$

Proof: The theorem follows immediately from the preceding comment and Theorem 8.7 when the values of each $f_n(P)$, $f(P)$, and $g(P)$ are changed to zero over all the sets of measure zero mentioned in the theorem.

Definition 8.4: *Let $f(P)$ be any real-valued function and let $f(P) = f_1(P) - f_2(P)$, where $f_1(P)$ and $f_2(P)$ are real and non-negative. If such a pair of functions $f_1(P)$ and $f_2(P)$ exist which are summable, then $f(P)$ is called summable; if such a pair of functions exist which are measurable, then $f(P)$ is called measurable and $\int f(P)\, dv_P$ is taken to be $\int f_1(P)\, dv_P - \int f_2(P) dv_P$.*

It follows from Theorem 8.10 that the value of $\int f(P)\, dv_P$ is independent of the manner in which $f(P)$ is resolved into $f_1(P)$ and $f_2(P)$ so long as the difference $\int f_1(P)\, dv_P - \int f_2(P)\, dv_P$ has sense.

Let $f(P)$ be any real-valued function and let

$$f_1^o(P) = \begin{cases} f(P) & \text{if } f(P) \geqq 0, \\ 0 & \text{if } f(P) < 0, \end{cases} \qquad f_2^o(P) = \begin{cases} 0 & \text{if } f(P) \geqq 0, \\ -f(P) & \text{if } f(P) < 0. \end{cases}$$

THEOREM 8.13: *A necessary and sufficient condition that a real-valued function $f(P)$ be measurable (summable) is that $f_1^o(P)$ and $f_2^o(P)$ be measurable (summable).*

Proof: That the condition is sufficient is evident. To prove that the condition of measurability is necessary, it is necessary to show that $S_P[f_1^o(P) > a]$ and $S_P[f_2^o(P) > a]$ are measurable for every real <u>a</u>. Since f(P) is measurable, the functions $f_1(P)$ and $f_2(P)$ of Definition 8.4 exist and $f_1^o(P) = \max\, [f_1(P) - f_2(P), 0]$. $S_P[f_1^o(P) > a]$ is obviously R_n if $a < 0$, and therefore it is measurable. If $a \geqq 0$, the condition $\max[f_1(P) - f_2(P), 0] > a$ is equivalent to the condition $f_1(P) - f_2(P) > a$, that is, to the condition $f_1(P) > f_2(P) + a$. But $S_P[f_1(P) > f_2(P) + a] = \sum_{\rho \text{ rational}} S_P[f_1(P) > \rho > f_2(P) + a]$, and $S_P[f_1(P) > \rho > f_2(P) + a] = S_P[f_1(P) > \rho] \cdot S_P[f_2(P) < \rho - a]$. Since the two sets in this product are measurable, $f_1^o(P)$ is measurable; $f_2^o(P)$ may be shown to be measurable in a similar way. That the condition of summability is necessary follows from the fact that $f_1^o(P) \leqq f_1(P)$ and $f_2^o(P) \leqq f_2(P)$.

THEOREM 8.14: <u>A necessary and sufficient condition that a real-valued function</u> f(P) <u>be measurable is that the sets</u> $S_P[f(P) > a]$ <u>be measurable for an everywhere dense set of values of</u> a. (Cf. Theorem 8.3).

Proof: The fact that $S_P[f(P) > a]$ is measurable for all real <u>a</u> if measurable for an everywhere dense set of values of <u>a</u> follows as in the proof of Theorem 8.3. Since $S_P[f_1^o(P) > a]$ is R_n if $a < 0$ and $S_P[f_1^o(P) > a] = S_P[f(P) > a]$ if $a \geqq 0$, and since similar relations hold for $f_2^o(P)$, it follows by Theorems 8.3 and 8.12 that the condition is sufficient. Since $S_P[f(P) > a] = S_P[f_1^o(P) > a]$ if $a \geqq 0$ and $S_P[f(P) > a] = S_P[f_2^o(P) < -a]$ if $a < 0$, it follows by Theorem 8.12 and 8.2 that the condition is necessary.

<u>Corollary: In Theorem 8.13 the condition</u> $f(P) > a$ <u>may be replaced by any of the conditions</u> $\geqq a$, $< a$, $\leqq a$.

Proof: This may be proved in the same way as the Corollary of Theorem 8.3.

THEOREM 8.15: *A necessary and sufficient condition that the measurable function* $f(P)$ *be summable is that* $|f(P)|$ *be summable.*

Proof: This follows immediately from the resolution $|f(P)| = f_1^o(P) + f_2^o(P)$.

It is readily seen that Postulates 1-5 are satisfied by the integral of Definition 8.4 and that Theorem 8.12 generalizes to sequences of real-valued functions when the condition $f_n(P) \leqq g(P)$ is replaced by the condition $|f_n(P)| \leqq g(P)$.

Definition 8.5: *If* $f(P) = g(P) + ih(P)$, *where* $g(P)$ *and* $h(P)$ *are real-valued functions, then* $\int f(P)\, dv_P$ *is taken to be* $\int g(P)\, dv_P + i \int h(P)\, dv_P$. *Thus* $f(P)$ *is measurable (summable) if* $g(P)$ *and* $h(P)$ *are both measurable (summable).*

Let $f_M(P) = \begin{cases} f(P) & \text{for } P \in M, \\ 0 & \text{for } P \in -M, \end{cases}$ where $f(P)$ is any complex function and M is any measurable point set.

Definition 8.6: $\int_M f(P)\, dv_P$ *is taken to be* $\int f_M(P)\, dv_P$, *where* $f_M(P)$ *is as defined in the preceding paragraph and* $\int f_M(P)\, dv_P$ *is the integral of Definition 8.5.*

THEOREM 8.16: $\int_M f(P)\, dv_P$ *has the following properties, where* M *is a measurable point set,* $f(P)$ *is a measurable complex function, and* $\int_M f(P)\, dv_P$ *is the integral of Definition 8.6:*

1) $\int_M c\, f(P)\, dv_P = c \int_M f(P)\, dv_P$, *where* c *is any constant.*

2) $\int_M [f(P) + g(P)]\, dv_P = \int_M f(P)\, dv_P + \int_M g(P)\, dv_P$.

3) $\int_M 1\, dv_P = \mu(M)$.

4) $\int_M f(P)\, dv_P \geqq 0$ if $f(P)$ *is real and* $\geqq 0$ *for all* P *in* M.

5) $\int_{T^{-1}(M)} f(TP)\, dv_P = \frac{1}{|D|} \int_M f(P)\, dv_P$, where T is any linear transformation of determinant D.

6) If $f(P) = g(P) + ih(P)$, $f(P)$ is measurable when and only when $S_P[g(P) > a]$ and $S_P[h(P) > a]$ are measurable for all real a, and $f(P)$ is summable when and only when $|f(P)|$ is summable.

7) $\int_M f(P)\, dv_P + \int_N f(P)\, dv_P = \int_{M+N} f(P)\, dv_P$ if $MN = 0$.

8) If $f(P)$ is measurable and if $M_i M_j = 0$ for $i \neq j$, then the following statements hold: $f(P)$ is summable over $\sum M_i$ when and only when it is summable over each M_i and $\sum \int_{M_i} |f(P)|\, dv_P < \infty$, and in this case we have $\int_{\sum M_i} f(P)\, dv_P = \sum \int_{M_i} f(P)\, dv_P$.

9) If $f_1(P), f_2(P), \ldots$ is a sequence of measurable complex functions which approach a limit function $f(P)$ over a measurable set M except for a set of measure zero, and if there exists a real summable function $g(P)$ such that $|f_n(P)| \leq g(P)$ over M for each n except for a set of measure zero, then $f(P)$ is measurable and $\lim_{n\to\infty} \int_M f_n(P) dv_P = \int_M f(P)\, dv_P$.

Proof: The proofs of all parts of this theorem are apparent.

Definition 8.7a: The set of continuous functions is called the class C_1 of Baire functions; any function $f(P)$ which is the limit of a sequence of functions $f_1(P), f_2(P), \ldots$, where each $f_n(P)$ is the class C_k of Baire functions, is in the class C_{k+1} of Baire functions provided it is not in any class C_i, $1 \leq i \leq k$.

Definition 8.7b: Let I be an interval and let $1_I(P) = \begin{cases} 1 \text{ if } P \in I, \\ 0 \text{ if } P \in -I. \end{cases}$

The set of functions $a_1 1_{I_1}(P) + \ldots + a_k 1_{I_k}(P)$, where each a_i and I_i is rational, is called the class C_1 of Baire functions; succeeding classes of Baire functions arise from C_1 as in Definition 8.7a.

Definition 8.7c: The set of polynomials $\sum a_{i\ldots k} x_1^i \ldots x_n^k$ is called the class C_1 of Baire functions; succeeding classes of Baire functions arise from C_1 as in Definition 8.7a.

These three definitions are equivalent in the sense that they all lead to the same set of functions contained in the sum of all the classes C_i. This may be seen by noting that the functions of class C_1 of each of our Definitions 8.7a,b, c belong to some class C_m of the next Definition (b,c,a, respectively) and therefore each C_p of these Definitions belongs to some C_q of the next one also. To prove statement concerning C_1, consider a, b, and c in turn.

A continuous function f(P) is the limit of a suitably chosen sequence of functions of the form $a_1 1_{I_1}(P) + \ldots + a_K 1_{I_K}(P)$. Let K_N be the cube $-N \leq x_i \leq N$, $i = 1, \ldots, n$. It is sufficient to find such a function that $|a_1 1_{I_1}(P) + \ldots + a_K 1_{I_K}(P) - f(P)| \leq \frac{1}{N}$ over K_N. Finding such a function for each $N = 1, 2, \ldots$ leads to the desired result. As f(P) is uniformly continuous in K_N, this can be done by a simple construction which may be left to the reader. Thus C_1 of 8.7a is part of C_2 of 8.7b.

In order to see that every function $a_1 1_{I_1}(P) + \ldots + a_K 1_{I_K}(P)$ is obtainable by taking successive limits of sequences of polynomials, it is sufficient to prove this for a single interval function $1_I(P)$, where I is the interval $a_i < x_i < b_i$, $i = 1, \ldots, n$. Now $1_I(P)$ is obviously the limit of the sequence of functions $g_N(x_1, \ldots, x_n) = f(N(a_1 - x_1)(x_1 - b_1)) \cdot \ldots \cdot f(N(a_n - x_n)(x_n - b_n))$, where $f(u) = \frac{e^u(e^u - 1)}{(e^u + 1)^2}$. (This follows from the fact that $(a_i - x_i)(x_i - b_i) \gtreqless 0$ according as $x_i \begin{cases} > a_i \text{ or } < b_i, \\ = a_i \text{ or } = b_i, \\ < a_i \text{ or } > b_i, \end{cases}$ respectively, thus

$\lim_{N\to\infty} N(a_i - x_i)(x_i - b_i) = \begin{Bmatrix} +\infty \\ 0 \\ -\infty \end{Bmatrix}$ in these cases respectively, and $\lim f(u) = \begin{Bmatrix} 1 \\ 0 \\ 0 \end{Bmatrix}$ according as $\lim u = \begin{Bmatrix} +\infty \\ 0 \\ -\infty \end{Bmatrix}$.) Each $g_N(x_1, \ldots, x_n)$ is analytical in $x_1, \ldots, x_n$ and therefore the limit of a sequence of polynomials. Thus C_1 of 8.7b is part of C_3 of 8.7c (one could even replace C_3 by C_2, but this is unimportant to us).

Finally it is obvious that C_1 of 8.7c is part of C_1 of 8.7a.

It ought to be mentioned that the classes C_m of Baire functions can be continued beyond the finite numbers m = 1, 2, ... to all elements of the so-called "Cantor's second class of ordinal numbers", but it is not desirable to go into the details of this problem here.

<u>THEOREM 8.17</u>: <u>If</u> $f(x_1, \ldots, x_m)$ <u>and</u> $g_1(x_1, \ldots, x_n), \ldots, g_m(x_1, \ldots, x_n)$ <u>are Baire functions, then</u> $h(x_1, \ldots, x_n) = f(g_1(x_1, \ldots, x_n), \ldots, g_m(x_1, \ldots, x_n))$ <u>is also a Baire function.</u>

Proof: Consider Definition 8.7a. Assume first that f belongs to C_1, i.e., that it is continuous. Then the theorem is obvious for $g_1, \ldots, g_m$ in C_1, i.e., for continuous functions, and it follows by induction for $g_1, \ldots, g_m$ in any other classes C_μ. Thus it is proved for f in C_1 and arbitrary $g_1, \ldots, g_m$. An obvious induction extends it to f in any class C_ν and arbitrary $g_1, \ldots, g_m$.

Hence it is possible to choose any continuous function for $f(x_1, \ldots, x_n)$, for instance, $\max(x_1, \ldots, x_n)$, $\min(x_1, \ldots, x_n)$ or any polynomial; or again the Baire function $f_N(x) = \begin{cases} x & \text{if } |K| \leq N, \\ 0 & \text{otherwise,} \end{cases}$ thus obtaining from a Baire function $g(x_1, \ldots, x_n)$ another Baire function $g_N(x_1, \ldots, x_n) =$

$$= f_N(g(x_1, \ldots, x_n)) = \begin{cases} g(x_1, \ldots, x_n) & \text{if } |g(x_1, \ldots, x_n)| \leq N, \\ 0 & \text{otherwise.} \end{cases}$$

<u>THEOREM 8.18</u>: <u>If for the real Baire functions</u> $f_1(P), f_2(P), \ldots,$

$\lim_{m\to\infty} \left\{\begin{matrix}\sup \\ \inf\end{matrix}\right\} f_m(P)$ exists everywhere, then this limit is a Baire function. This limit certainly exists if $f_1(P)$, $f_2(P)$, ... are uniformly bounded $\left\{\begin{matrix}\text{above} \\ \text{below}\end{matrix}\right\}$.

Proof: It is sufficient to consider $\liminf_{m\to\infty} f_m(P)$. This limit can be expressed by using only the operations "min" (for a finite number of functions) and "lim" (for everywhere convergent sequences) as can be seen in the beginning of the proof of Theorem 8.7. This, together with the remarks preceding the theorem, completes the proof.

THEOREM 8.19: If $f_1(P)$, $f_2(P)$, ... is a sequence of Baire functions, there exists another Baire function $f(P)$ such that, whenever $\lim_{m\to\infty} f_m(P)$ exists, this limit is $f(P)$.

Proof: It may be assumed that $f_1(P)$, $f_2(P)$, ... are all real, as otherwise the real and imaginary parts could be considered separately. Let

$$f_{Nm}(P) = \begin{cases} f_m(P) & \text{if } |f_m(P)| \leqq N, \\ 0 & \text{otherwise.}\end{cases}$$

If $f_N(P) = \liminf_{m\to\infty} f_{Nm}(P)$, and if $f(P) = \lim_{N\to\infty} f_N(P)$, then $f(P)$ meets the requirements of the theorem.

After these general theorems on Baire functions, it is desirable to investigate the connection between measurable and Baire functions.

THEOREM 8.20: Every Baire function $f(P)$ is measurable.

Proof: This is obvious for the class C_1 (using Definition 8.7b), and it follows by induction (by Theorem 8.7) for all classes C_m.

THEOREM 8.21: A function $f(P)$ is measurable when and only when it is everywhere equal to a Baire function except over a set of points P of measure zero.

Proof: That the condition is sufficient for the measurability of $f(P)$ follows from Theorem 8.20. Therefore only the necessity of the condition must be proved. Suppose that f(P) is measurable. f(P) may be assumed real, as otherwise the real and imaginary parts could be considered separately. It may even be assumed non-negative, as otherwise it is the difference of two such functions ($f_1^o(P)$ and $f_2^o(P)$, Definition 8.4). By Definition 8.3 and Theorem 8.9, f(P) is the limit of a sequence of finitely valued functions, so that, by Theorem 8.19, it is sufficient to assume that f(P) is of the form $v_1 1_{M_1}(P) + \dots + v_n 1_{M_n}(P)$, where $1_M(P) = \begin{cases} 1 \text{ if } P \in M, \\ 0 \text{ if } P \in -M, \end{cases}$ and where each set M_i is measurable and of finite measure. But such a function is a Baire function if any function $f(P) = 1_M(P)$, where M is of finite measure, is a Baire function. By Theorem 3.17, M is a Borel set of type C_σ except for a set of measure zero, so that it is sufficient to consider the case where M is a set of the type C_σ. But, in this event, f(P) is obviously a Baire function.

Definition 8.8: *A sequence $f_1(P)$, $f_2(P)$, ... of functions is said to approach a limit function f(P) uniformly if, corresponding to any positive number δ, there exists an integer N such that $|f_n(P) - f(P)| \leq \delta$ for all P and for all $n \geq N$. A sequence of functions is said to approach a limit function essentially uniformly if, for every $\varepsilon > 0$, there exists a set of measure $< \varepsilon$ such that over its complementary set the approach is uniform.*

THEOREM 8.22: *If $f_1(P)$, $f_2(P)$, ... is a sequence S of functions summable over a point set M of finite measure, and if S has a limit function f(P), then the approach of S to f(P) is essentially uniform.*

Proof: Corresponding to any $\delta > 0$ let $N_{n,\delta}$ be the set of points P such that $|f_n(P) - f(P)| > \delta$. It follows that $\prod_{n=n_0}^{\infty} \sum_{\ell=0}^{\infty} MN_{n_0+\ell,\delta} = 0$ since, for any fixed point P, $|f_n(P) - f(P)| < \delta$ for all sufficiently large n. But $\sum_{\ell=0}^{\infty} MN_{n_0+\ell,\delta} \supset \sum_{\ell=0}^{\infty} MN_{n_0+1+\ell,\delta} \supset \dots$. Each $N_{n,\delta}$ is measurable since each $f_i(P)$ is measurable. Hence, by the Corollary of Lemma 2 of Chapter III, for $\delta > 0$ and $\eta > 0$ there exists a value n' of n such that $\mu[\sum_{\ell=0}^{\infty} MN_{n'+\ell,\delta}] \leq \eta$. Let δ have the sequence of values 1, 1/2, 1/3, ... and let η have the values $\frac{\varepsilon}{2}, \frac{\varepsilon}{2^2}, \frac{\varepsilon}{2^3}, \ldots$; if n_k is the integer n' corresponding to $\delta = \frac{1}{k}$ and $\eta = \frac{1}{2^k}$, then $\mu[\sum_{\ell=0}^{\infty} MN_{n_k+\ell,\frac{1}{k}}] \leq \frac{\varepsilon}{2^k}$ and $\mu[\sum_{k=1}^{\infty} \sum_{\ell=0}^{\infty} MN_{n_k+\ell,\frac{1}{k}}] = \mu(E) \leq \varepsilon$. Hence the sequence $f_i(P)$ approaches f(P) uniformly over the point set M - E.

THEOREM 8.23: If f(P) is defined and measurable over a set M, then it is possible to exclude from M a set M_ε of arbitrarily small positive measure ε so that, when f(P) is defined over only $M-M_\varepsilon$, f(P) is everywhere continuous over $M-M_\varepsilon$.

Proof: Suppose that M is of finite measure. By Theorem 8.21 it is sufficient to consider the case where f(P) is a Baire function of some class C_m. If f(P) is of class C_1, then the theorem is apparent. Suppose the theorem proved for Baire functions of class C_k. If f(P) is of class C_{k+1}, then f(P) is the limit of a sequence S of functions $f_i(P)$ of class C_k. By hypothesis, $f_i(P)$ is continuous over $M-M_i$, where $\mu(M_i) \leq \frac{\varepsilon}{2^{i+1}}$. By Theorem 8.22, the functions $f_i(P)$ approach f(P) uniformly over M-E, where $\mu(E) \leq \frac{\varepsilon}{2}$. Hence the functions $f_i(P)$ are continuous over the set M-T, where $T = E + \sum_{i=1}^{\infty} M_i$ and where $\mu(T) \leq \varepsilon$, and S approaches f(P) uniformly over M-T. Thus f(P) is continuous over M-T. If M is of infinite measure, the proof is obtained by considering the part M^N of M lying in the cube $-N \leq x_i \leq N$, i = 1, ..., n, constructing the exceptional set M_i^N for N = 1, 2, ..., where $\mu(M_i^N) \leq \frac{\varepsilon}{2^{i+1}2^N}$, putting $M_i = M_i^1 + M_i^2 + \ldots$, and replacing E by $\sum_{N=1}^{\infty} E^N$, where $\mu(E^N) \leq \frac{\varepsilon}{2^{N+1}}$.

It is desirable to exhibit a measurable function f(P) such that ε in the preceding theorem cannot be zero. Let f(P) be $I_M(P)$, where M is a measurable set defined as follows: let all rational open intervals $a < x < b$ be ordered into a sequence $I_1, I_2, \ldots$. Define a sequence of intervals $J_1, J_2, \ldots$ in such a manner that $\mu(J_\lambda) \leqq \frac{1}{3}\mu(J_{\lambda-1})$ for every λ and such that $J_{2\nu-1}$ and $J_{2\nu}$ are in I_ν for every ν. Consider those points x which belong to only a finite positive number of the intervals J_λ. For each such point x there is a last interval J_λ which contains x and its index $\lambda = \lambda(x)$ is a function of x. Let M be the set of all such points x for which λ is even, and let N be the set of all such points x for which λ is odd. Let $K_\lambda = J_\lambda - J_\lambda(\sum_{i=1}^{\infty} J_{\lambda+i})$. Then $\mu(K_\lambda) \geqq \mu(J_\lambda) - \mu[J_\lambda(\sum_{i=1}^{\infty} J_{\lambda+i})] \geqq \mu(J_\lambda) - \sum_{i=1}^{\infty}\mu(J_{\lambda+i}) \geqq$

$\geqq \mu(J_\lambda)(1 - \frac{1}{3} - \frac{1}{9} - \ldots) = \frac{1}{2}\mu(J_\lambda)$. Since $M = K_2 + K_4 + \ldots$ and $N = K_1 + K_3 + \ldots$ it follows that M and N are measurable. Furthermore $I_\nu \supset K_{2\nu}$ and $I_\nu \supset K_{2\nu-1}$, so that in every interval there is a part of M and a part of N and each such part is measurable and of positive measure. Hence in every interval f(P) assumes both of the values 0 and 1 over sets of positive measure, and ε must be positive.

Definition 8.9: If for each $\varepsilon > 0$ and each θ, $0 < \theta < 1$, there exists a $\delta > 0$ such that, for every cube C: $x_i^{(o)} - \frac{a}{2} < x_i < x_i^{(o)} + \frac{a}{2}$ $(i = 1, \ldots, n)$ with edge $a < \delta$, the set of points P in C for which $|f(P^o) - f(P)| \leqq \varepsilon$ is of measure $\geqq \theta a^n$, then f(P) is called approximately continuous at P^o.

THEOREM 8.24. If f(P) is defined and measurable over a set M, then f(P) is approximately continuous over M except for at most a set of measure zero.

Proof: Let R_2 be the complex plane so that, for any point P, $f(P) \in R_2$. Let all rational intervals in R_2 be ordered in a sequence $I_1, I_2, \ldots$. In the notation of Theorem 8.3, each set $S^\nu = S_P[f(P) \in I_\nu]$ is measurable. By Theorem 6.5, for every point P, except for a set Z of measure zero, it follows that all S^ν containing P have unit density at P. Let P be a fixed point in

R_n but not in Z. Let C be the circle in R_2 with center f(P) and radius ε. Inside C there exists a square I_ν containing f(P). Since the corresponding set S^ν is of unit density, the condition for approximate continuity at P is satisfied.

CHAPTER IX.

MONOTONIC FUNCTIONS

Definition 9.1: In the space R_1 of real numbers, let $f(x)$ be real and defined over any finite or infinite interval. $f(x)$ is called

increasing if $f(b) > f(a)$ whenever $b > a$,

decreasing if $f(b) < f(a)$ whenever $b > a$,

monotonically increasing if $f(b) \geqq f(a)$ whenever $b > a$,

monotonically decreasing if $f(b) \leqq f(a)$ whenever $b > a$.

It is apparent that the third class includes the first and that the fourth includes the second; if $f(x)$ is in the fourth class, $-f(x)$ is in the third. Hence it is sufficient to consider only monotonically increasing (m.i.) functions in developing the properties of these various types of functions.

Definition 9.2: If $f(x)$ is m.i., then the least upper bound of $f(x)$ for $x < x_o$ is denoted by $f_-(x_o)$ and the greatest lower bound of $f(x)$ for $x > x_o$ is denoted by $f_+(x_o)$.

THEOREM 9.1: If $f(x)$ is m.i., then $\lim_{x \underset{<}{\to} x_o} f(x) = f_-(x_o)$ and $\lim_{x \underset{>}{\to} x_o} f(x) = f_+(x_o)$.

Proof: There exists a number $x_1 < x_o$ such that $f(x_1) > f_-(x_o) - \epsilon$. Hence, for all x such that $x_1 < x < x_o$, $f_-(x_o) - \epsilon < f(x) \leqq f_-(x_o)$, and the first part of the theorem is immediate; the second part may be treated analogously.

THEOREM 9.2: If $f(x)$ is m.i. and if $x_o < x_1$, then $f_+(x_o) \leqq f_-(x_1)$.

Proof: Let x_2 be such that $x_o < x_2 < x_1$. Then, by Definition 9.2, $f_+(x_o) \leqq f(x_2) \leqq f_-(x_1)$.

THEOREM 9.3: *If $f(x)$ is m.i., then $f_-(x_0) \leq f(x_0) \leq f_+(x_0)$.*

The proof is trivial.

Definition 9.3: *If $f(x)$ is m.i., the difference $f_+(x_0) - f_-(x_0)$ is called the oscillation, osc $f(x_0)$, of $f(x)$ at x_0.*

It follows from Theorem 9.3 that osc $f(x_0) \geq 0$. It is apparent that osc $f(x_0) = 0$ when and only when $f(x)$ is continuous at x_0.

THEOREM 9.4: *If $f(x)$ is m.i., then the number of points at which it is discontinuous is at most countable.*

Proof: By Theorem 9.2, for $x_0 \neq x_1$ the intervals $f_-(x_0) \leq y \leq f_+(x_0)$ and $f_-(x_1) \leq y \leq f_+(x_1)$ are non-overlapping except possibly for a common end point. The interval $-\infty < y < +\infty$ can be divided up into only a countable number of non-overlapping intervals of length $\geq \varepsilon > 0$. Hence the number of points at which osc $f(x) \geq \varepsilon > 0$ is at most countable. If ε is given the values 1/2, 1/3, 1/4, ... , the oscillation of $f(x)$ at each point of discontinuity is greater than almost all of these values of ε. Hence the number of points of discontinuity is at most countable.

THEOREM 9.5: *If $f(x)$ is m.i., the set of points of discontinuity of $f(x)$ may be everywhere dense over R_1.*

Proof: By Theorems 1.8 and 1.9 the set of rational points in R_1 is everywhere dense in R_1 and may be ordered in a sequence S: $x_1, x_2, \ldots$. Let x be a point of R_1 and let $x_{n_1}, x_{n_2}, \ldots$ be the elements of S less than x. Then $f(x) = \frac{1}{2^{n_1}} + \frac{1}{2^{n_2}} + \ldots$ is m.i. and is discontinuous at each rational point of R_1.

Definition 9.4: *If $f(x)$ and $g(x)$ are m.i, and if the conditions $f_-(x_0) = g_-(x_0)$ and $f_+(x_0) = g_+(x_0)$ hold at every point x_0, then $f(x)$ and $g(x)$ are called equivalent.*

THEOREM 9.6: *If $f(x)$ and $g(x)$ are m.i., a necessary and sufficient condition that f and g be equivalent is that* $\left\{\begin{matrix} f(x_1) \\ \text{and} \\ g(x_1) \end{matrix}\right\} \leq \left\{\begin{matrix} f(x_2) \\ \text{and} \\ g(x_2) \end{matrix}\right\}$ *whenever* $x_1 < x_2$.

Proof: The condition is necessary, for $f(x_1) \leq f_-(x_2) = g_-(x_2) \leq g(x_2)$. Similarly, $g(x_1) \leq g_-(x_2) = f_-(x_2) \leq f(x_2)$. By Definition 9.1, $f(x_1) \leq f(x_2)$ and $g(x_1) \leq g(x_2)$. The condition is sufficient, for let x_o be any real number. For all $x < x_o$, $f(x) \leq g(x') \leq g_-(x_o)$ where $x < x' < x_o$. Hence $f_-(x_o) \leq g_-(x_o)$. By a similar argument $g_-(x_o) \leq f_-(x_o)$, so that $f_-(x_o) =$ $= g_-(x_o)$. It may be shown in the same way that $f_+(x_o) = g_+(x_o)$.

The preceding proof shows by its construction that either one of the conditions $f_-(x_o) = g_-(x_o)$ and $f_+(x_o) = g_+(x_o)$ in Definition 9.4 is sufficient for the equivalence of $f(x)$ and $g(x)$.

THEOREM 9.7: *If the m.i. functions $f(x)$ and $g(x)$ are equivalent, then $f_-(x_o) \leq g(x_o) \leq f_+(x_o)$ and $g_-(x_o) \leq f(x_o) \leq g_+(x_o)$ for all x_o; conversely, if either of these conditions obtains, then f and g are equivalent.*

Proof: The proof is apparent.

It follows from this theorem that two equivalent functions $f(x)$ and $g(x)$ have the same points of continuity and are equal at each point of continuity.

THEOREM 9.8: *A necessary and sufficient condition that the m.i. functions $f(x)$ and $g(x)$ be equivalent is that they be equal over an everywhere dense set of real numbers.*

Proof: The proof is apparent, as is also the following

Corollary: *If $f(x)$ and $g(x)$ are equivalent, there exists a number x_2 between any two given numbers x_o and x_1 such that $f(x_2) = g(x_2)$.*

It is readily seen that $f_-(x)$, $f_+(x)$, and $\dfrac{f_-(x)+f_+(x)}{2}$ are equivalent to $f(x)$; also that $f_-(x)$ is identical with $f_{--}(x)$ and that $f_+(x)$ is identical with $f_{++}(x)$; also that if $f(x)$ is identical with $f_-(x)$, $f(x)$ is left continuous, and that if $f(x)$ is identical with $f_+(x)$, $f(x)$ is right continuous.

The sum of two m.i. functions is m.i., but the difference between two m.i. functions may not be m.i.

Definition 9.5: If $f(x)$ is any real-valued function defined over the interval $a \leq x \leq b$, and if $x_1, \ldots, x_n$ are any n points in this interval such that $a = x_1 \leq x_2 \leq \ldots \leq x_n = b$, then the least upper bound of $\sum_{i=1}^{n} |f(x_i) - f(x_{i-1})|$ is called the variation, $\text{var}_{ab}f(x)$, of $f(x)$ in this interval. If $\text{var}_{ab}f(x)$ is finite, $f(x)$ is said to be of bounded variation (b.v.) in this interval.

THEOREM 9.9: If $f(x)$ and $g(x)$ are functions of b.v. over an interval $a \leq x \leq b$, then 1) $cf(x)$ is of b.v., where c is a constant, 2) $f(x) \pm g(x)$ and $f(x)g(x)$ are of b.v., 3) $f(x)$ is of b.v. over any interval $a' \leq x \leq b'$, where $a \leq a' \leq b' \leq b$, and $\text{var}_{a'b'}f(x) \leq \text{var}_{ab}f(x)$, 4) if $f(x)$ is of b.v. also over the interval $b \leq x \leq c$, then $f(x)$ is of b.v. over the interval $a \leq x \leq c$, and $\text{var}_{ac}f(x) = \text{var}_{ab}f(x) + \text{var}_{bc}f(x)$.

THEOREM 9.10: Every function with a continuous bounded derivative is of b.v., every monotonic function is of b.v., and the difference between two monotonic functions is of b.v.

Proofs: The proofs of these two theorems are apparent.

If $f(x)$ fails to have a right-hand limit at any point P, it is not of b.v. in any interval containing P. The function $f(x) = \begin{cases} x \sin \frac{1}{x}, & x \neq 0, \\ 0, & x = 0 \end{cases}$ is of unbounded variation near the origin even though it is continuous.

Definition 9.6: If $f(x)$ is defined over $a \leq x \leq b$ and if $x_1, \ldots, x_n$ are any n points in this interval such that $a = x_1 \leq x_2 \leq \ldots \leq x_n = b$, then

$$\text{var}^+_{ab}f(x) = \text{l.u.b.} \sum_{i=1}^{n} \max\{[f(x_i) - f(x_{i-1})], 0\}$$

and

$$\text{var}^-_{ab}f(x) = \text{l.u.b.} \sum_{i=1}^{n} \max\{[-f(x_i) + f(x_{i-1})], 0\}.$$

THEOREM 9.11: $\text{var}^+_{ab}f(x) + \text{var}^-_{ab}f(x) = \text{var}_{ab}f(x)$.

Proof: It is apparent that the left member of this equation is not less than the right member since, for any particular set of x's,

$$\sum_{i=1}^{n} \max\{[f(x_i)-f(x_{i-1})], 0\} + \sum_{i=1}^{n} \max\{[-f(x_i)+f(x_{i-1})], 0\} = \sum_{i=1}^{n} |f(x_i)-f(x_{i-1})|$$

and since $\text{var}^+_{ab}f(x) + \text{var}^-_{ab}f(x) \geqq$

$\geqq \text{l.u.b.}[\sum_{i=1}^{n} \max\{[f(x_i)-f(x_{i-1})], 0\} + \sum_{i=1}^{n} \max\{[-f(x_i)+f(x_{i-1})], 0\}]$. (The inequality is necessary inasmuch as the same set of x's is used in the two sums, whereas $\text{var}^+_{ab}f(x)$ and $\text{var}^-_{ab}f(x)$ must be approximated independently.) On the other hand, there exists a set S_1 of x's, $a = x_1^{(1)} \geqq \dots \geqq x_m^{(1)} = b$, such that $\sum_{i=1}^{m} \max\{[f(x_i^{(1)}) - f(x_{i-1}^{(1)})], 0\} > \text{var}^+_{ab}f(x) - \frac{\varepsilon}{2}$, and the left member of this inequality is not decreased if more x's are added to S_1. Likewise there exists a set S_2 of x's, $a = x_1^{(2)} \geqq \dots \geqq x_n^{(2)} = b$, such that $\sum_{i=1}^{n} \max\{[-f(x_i^{(2)}) + f(x_{i-1}^{(2)}), 0\} > \text{var}^-_{ab}f(x) - \frac{\varepsilon}{2}$, and the left member of this inequality is not decreased if more x's are added to S_2. Let S_1 and S_2 be combined into a single set S whose elements (after proper reordering) may be denoted by $x_1, \dots, x_k$. Then $\text{var}_{ab}f(x) \geqq \sum_{i=1}^{k} |f(x_i) - f(x_{i-1})| =$

$= \sum_{i=1}^{k} \max\{[f(x_i) - f(x_{i-1})], 0\} + \sum_{i=1}^{k} \max\{[-f(x_i) + f(x_{i-1})], 0\} > \text{var}^+_{ab}f(x) +$

$+ \text{var}^-_{ab}f(x) - \varepsilon$. Hence $\text{var}_{ab}f(x) \leqq \text{var}^+_{ab}f(x) + \text{var}^-_{ab}f(x)$ and the proof is completed.

THEOREM 9.12: $\mathrm{var}^{+}_{ab} f(x) - \mathrm{var}^{-}_{ab} f(x) = f(b) - f(a)$.

Proof: Let S_1, S_2, and S be the sets of x's occurring in the proof of the preceding theorem. Then

$$0 \leqq \left\{ \begin{array}{l} \mathrm{var}^{+}_{ab} f(x) - \sum_{i=1}^{k} \max\{[f(x_i) - f(x_{i-1})],\ 0\} \\ \\ \mathrm{var}^{-}_{ab} f(x) - \sum_{i=1}^{k} \max\{[-f(x_i) + f(x_{i-1})],\ 0\} \end{array} \right\} \leqq \frac{\varepsilon}{2} .$$

Hence

$$\left| \mathrm{var}^{+}_{ab} f(x) - \mathrm{var}^{-}_{ab} f(x) - \sum_{i=1}^{k} \{f(x_i) - f(x_{i-1})\} \right| =$$

$$= \left| \mathrm{var}^{+}_{ab} f(x) - \mathrm{var}^{-}_{ab} f(x) - \{f(b) - f(a)\} \right| \leqq \frac{\varepsilon}{2} ,$$

and the theorem is immediate.

THEOREM 9.13: *Every function $f(x)$ of b.v. over $a \leqq x \leqq b$ can be represented as the difference between two m.i. functions.*

Proof: Let $\bar{g}(x) = \mathrm{var}^{+}_{ax} f(x)$ and let $\bar{h}(x) = \mathrm{var}^{-}_{ax} f(x)$. It is obvious that $\bar{g}$ and $\bar{h}$ are m.i. By Theorem 9.12, $\bar{g}(x) - \bar{h}(x) = f(x) - f(a)$, so that $f(x) = [f(a) + \bar{g}(x)] - \bar{h}(x)$.

The resolution of $f(x)$ just given is of interest in that $g(a) \quad h(a) = 0$.

THEOREM 9.14: *If $f(x) = g(x) - h(x)$ is any resolution of a function of b.v. over $a \leqq x \leqq b$, where $g(x)$ and $h(x)$ are m.i., then, in the notation of the proof of the preceding theorem, $g(x) = f(a) + \bar{g}(x) + \varphi(x)$ and $h(x) = \bar{h}(x) + \varphi(x)$ where $\varphi(x)$ is m.i.*

Proof: Let $g(x) - \bar{g}(x) = f(a) + \varphi(x)$. Then $h(x) - \bar{h}(x) = \varphi(x)$. But if $a \leqq u \leqq v \leqq b$, then $\mathrm{var}^{+}_{uv} f(x) = \mathrm{var}^{+}_{uv}[g(x) - h(x)] \leqq \mathrm{var}^{+}_{uv} g(x) = g(v) - g(u)$ since $g(x)$ is m.i. Inasmuch as $\mathrm{var}^{+}_{uv} f(x) = \mathrm{var}^{+}_{av} f(x) - \mathrm{var}^{+}_{au} f(x)$, it follows that $g(v) - \mathrm{var}^{+}_{av} f(x) \geqq g(u) - \mathrm{var}^{+}_{au} f(x)$, that is, $g(v) - \bar{g}(v) \geqq g(u) - \bar{g}(u)$, so that $\varphi(v) \geqq \varphi(u)$ and $\varphi(x)$ is m.i.

Suppose $f(x)$ is m.i. over $a \leqq x \leqq b$. Then, for a fixed x_0, $f_-(x_0) - \sum_{x<x_0} \operatorname{osc} f(x) = f_+(x_0) - \sum_{x \leqq x_0} \operatorname{osc} f(x)$. Each term of the left member of this equation is left continuous and each term of the right member is right continuous. Hence the two members of the equation are continuous.

Definition 9.7: *If $f(x)$ is m.i., then $f_-(x) - \sum_{x'<x} \operatorname{osc} f(x')$ is called the continuous part, $f_c(x)$, of $f(x)$. The difference $f_d(x) = f(x) - f_c(x)$ is called the discontinuous part of $f(x)$.*

THEOREM 9.15: *$f_c(x)$ and $f_d(x)$ are m.i.*

Proof: Suppose that $x_2 > x_1$. Then $f_c(x)$ is m.i. since $f_c(x_2) - f_c(x_1) = [f_-(x_2) - f_-(x_1)] - \sum_{x_1 \leqq x' < x_2} \operatorname{osc} f(x')$ and in the right member of this equation the first term is positive and the second term is not greater than the first inasmuch as $\operatorname{osc} f(x) = \operatorname{osc} f_-(x)$. Again, $f_d(x)$ is m.i. since

$$f_d(x_2)-f_d(x_1) = [f(x_2)-f(x_1)] - [f_-(x_2) - f_-(x_1)] + \sum_{x_1 \leqq x' < x_2} \operatorname{osc} f(x') =$$

$$= [f(x_2)-f_-(x_2)] - [f(x_1)-f_-(x_1)] + \sum_{x_1 \leqq x' < x_2} \operatorname{osc} f(x') =$$

$$= [f(x_2)-f_-(x_2)] + [f_+(x_1)- f(x_1)] + \sum_{x_1 < x' < x_2} \operatorname{osc} f(x'),$$

and all the terms in this last expression are non-negative.

The following development is preparatory to proving Theorem 9.18, the well known theorem of Lebesgue on the differentiability of monotonic functions.

Definition 9.8: *$f(x)$ is said to be upper semi-continuous (u.s.c.) at $x = x_0$ if, corresponding to each $\varepsilon > 0$, there exists a $\delta > 0$ such that $f(x) \leqq f(x_0) + \varepsilon$ whenever $x_0 - \delta \leqq x \leqq x_0 + \delta$.*

THEOREM 9.16: If $f(x)$ is u.s.c. over the interval $a < x \leqq b$, if $\bar{x}$ is a point in this interval corresponding to which there exists another point x' in this interval such that $f(x') > f(\bar{x})$, and if $\emptyset$ is the set of all points $\bar{x}$ in this interval, then 1) $\emptyset$ is the sum of a sequence (finite or infinite) of non-overlapping open intervals I_n: $a_n < x < b_n$, 2) if x_o is any point of I_n, $f(x_o) \leqq f(b_n)$, 3) $\limsup_{x \to a_n,\, x > a_n} f(x) \leqq f(b_n)$.

Proof: Part 1) If $\bar{x} \in \emptyset$ and if $f(x') > f(\bar{x})$, then, since $f(x)$ is u.s.c., there exists an open interval I: $\bar{x} - \delta < x < \bar{x} + \delta$ such that every $x \in I$ satisfies the condition $f(x') > f(x)$. But I may be chosen so that it lies in the interval $a < x < b$, and so that $x < x'$ for all x in I. Hence $I \subset \emptyset$. As this holds for every $\bar{x} \in \emptyset$, $\emptyset$ is an open set. Therefore $\emptyset$ is the sum of a countable set of non-overlapping intervals.

Part 2) Let x_o be any point in the interval $a_n < x < b_n$. Let C be the set of all points $\bar{\bar{x}}$ in the interval $x_o \leqq \bar{\bar{x}} \leqq b_n$ for which $f(x_o) \leqq f(\bar{\bar{x}})$. ($C$ is not empty since $x_o \in C$.) C is closed (since $f(x)$ is u.s.c.), and therefore C contains a maximum x_1. If $x_1 = b_n$, then $f(x_o) \leqq f(b_n)$. If $x_1 < b_n$, then $x_1 \in \emptyset$, and there would exist an $x' > x_1$ such that $f(x') > f(x_1)$. The condition $x' \leqq b_n$ would imply that $x' \in C$, so that $x' < x_1$. Therefore $x' > b_n$. If $f(x')$ were $> f(b_n)$, b_n would be in $\emptyset$. Since it is not, $f(x_o) \leqq f(x_1) < f(x') \leqq f(b_n)$.

Part 3) follows from Part 2).

If $f(x)$ is m.i. over the interval $a \leqq x \leqq b$, it is obvious that $f_+(x)$ is u.s.c. over this interval. Hence $f_+(-x)$ is u.s.c. over the interval $-b \leqq x \leqq -a$, and therefore the functions $f_+(x) - Ax$ and $f_+(-x) + Ax$ are u.s.c., where A is an arbitrary constant. (The notation $f_+(-x)$ is ambiguous in that it makes a difference whether the operation "+" or "-" is applied first to $f(x)$.

It is to be understood here that the operation "+" is applied first.)

If $a_n < x < b_n$ is one of the intervals I_n of the preceding theorem when applied to the functions $f_+(x) - Ax$ or $f_+(-x) + Ax$ respectively, then Part 3) assumes the form $f_+(a_n) - Aa_n \leqq f_+(b_n) - Ab_n$ or $f_-(-a_n) + Aa_n \leqq f_+(-b_n) + Ab_n$; that is, $f_+(b_n) - f_+(a_n) \geqq A(b_n - a_n)$ or $f_-(-a_n) - f_+(-b_n) \leqq A(b_n - a_n)$. In the first case it should be remarked, however, that if $b_n = b$, then $f_+(b_n)$ may be replaced by $f_-(b_n)$ because $f(x)$ may be redefined for $x \geqq b$ as $f(x) = f_-(b) + \varepsilon$ without changing anything in the above discussion, where ε is any constant > 0, and where ε is then allowed to approach zero.

<u>THEOREM 9.17: In the preceding notation (with regard to the function $f_+(x) - Ax$ in the following parts 1) and 2), and with regard to the function $f_+(-x) + Ax$ in parts 3) and 4)),</u> 1) $b_n - a_n \leqq \frac{1}{A}[f_+(b_n) - f_+(a_n)]$ <u>(where $f_+(b_n)$ may be replaced by $f_-(b_n)$ if $b_n = b$),</u> 2) $\sum(b_n - a_n) \leqq \frac{1}{A}[f_-(b) - f_+(a)]$, 3) $f_-(-a_n) - f_+(-b_n) \leqq A(b_n - a_n)$, <u>and</u> 4) $\sum[f_-(-a_n) - f_+(-b_n)] \leqq A(b-a)$.

Proof: Parts 1) and 3) have been proved above, and parts 2) and 4) follow by summing parts 1) and 3) over $n = 1, \ldots, N$, and then allowing $N \to \infty$.

If $f(x)$ has a derivative at $x = x_0$, the derivative will be denoted by $f'(x_0)$.

<u>THEOREM 9.18: If $f(x)$ is m.i. over $a \leqq x \leqq b$, then $f'(x)$ exists and is finite over this interval except for a set of measure zero.</u>

Proof: Let $d(x) = \dfrac{f(x) - f(x_0)}{x - x_0}$, and let $D_r^+ f(x_0) = \limsup_{x \to x_0,\, x > x_0} d(x)$,

$D_r^- f(x_0) = \liminf_{x \to x_0,\, x > x_0} d(x)$, $D_\ell^+ f(x_0) = \limsup_{x \to x_0,\, x < x_0} d(x)$, and $D_\ell^- f(x_0) = \liminf_{x \to x_0,\, x < x_0} d(x)$.

It will be shown that 1) $D_r^- f(x) = \infty$ over a set of measure zero, 2) $D_\ell^- f(x) = \infty$ over a set of measure zero, 3) the condition $D_\ell^- f(x) < C < D < D_r^+ f(x)$ holds over a set of measure zero (where C and D are any two numbers such that $C < D$), and 4) the condition $D_r^- f(x) < C < D < D_\ell^+ f(x)$ holds over a set of measure zero.

Suppose that these four assertions have been proved. Let I_i: $c_i \leqq x \leqq d_i$ (where $c_i < d_i$ and $i = 1,2,\ldots$) be a sequence of all rational intervals in $0 \leqq x < \infty$. Then, except over a set of measure zero, all $D^{+}_{r,\ell} f(x)$ are finite and no relation $D_\ell^- f(x) < c_i < d_i < D_r^+ f(x)$ or $D_r^- f(x) < c_i < d_i < D_\ell^+ f(x)$ holds. Therefore $D_\ell^- f(x) \geqq D_r^+ f(x)$ and $D_r^- f(x) \geqq D_\ell^+ f(x)$. Since $D_r^+ f(x) \geqq D_r^- f(x)$ and $D_\ell^+ f(x) \geqq D_\ell^- f(x)$, it follows that $D_\ell^- f(x) = D_\ell^+ f(x) = D_r^- f(x) = D_r^+ f(x) < \infty$. At such a point x, f(x) is differentiable and the theorem is proved. Now assertions 1) and 3) become 2) and 4) respectively if f(x) is replaced by $-f(-x)$. Thus only 1) and 3) need be proved. As the points of discontinuity of f(x) form a set of measure zero (it is countable by Theorem 9.4), it is necessary to discuss only the points of continuity of f(x).

Proof of 1): Let C be a positive number. If $D_r^- f(x_0) > C$, then there exists a point $x' > x_0$ such that $\frac{f(x') - f(x_0)}{x' - x_0} > C$, so that $\frac{f_+(x') - f_+(x_0)}{x' - x_0} > C$. ($x_0$ being point of continuity); that is, $g(x') > g(x_0)$, where $g(x) = f_+(x) - Cx$. Hence Theorems 9.16 and 9.17 apply. x_0 is in the set $\emptyset$, that is, in one of the intervals $a_n < x < b_n$. By Theorem 9.17, part 2), the sum of all such intervals may be made arbitrarily small by taking C sufficiently large. This completes the proof of 1).

Proof of 3): Suppose $D_\ell^- f(x_0) < C < D < D_r^+ f(x_0)$. Then there exists a point $x' < x_0$ for which $\frac{f(x') - f(x_0)}{x' - x_0} < C$; therefore $\frac{f_+(x') - f_+(x_0)}{x' - x_0} < C$ (x_0 being a point of continuity) so that $f_+(x_0) - Cx_0 < f_+(x') - Cx'$. Let

$y_o = -x_o$ and let $y' = -x'$. Then $y' > y_o$ and $f_+(-y_o) + Cy_o > f_+(-y') + Cy'$. Again, there exists a point $x'' > x_o$ for which $\frac{f(x'') - f(x_o)}{x'' - x_o} > D$, so that $\frac{f_+(x'') - f_+(x_o)}{x'' - x_o} > D$ and $f_+(x_o) - Dx_o < f_+(x'') - Dx''$. (In the preceding discussion x' and x'' can be arbitrarily close to x_o.) By Theorem 9.17, part 4), where $A = C$, if $y_o = -x_o$ is regarded as lying in some interval I: $a' < y < b'$, then y_o lies in one of the intervals $a_n < y < b_n$ and $\sum_n [f_-(-a_n) - f_+(-b_n)] \leqq C(b' - a')$. Suppose y_o lies in the interval $a_{n'} < y < b_{n'}$. Then $-b_{n'} < x_o < -a_{n'}$. By Theorem 9.17, part 2), where $A = D$, it follows that x_o lies in one of the intervals $-b_{n'm} < x < -a_{n'm}$ and $\sum_m (-a_{n'm} + b_{n'm}) \leqq \frac{1}{D}[f_-(-a_{n'}) - f_+(-b_{n'})]$. Hence $\sum_{m,n} (b_{nm} - a_{nm}) \leqq \frac{C}{D}(b' - a')$ and every y_o in I lies in some interval $a_{nm} < y < b_{nm}$. Thus for every interval I the part of the set $S_x[D_\ell^- f(x) < C < D < D_r^+ f(x)]$ lying in I has measure $\leqq \frac{C}{D}\mu(I)$, where $\frac{C}{D}$ is a fixed number $\geqq 0$ and < 1. By Theorem 6.4, the set S has measure zero. (The use of Theorem 6.4 may be avoided by the following argument: if the measure of the set is μ, then it can be covered by intervals I of total length $< \mu + \varepsilon$. Hence the measure of S is $\leqq \frac{C}{D}(\mu + \varepsilon)$. If $\varepsilon \to 0$, then $\mu \leqq \frac{C}{D}\mu$ and $\mu = 0$.)

<u>Corollary: Every function of b.v. is everywhere differentiable except for a set of measure zero.</u>

Proof: This follows immediately from Theorem 9.13.

<u>Lemma: If $\varphi_1(x)$, $\varphi_2(x)$, ... is a sequence of m.i. functions approaching a limit function $\varphi(x)$ (which is necessarily m.i.) over the interval $a \leqq x \leqq b$, and if either $\left\{\begin{matrix}\varphi_n(x) - \varphi_{n+1}(x)\\ \varphi_{n+1}(x) - \varphi_n(x)\end{matrix}\right\}$ is m.i. for each n, then</u>

$\lim_{n\to\infty} \varphi_n'(x) = \varphi'(x)$ <u>for $a \leqq x \leqq b$ except for a set of measure zero.</u>

Proof: If $m > n$, $\left\{\begin{matrix}\varphi_n(x) - \varphi_m(x)\\ \varphi_m(x) - \varphi_n(x)\end{matrix}\right\}$ is m.i. and $\lim_{m\to\infty}\left\{\begin{matrix}\varphi_n(x) - \varphi_m(x)\\ \varphi_m(x) - \varphi_n(x)\end{matrix}\right\} =$ $= \left\{\begin{matrix}\varphi_n(x) - \varphi(x)\\ \varphi(x) - \varphi_n(x)\end{matrix}\right\} = \psi_n(x)$ is m.i. In either of these cases, $\lim_{n\to\infty} \psi_n(x) = 0$ and $\psi_n(x) - \psi_m(x)$ is m.i. for $m > n$. Hence it is sufficient to prove the theorem for the sequence $\psi_n(x)$. If there exists a sequence of values $n_1, n_2, \ldots$ of n for which the theorem holds, then the theorem holds for the sequence $\psi_n(x)$, for if $m > n$, $\dfrac{\psi_n(x) - \psi_n(y)}{x - y} \geqq \dfrac{\psi_m(x) - \psi_m(y)}{x - y} \geqq 0$ and $\psi_n'(y) \geqq \psi_m'(y) \geqq 0$.

There is no loss of generality in assuming that $\psi_n(a) = 0$ for all n. Since $\lim_{n\to\infty} \psi_n(b) = 0$, there exists a subsequence $\psi_{n_1}(x), \psi_{n_2}(x), \ldots$ such that $\sum_{i=1}^{\infty} \psi_{n_i}(b)$ is finite. Let this subsequence be denoted by $w_1(x), w_2(x), \ldots$.

Then $w_1(x) + \ldots + w_\nu(x) + \sum_{\mu=\nu+1}^{\infty} w_\mu(x) = \sum_{\mu=1}^{\infty} w_\mu(x)$. Since the right member is bounded, each term in this equation is differentiable over $a \leqq x \leqq b$ except for a set of measure zero. Then $\dfrac{w_1(x) - w_1(y)}{x - y} + \ldots + \dfrac{w_\nu(x) - w_\nu(y)}{x - y} +$ $+ \sum_{\mu=\nu+1}^{\infty} \dfrac{w_\mu(x) - w_\mu(y)}{x - y} = \sum_{\mu=1}^{\infty} \dfrac{w_\mu(x) - w_\mu(y)}{x - y}$ and the sum of the first ν terms is not greater than the right member of this equation. Let y be constant and let x approach y. The limit of the right member exists (except over a set of measure zero) and is independent of ν. Hence $\sum_{\mu=1}^{\infty} w_\mu'(y)$ is bounded and convergent (except over a set of measure zero) and $\lim_{u\to\infty} w_\mu'(y) = 0$.

<u>THEOREM 9.19: If $f(x)$ is real and summable over $a \leqq x \leqq b$ and if $\varphi(x) = \int_a^x f(y)dy$, then 1) $\varphi(x)$ is continuous and of b.v., 2) $\varphi(x)$ is differentiable over $a \leqq x \leqq b$ except for a set of measure zero., and 3) $\varphi'(x) = f(x)$ except over a set of measure zero.</u>

Proof: It is sufficient to consider the case where $f(x) \geqq 0$, for any real function $f(x) = \frac{|f(x)| + f(x)}{2} - \frac{|f(x)| - f(x)}{2}$; each of these fractions is always $\geqq 0$, and it is evident that the theorem extends to the difference between two non-negative functions. Since $f(x) \geqq 0$, $\varphi(x)$ is m.i.; therefore $\varphi(x)$ is of b.v. and, by Theorem 9.18, part 2) obtains. Let $x_1, x_2, \dots$ be a sequence of values of x with limit x'. Let $f_x(y) = \begin{cases} f(y) \text{ for } a \leqq y \leqq x \leqq b, \\ 0 \text{ elsewhere.} \end{cases}$ Then $\varphi(x) = \int_a^x f(y)dy = \int_a^b f_x(y)dy$. Since $|f_x(y)| \leqq |f(y)|$ for all x, and since $\lim_{n\to\infty} f_{x_n}(y) = f_{x'}(y)$, Theorem 8.16 is applicable and it follows that $\lim_{n\to\infty} \varphi(x_n) = \varphi(x')$, so that part 1) obtains. It remains to prove part 3). If $f(x)$ is not bounded, let $f_N(x) = \begin{cases} f(x), & f(x) \leqq N, \\ 0, & f(x) > N, \end{cases}$ $N = 1, 2, \dots$. Then $\varphi_N(x) = \int_a^x f_N(y)dy$ is m.i., $\varphi_{N+1}(x) - \varphi_N(x)$ is m.i., and $\lim_{N\to\infty} \varphi_N(x) = \varphi(x)$. The preceding lemma shows that it is sufficient to consider $f_N(x)$, that is, bounded functions. In Definition 8.9 let $\theta = 1 - \varepsilon$, make $|h| < \frac{\delta}{2}$, and take $a = 2|h|$. Let M be the set of points such that $|f(x) - f(x_0)| < \varepsilon$, let N be the complement of M in $a \leqq x \leqq b$, let M_1 be the part of M in the interval $x_0 \leqq x \leqq x_0+h$ or in the interval $x_0 \geqq x \geqq x_0+h$ (corresponding to the cases $h > 0$ and $h < 0$), and let N_1 be the complement of M_1 in this interval. Then

$$\frac{\varphi(x_0+h) - \varphi(x_0)}{h} - f(x_0) = \frac{1}{h}\int_{x_0}^{x_0+h} f(x)dx - f(x_0) = \frac{2}{a}\left(\int_{M_1} f(x)dx + \int_{N_1} f(x)dx\right) - f(x_0) =$$

$$= \frac{2}{a}\left[\int_{M_1} [f(x)-f(x_0)]dx + \int_{N_1} [f(x)-f(x_0)]dx\right].$$

Because of the approximate continuity of $f(x)$, the first integral is $\leqq \frac{\varepsilon a}{2}$. The second integral is $\leqq 2A(a\varepsilon)$, where A is a bound of $f(x)$. Hence the entire expression is $\leqq (4A+1)\varepsilon$, and part 3) is immediate.

If $\varphi(x)$ is m.i. and if $f(x) = \varphi'(x)$ (where it exists), the assertion that $\int_a^x f(y)dy = \varphi(x)$ is obviously false because $\int_a^x f(y)dy$ is continuous, while $\varphi(x)$ need not be continuous. It will be shown in the sequel that this assertion is generally false even if $\varphi(x)$ is m.i. and continuous.

Let f(x) be summable, unbounded, and non-negative over $a \leqq x \leqq b$; let $f_N(x) = \begin{cases} f(x) & \text{if } f(x) \leqq N, \\ 0 & \text{if } f(x) > N. \end{cases}$ Then $\lim_{N\to\infty} \int_a^b f_N(x)dx = \int_a^b f(x)dx$. Hence there exists an N_o sufficiently large such that $\int_a^b f_{N_o}(x)dx \geqq \int_a^b f(x)dx - \varepsilon$. Thus $\varepsilon \geqq \int_a^b [f(x) - f_{N_o}(x)]dx = \int_S f(x)dx \geqq 0$, where $S = S_x[f(x) > N_o]$. This proves

<u>THEOREM 9.20: If f(x) is summable, unbounded, and non-negative over $a \leqq x \leqq b$, then it is possible to integrate it over such a set S in this interval that its integral is arbitrarily small while f(x) is bounded in the set complementary to S.</u>

<u>Definition 9.9: A function f(x) defined over an interval $a \leqq x \leqq b$ is said to be absolutely continuous over $a \leqq x \leqq b$ if, corresponding to each $\varepsilon > 0$, there exists a $\delta > 0$ such that whenever $\sum_{i=1}^{n} (b_i - a_i) < \delta$, where</u> $a \leqq a_1 \leqq b_1 \leqq \dots \leqq a_n \leqq b_n \leqq b$, $\sum_{i=1}^{n} | f(b_i) - f(a_i) | < \varepsilon$. (This definition is obviously unaltered if n is allowed to become infinite.)

It is desirable at this point to indicate the object of the discussion in the next few pages. Suppose $\varphi(x)$ is m.i. over the interval $a \leqq x \leqq b$. Other possible properties of $\varphi(x)$ are 1) $\varphi(x) - \varphi(a)$ is the integral (from a to x) of some function f(x), 2) $\varphi(x) - \varphi(a) = \int_a^x \varphi'(y)dy$, 3) $\varphi(x)$ represents sets of measure zero on sets of measure zero, 4) $\varphi(x)$ represents measurable sets on measurable sets, and 5) $\varphi(x)$ is absolutely continuous. It will be shown that if $\varphi(x)$ has any one of these properties, then it has all of them. The equivalences will be proved in the following order: $1\to5$, $5\to3$, $3\to2\to1$,

$3 \rightarrow 4 \rightarrow 3$.

THEOREM 9.21: *If* $\varphi(x) - \varphi(a) = \int_a^x f(y)dy$, *then* $\varphi(x)$ *is absolutely continuous.*

Proof: It is sufficient to consider the case where $f(x) \geqq 0$. Let ε be a given positive number. Let $S = S_x[f(x) > N]$, where $a \leqq x \leqq b$. Then, by Theorem 9.20 there exists a sufficiently large N such that $\int_S f(x)dx < \frac{\varepsilon}{2}$. Let M be any subset of the interval $a \leqq x \leqq b$ of measure $< \delta$. Let $M_1 = M - MS$. Then $\int_M f(x)dx = \int_{M_1} f(x)dx + \int_{MS} f(x)dx < N\delta + \frac{\varepsilon}{2}$. If δ is taken to be $\frac{\varepsilon}{2N}$, then $\int_M f(x)dx < \varepsilon$. Now let M be the sum of all intervals $a_i \leqq x \leqq b_i$, where $a \leqq a_1 \leqq b_1 \leqq \ldots \leqq a_n \leqq b_n \leqq b$, and where $\mu(M) = \sum_{i=1}^{n} (b_i - a_i) < \delta$. Then $\int_M f(x)dx = \sum_{i=1}^{n}[\int_a^{b_i} f(x)dx - \int_a^{a_i} f(x)dx] = \sum_{i=1}^{n}[\varphi(b_i) - \varphi(a_i)]$. There exists a $\delta > 0$ such that this last sum is always $< \varepsilon$. Since $\varphi(x)$ is m.i., this last sum $= \sum_{i=1}^{n}|\varphi(b_i) - \varphi(a_i)|$ and $\varphi(x)$ is absolutely continuous.

THEOREM 9.22: *If* $\varphi(x)$ *is m.i. and absolutely continuous, then the transformation* $x' = \varphi(x)$ *maps sets of measure zero on sets of measure zero.*

Proof: Let M be a set of measure zero in R_1 and let N be its image under the transformation $x' = \varphi(x)$. Let M be covered by a sequence of non-overlapping intervals $I_1, I_2, \ldots$ such that $\sum_{i=1}^{\infty} \mu(I_i) < \delta$. If I_i is the interval $a_i \leqq x \leqq b_i$, then δ can be taken sufficiently small so that $\sum_{i=1}^{\infty}[\varphi(b_i) - \varphi(a_i)] < \varepsilon$ because of the absolute continuity of $\varphi(x)$. (The absolute continuity gives this directly only for $\sum_{i=1}^{n}$, but n may be allowed to become infinite.) But the set of intervals $\varphi(a_i) \leqq y \leqq \varphi(b_i)$ covers N. Hence $\mu^*(N) < \varepsilon$, that is, $\mu(N) = 0$.

It is desirable to show that the condition of continuity alone is insufficient in the preceding theorem. This will be done by exhibiting a m.i. continuous function which does not map a set of measure zero on a set of measure zero. Let $\varphi(x)$ be defined over the interval $0 \leqq x \leqq 1$ as follows: let M_1 be the point $x = \frac{1}{2}$ and let M'_1 be the point $\varphi(x) = \frac{1}{2}$. Let P_1Q_1 be a subinterval of $0 \leqq x \leqq 1$ with center M_1 and length α_1, $0 < \alpha_1 < 1$, and let $P'_1\,Q'_1$ be a subinterval of $0 \leqq \varphi(x) \leqq 1$ with center M'_1 and length β_1, $0 < \beta_1 < 1$. Let M_{11} and M_{12} be the centers of the remaining intervals OP_1 and $Q_1 1$, and let M'_{11} and M'_{12} be the centers of the remaining intervals OP'_1 and $Q'_1 1$. Let

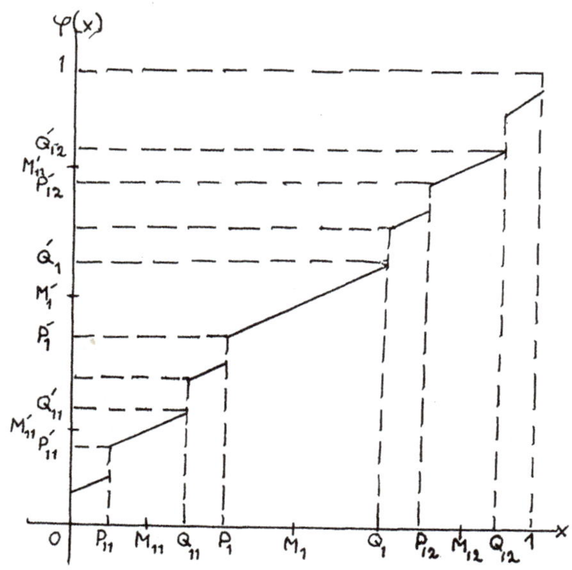

$P_{11}Q_{11}$ and $P_{12}Q_{12}$ be subintervals of OP_1 and $Q_1 1$ with centers M_{11} and M_{12} and length ℓ_2, where $\frac{2\ell_2}{1-\alpha_1} = \alpha_2$, $0 < \alpha_2 < 1$. Let $P'_{11}Q'_{11}$ and $P'_{12}Q'_{12}$ be subintervals of OP'_1 and $Q'_1\,1$ with centers M'_{11} and M'_{12} and length ℓ'_2, where $\frac{2\ell'_2}{1-\beta_1} = \beta_2$, $0 < \beta_2 < 1$. Let this process of selecting intervals in the "middle" of the remaining intervals be continued indefinitely, where all the intervals $P_{\ldots i}\;Q_{\ldots i}$ are of equal length, where the ratio of their total length to the total length of the remaining intervals in which they are located is α_i, $0 < \alpha_i < 1$, where a similar statement applies to the intervals $P'_{\ldots i}Q'_{\ldots i}$, and where $0 < \beta_i < 1$. If x is either a point P or a point Q, then $\varphi(x)$ is represented by the corresponding point P' or Q'. $\varphi(x)$ is defined to be linear over each closed interval PQ, linear over the complement M of the sum of the intervals PQ, and continuous over the interval $0 \leqq x \leqq 1$. N is then the complement of the sum of the intervals P'Q'. $\mu(M) = \prod_{i=1}^{\infty}(1-\alpha_i)$ and $\mu(N) = \prod_{i=1}^{\infty}(1-\beta_i)$. If $\alpha_i = \frac{1}{2}$ for each i and if $\beta_i = \frac{1}{2^i}$, then $\mu(M) = 0$ and $\mu(N) > 0$.

LEMMA: _If $\varphi(x)$ is m.i. over J: $a \leqq x \leqq b$, and maps sets of measure zero on sets of measure zero, if M is any measurable point set in J such that $\varphi'(x)$ exists over M and is $< A$ (A being a positive constant), and if N is the image of M under the transformation $x' = \varphi(x)$, then $\mu(M) \geqq \frac{1}{A}\mu^*(N)$._

Proof: If $x \in M$, $\lim_{h\to 0} \frac{\varphi(x+h) - \varphi(x)}{h} < A$ and $\lim_{h\to 0} \frac{\varphi(x) - \varphi(x-h)}{h} < A$. Therefore $\lim_{h\to 0} \frac{\varphi(x+h) - \varphi(x-h)}{2h} < A$. Hence, for each point $x \in M$, there exists an interval I_x: $a_x \leqq x \leqq b_x$ containing x such that $\varphi(b_x) - \varphi(a_x) < A(b_x - a_x)$. Let I_x^n be a sequence of such intervals containing the point x such that $\lim_{n\to\infty} \mu(I_x^n) = 0$. By Theorem 6.2, there exists a sequence S: $I_{x_1}^{n_1}, I_{x_2}^{n_2}, \ldots$ of these intervals such that 1) no two intervals of S have a common point, 2) M is covered by S except for a set of measure zero, and 3) $\sum_{i=1}^{\infty} \mu(I_{x_i}^{n_i}) \leqq$ $\leqq \mu(M) + \varepsilon$. Let M_1 be the subset of M covered by S and let N_1 be the image of M_1 under the transformation $x' = \varphi(x)$. N_1 is covered by the intervals $I'^{n_i}_{x_i} = [\varphi(a_{x_i}^{n_i}) \leqq y \leqq \varphi(b_{x_i}^{n_i})]$ since $\varphi(x)$ is m.i. Hence $\mu^*(N_1) \leqq \sum_{i=1}^{\infty}\mu(I'^{n_i}_{x_i}) <$ $< \sum_{i=1}^{\infty} A(b_{x_i}^{n_i} - a_{x_i}^{n_i}) < A[\mu(M) + \varepsilon]$, that is, $\mu^*(N_1) \leqq A\,\mu(M)$. By assumption the image of $M-M_1$ is a set of measure zero. Hence $\mu^*(N) \leqq A\,\mu(M)$.

THEOREM 9.23: _If $\varphi(x)$ is m.i. over $a \leqq x \leqq b$ and maps sets of measure zero on sets of measure zero, then $\varphi(b) - \varphi(a) = \int_a^b \varphi'(x)dx$ (where $\varphi'(x)$ is the derivative of $\varphi(x)$ where it exists and where $\varphi'(x)$ may be defined arbitrarily over the set of measure zero where the derivative fails to exist)._

Proof: Let $\varphi(x)$ be extended continuously over $x \geqq b$, for example, by setting $\varphi(x) \equiv \varphi(b)$ for $x \geqq b$. Since $\varphi(x)$ is m.i. $\lim_{n\to\infty} \frac{\varphi(x+1/n) - \varphi(x)}{1/n} =$ $= \varphi'(x)$ except for a set of measure zero. But $\int_a^b n[\varphi(x+1/n) - \varphi(x)]\,dx =$ $= n[\int_{a+1/n}^{b+1/n} \varphi(x)dx - \int_a^b \varphi(x)dx] = n[\int_b^{b+1/n} \varphi(x)dx - \int_a^{a+1/n} \varphi(x)dx]$ and the limit of this expression as $n\to\infty$ is $\varphi(b) - \varphi(a)$. In the proof of Theorem 8.7 it was shown that $\liminf_{n\to\infty} \int_a^b n[\varphi(x+1/n) - \varphi(x)]dx \geqq \int_a^b \varphi'(x)dx$. Hence $\varphi(b) - \varphi(a) \geqq \int_a^b \varphi'(x)dx$.

Now let $0 = \alpha_0 < \alpha_1 < \alpha_2 < \dots$ be a sequence of numbers such that $\lim_{i\to\infty} \alpha_i = \infty$. Let M_i be the set of points in $a \leqq x \leqq b$ where $\varphi'(x)$ exists and satisfies the condition $\alpha_{i-1} \leqq \varphi'(x) < \alpha_i$ and let M'_i be the image of M_i under the transformation $x' = \varphi(x)$. By Theorem 9.18, $\sum_{i=1}^{\infty} \mu(M_i) = b-a$. But $\sum_{i=1}^{\infty} \mu^*(M'_i) \geqq \mu^*[\sum_{i=1}^{\infty}(M'_i)] \geqq \varphi(b) - \varphi(a)$ since $\varphi(x)$ is continuous, and, by the preceding lemma, $\sum_{i=1}^{\infty} \mu^*(M'_i) \leqq \sum_{i=1}^{\infty} \alpha_i \mu(M_i)$. If $\delta = \max(\alpha_{i+1} - \alpha_i) \to 0$, then $\lim_{\delta\to 0} \sum_{i=1}^{\infty} \alpha_i\, \mu(M_i) = \int_a^b \varphi'(x)dx$. Hence $\int_a^b \varphi'(x)dx \geqq \varphi(b) - \varphi(a)$.

Thus $\varphi(b) - \varphi(a) = \int_a^b \varphi'(x)dx$, and the proof is complete.

THEOREM 9.24: *If $\varphi(x)$ is m.i. over J: $a \leqq x \leqq b$ and maps sets of measure zero on sets of measure zero, then $\varphi(x)$ maps measurable sets on measurable sets.*

Proof: Let P be a measurable set in J. If, in the proof of the preceding theorem, the sets M_i are replaced by PM_i and M'_i by QM'_i (where Q is the image of P), then $\int_P \varphi'(x)dx \geqq \mu^*(Q)$. If K is the interval $\varphi(a) \leqq y \leqq \varphi(b)$, then, in the same way, $\int_{J-P} \varphi'(x)dx \geqq \mu^*(K-Q)$. Hence $\mu(K) = \int_J \varphi'(x)dx \geqq$ $\geqq \mu^*(Q) + \mu^*(K-Q)$, so that $\mu(K) = \mu^*(Q) + \mu^*(K-Q)$. By Theorem 4.7, $\mu(K) =$ $= \mu_*(Q) + \mu^*(K-Q)$. Hence $\mu_*(Q) = \mu^*(Q)$ and, by Theorem 4.3, Q is measurable. Thus $\int_P \varphi'(x)dx = \mu(Q)$.

THEOREM 9.25: *If $\varphi(x)$ is m.i. over $a \leqq x \leqq b$ and if M is a set of measure zero whose image M' under the transformation $x' = \varphi(x)$ is not of measure zero, then $\varphi(x)$ maps a measurable part of M on a non-measurable set.*

Proof: The theorem is immediate if M' itself is non-measurable. If M' is measurable, let A and B be the two non-measurable sets of Theorem 7.11. Then AM' and BN' are not measurable and are the maps of measurable parts of M.

It is now possible to prove the equivalence of the various conditions enumerated in the paragraph just preceding Theorem 9.21.

THEOREM 9.26: *The five conditions enumerated in the paragraph preceding Theorem 9.21 are all equivalent.*

Proof: 2) obviously implies 1), 1) implies 5) by Theorem 9.21, 5) implies 3) by Theorem 9.22, 3) implies 2) by Theorem 9.23. Thus 1), 2), 3), and 5) are equivalent. 3) implies 4) by Theorem 9.24, 4) implies 3) by Theorem 9.25. Thus 3) and 4) are also equivalent and the proof is complete.

These results can be applied to the problem of determining the measurability of a function $f(g(x))$ from properties of $f(x)$ and $g(x)$.

THEOREM 9.27: *If $\varphi(x)$ is increasing, then $f(\varphi(x))$ will be measurable for every measurable function $f(y)$ when and only when $\varphi^{-1}(x)$ is absolutely continuous (where $\varphi^{-1}(x)$ is the inverse of $\varphi(x)$).*

Proof: Let $\psi(x) = f(\varphi(x))$. The measurability of $f(y)$ means that every set $M_c = S_y[f(y) > c]$ is measurable (c any real number), while the measurability of $f(\varphi(x))$ means that every set $M'_c = S_x[f(\varphi(x)) > c]$ is measurable. But M_c is a perfectly arbitrary measurable set, and M'_c is its image under the transformation $x = \varphi^{-1}(y)$. Thus the condition is really condition 4) in Theorem 9.27, and is equivalent to absolute continuity.

THEOREM 9.29: *f(g(x)) is measurable when g(x) is measurable and f(y) is a Baire function.*

Proof: This follows immediately from Definition 8.7c, Theorem 8.16, and the fact that the product of two measurable functions is measurable.

CHAPTER X

GENERAL MEASURE FUNCTIONS AND OUTER MEASURES

§1. Elementary properties of measure functions

In this section we shall deal with certain collections of point sets, which collections we shall distinguish variously as half-rings, rings, restricted Borel-rings, Borel-rings. These are generalizations of certain elementary collections of sets which play an important part in the treatment of the Lebesgue measure and of the more general additive functions of point sets in a Euclidean space. An instance of what we shall call a half-ring is the set of all half open intervals in the Euclidean space. The set of all finite sums of half open intervals is an instance of what we shall call a ring; the set of all bounded Borel sets an instance of a restricted Borel-ring; the set of all Borel sets an instance of a Borel-ring. We shall be interested in generalizing only those properties of these elementary collections which make it possible to define additive functions of point sets over them. These properties, as it turns out, are non-topological. Hence the topology of the space in which the point sets lie - even if it exists - is irrelevant. Some use of topology will be made in section 5 when introducing the general notion of Stieltjes-Radon-Lebesgue measure. The considerations needed for this purpose extend from Definition 1.15 to Theorem 1.18. Lebesgue measure itself can be discussed without any use of topology, cf. §5.

Let S be an arbitrary class or space, fixed for the entire discussion; let $\mathcal{P}(S)$ be the power set of S, i.e., the totality of subsets of S.

Definition 10.1.1. A collection of sets $\mathcal{R} \subset \mathcal{P}(S)$ is called a Borel-

ring where:

(α_1) *If each set M_i of a sequence $\{M_i\}$ belongs to $\mathcal{R}$, then so does $\sum_i M_i$.*

(β_1) *If each set M_i of a sequence $\{M_i\}$ belongs to $\mathcal{R}$, then so does $\prod_i M_i$.*

(γ_1) *If M and N belong to $\mathcal{R}$, then so does M - MN.*

Definition 10.1.2. A collection of sets $\mathcal{R} \subset \mathcal{P}(S)$ is called a restricted Borel-ring when

(α_2) *If M and N belong to $\mathcal{R}$, then so does M + N.*

(β_2) *If each set M_i of a sequence $\{M_i\}$ belongs to $\mathcal{R}$, then so does $\prod_i M_i$.*

(γ_2) *If M and N belong to $\mathcal{R}$, then so does M - MN.*

Note that (α_2) is weaker than (α_1). However, if all M_i of a sequence M_i are subsets of a fixed $N \in \mathcal{R}$, that is if $M_i \in \mathcal{R}$, $M_i \subset N \in \mathcal{R}$, then we can affirm $\sum_i M_i \in \mathcal{R}$ on the basis of (β_2), (γ_2) alone, owing to $\sum_i M_i = N - \prod_i (N - M_i)$.

Definition 10.1.3. A collection of sets $\mathcal{R} \subset \mathcal{P}(S)$ is called a ring when:

(α_3) *If M and N belong to $\mathcal{R}$, then so does M + N.*

(β_3) *If M and N belong to $\mathcal{R}$, then so does MN.*

(γ_3) *If M and N belong to $\mathcal{R}$, then so does M - MN.*

We note that in Definitions 10.1.1 and 10.1.3 the condition (β_1), (β_3) respectively are superfluous, This is a consequence of the identity $\prod_i M_i = \sum_i M_i - \sum_j (\sum_i M_i - M_j)$.

Definition 10.1.4. Let $\mathcal{V}$ ($\mathcal{V} \subset \mathcal{P}(S)$) be any collection of sets. A set $M \in \mathcal{V}$ is said to be "immediately below" a set $N \in \mathcal{V}$ if: (1) $M \subset N$, *and* (2) $N - M \in \mathcal{V}$. *This relation is of course relative to the collection $\mathcal{V}$; it is denoted by* $M \Subset N$.

Definition 10.1.5. A collection of sets $\mathcal{R} \subset \mathcal{P}(S)$ is called a half-ring when

(β_4) If M and N belong to $\mathcal{R}$, so does MN.

(γ_4) If M and N belong to $\mathcal{R}$, and $M \supset N$, then there is a finite chain $N_o, N_1, \ldots, N_k$ of sets of $\mathcal{R}$ such that relatively to $\mathcal{R}$, $N = N_o \Subset N_1 \Subset N_2 \Subset \ldots \Subset N_k = M$.

The labeling (β_4) (γ_4) of the postulates in the definition of a half-ring is used to emphasize the analogies with the corresponding postulates for Borel-rings and rings. It is clear that the postulate (α_1) implies (α_2) which is equivalent to (α_3), that (β_1), which is equivalent to (β_2), implies (β_3) which is equivalent to (β_4), and that (γ_1), (γ_2) and (γ_3) are equivalent and imply (γ_4). Thus each one of the four properties of being a Borel-ring, a restricted Borel-ring, a ring, and a half-ring, implies all subsequent ones.

THEOREM 10.1.1: The intersection of any number of rings (respectively, restricted Borel-rings, Borel rings) contained in $\mathcal{P}(S)$ is again a ring (respectively, restricted Borel-ring, Borel-ring).

Proof: Let $\{\mathcal{R}', \mathcal{R}'', \ldots\}$ be any class of rings, and let $\mathcal{R}$ be the intersection. If M and N belong to $\mathcal{R}$ they belong to each $\mathcal{R}^{(\alpha)}$. Therefore M + N, MN, M - MN belong to each $\mathcal{R}^{(\alpha)}$, hence to $\mathcal{R}$. A similar proof applies to restricted Borel-rings and to Borel-rings.

A like theorem does not hold for half-rings. The reason that the theorem went through in the three other cases is that the postulates for these simply assert a kind of closure with respect to certain operations (addition, multiplication, and multiplication by the complement). The postulate (γ_4) is not of this type.

The collection $\mathcal{P}(S)$ is clearly a Borel-ring; therefore it has the three other properties too. Hence for any $\mathcal{T} \subset \mathcal{P}(S)$ there is a ring (restricted Borel-ring, Borel-ring) containing $\mathcal{T}$.

Definition 10.1.6. Assume that $\mathcal{T} \subset \mathcal{P}(S)$. We shall denote by $R(\mathcal{T})$ the smallest ring in $\mathcal{P}(S)$ containing $\mathcal{T}$; i.e., the intersection of all rings $\mathcal{R}$ for which $\mathcal{T} \subset \mathcal{R} \subset \mathcal{P}(S)$. Similarly $BR'(\mathcal{T})$ and $BR(\mathcal{T})$ are defined as the smallest restricted Borel-ring and the smallest Borel-ring containing $\mathcal{T}$.

Evidently we have $\mathcal{T} \subset R(\mathcal{T}) \subset BR'(\mathcal{T}) \subset BR(\mathcal{T})$.

THEOREM 10.1.2. Let $\mathcal{R}$ be a half-ring. Then $R(\mathcal{R})$ is the totality of sets of the form $\sum M_i$ where the sets M_i are finite in number, are mutually disjunct, and are contained in $\mathcal{R}$.

Proof: Denote by $\mathcal{T}$ the totality of the sets $\sum M_i$ where M_i are as described in the statement of the theorem. Since by an obvious induction applied to postulate (α_2) the sum of a finite number of sets from a ring belongs again to the ring, and since $R(\mathcal{R}) \supset \mathcal{R}$, it is evident that $\mathcal{T} \subset R(\mathcal{R})$. To prove the theorem it is sufficient simply to show that $\mathcal{T}$ is a ring, for than we shall have $\mathcal{T} \supset R(\mathcal{R})$, hence $\mathcal{T} = R(\mathcal{R})$. We make the following observations:

(i) If $M \in \mathcal{T}$ and $N \in \mathcal{T}$ then $MN \in \mathcal{T}$. For, let $M = \sum_i M_i$, $N = \sum_i N_i$ where the M_i and similarly the N_i are disjunct, finite in number, and belong to $\mathcal{R}$. Now $MN = \sum_i M_i \cdot \sum_j N_j = \sum_i \sum_j M_i N_j$. Since the sets $M_i N_j$ are also disjunct and contained in $\mathcal{R}$, $MN \in \mathcal{T}$. A similar remark follows inductively for the product of any finite number of sets from $\mathcal{T}$.

(ii) If $M \in \mathcal{T}$, $N \in \mathcal{T}$ and $MN = 0$, then $M + N \in \mathcal{T}$. For, again, let $M = \sum_i M_i$, $N = \sum_i N_i$ be representations of M and N as elements of $\mathcal{T}$ (i.e., be sums of sets of $\mathcal{R}$ of the type described in the definition of $\mathcal{T}$). Since $MN = 0$, $M_i N_j = 0$ and $M + N = \sum_i M_i + \sum_j N_j$ is a representation of $M + N$ as an element of $\mathcal{T}$. A similar remark follows inductively for the sum of a finite number of mutually disjunct addends from $\mathcal{T}$.

(iii) If $M \in \mathcal{R}$ and $N \in \mathcal{R}$ then $M - MN \in \mathcal{T}$. For, since $M \supset MN$ there is a chain $N_o, N_1, \ldots, N_k$ from $\mathcal{R}$ such that $MN = N_o \Subset N_1 \Subset N_2 \Subset \ldots \Subset N_k = M$. The sets $L_i \equiv N_i - N_{i-1}$ $(i=1, 2, \ldots, k)$ are mutually disjunct and belong to $\mathcal{R}$; furthermore, their sum is $M - MN$.

(iv) If $M \in \mathcal{T}$ and $N \in \mathcal{R}$ then $M - MN \in \mathcal{T}$. To prove this, represent M as an element of $\mathcal{T}$: $M = \sum_i M_i$. Let $L_i \equiv M_i - M_i N$. From (iii) the $L_i \in \mathcal{T}$. As parts of the M_i, the L_i have no points in common. It follows from (ii) that $M - MN = \sum_i (M_i - M_i N) = \sum_i L_i \in \mathcal{T}$.

(v) If $M \in \mathcal{T}$ and $N \in \mathcal{T}$ then $M - MN \in \mathcal{T}$. For let $N = \sum N_i$ be a representation of N as an element of $\mathcal{T}$. From (iv) $M - MN_i \in \mathcal{T}$; hence, using (i), $M - MN = \prod_i (M - MN_i) \in \mathcal{T}$.

(vi) If $M \in \mathcal{T}$ and $N \in \mathcal{T}$ then $M + N \in \mathcal{T}$. For, since $M + N = (M - MN) + N$, we can apply (v) and (ii).

(i), (v), and (vi) above, prove that $\mathcal{T}$ is a ring.

<u>THEOREM 10.1.3. Let $\mathcal{R}$ be a ring in $\mathcal{P}(S)$; let $\mathcal{R}'$ be such that $\mathcal{R} \subset \mathcal{R}' \subset \mathcal{P}(S)$ and such that</u>

(α) <u>If $\{M_i\}$ is a finite or denumerable sequence of mutually disjunct sets from $\mathcal{R}'$, then $\sum M_i \in \mathcal{R}'$.</u>

(γ) <u>If $M \in \mathcal{R}$ and $N \in \mathcal{R}'$, then $M - MN \in \mathcal{R}'$.</u>

<u>The theorem is to the effect that if $\mathcal{R}'$ is such a collection of sets, then $\mathcal{R}' \supset BR(\mathcal{R})$.</u>

Proof: Denote by $\mathcal{R}''$ the totality of sets $M \in \mathcal{R}'$ having the property that if $N \in \mathcal{R}'$, then $M - MN \in \mathcal{R}'$. By the definition of $\mathcal{R}''$, $\mathcal{R}' \supset \mathcal{R}''$; also by the condition (γ) it is clear that $\mathcal{R}'' \supset \mathcal{R}$. The theorem will be proved if it is shown that $\mathcal{R}''$ is a Borel-ring, for then we shall have $\mathcal{R}' \supset \mathcal{R}'' \supset BR(\mathcal{R})$. We proceed to show that $\mathcal{R}''$ is a Borel-ring.

(i) If $M \in \mathcal{R}''$, $N \in \mathcal{R}''$, then $M - MN \in \mathcal{R}''$. For, since $M \in \mathcal{R}''$, $N \in \mathcal{R}'' \subset \mathcal{R}'$, the definition of $\mathcal{R}''$ shows that $M - MN \in \mathcal{R}'$. We have thus to show that for any $L \in \mathcal{R}'$, $[(M - MN) - (M - MN)L] \in \mathcal{R}'$. Since $N \in \mathcal{R}''$, we have that if $L \in \mathcal{R}'$, then $N - NL \in \mathcal{R}'$, and hence by condition (α), $N + L = (N - NL) + L \in \mathcal{R}'$. Hence if $L \in \mathcal{R}'$, we have $[(M - MN) - (M - MN)L] = M - M(N + L) \in \mathcal{R}'$.

(ii) If $\{M_i\}$ is a sequence of disjunct sets belonging to $\mathcal{R}''$, then $\sum_i M_i \in \mathcal{R}''$. Let $N \in \mathcal{R}'$; we have to prove that $[\sum_i M_i - N\sum_i M_i] \in \mathcal{R}'$. Now $\sum_i M_i - N\sum_i M_i = \sum_i (M_i - M_iN)$. The sets $(M_i - M_iN)$ are disjunct, and since $M_i \in \mathcal{R}''$, they belong to $\mathcal{R}'$; by the condition (α) their sum does also.

(iii) If $\{M_i\}$ is a sequence of sets from $\mathcal{R}''$, there exists a sequence $\{N_i\}$ of disjunct sets from $\mathcal{R}''$ such that $\sum_i M_i = \sum_i N_i$. For, define the sets $N_{ij} (i \leq j)$ as follows:

$$N_{11} = M_1 ,$$
$$N_{12} = M_2, \; N_{22} = N_{12} - N_{12}M_1 ,$$
$$N_{13} = M_3, \; N_{23} = N_{13} - N_{13}M_1, \; N_{33} = N_{23} - N_{23}M_2 ,$$

etc., etc.

These recursion formulae and the observation (i) above, prove inductively that $N_{ij} \in \mathcal{R}''$. Now define $N_i \equiv N_{ii}$. Then

$$N_1 = M_1,$$
$$N_2 = M_2 - M_2M_1 ,$$
$$N_3 = M_3 - M_3(M_1 + M_2) ,$$

etc., etc.

The sets N_i have no common points, they belong to $\mathcal{R}''$ and their sum is $\sum_i M_i$.

The observations (ii) and (iii) together show that $\mathcal{R}''$ satisfies postulate (α_1) for Borel-rings; (i) shows that (γ_1) is satisfied. Since (α_1) and (γ_1) together imply (β_1), $\mathcal{R}''$ is a Borel-ring.

THEOREM 10.1.4. *If* $\mathcal{T} \subset \mathcal{P}(S)$, *then* $BR'(\mathcal{T})$ *is the set of all* $(M_1 + \dots + M_n)N$, *where* $M_1, \dots, M_n \in \mathcal{T}$, $N \in BR(\mathcal{T})$.

Proof: Denote the set in question by $\mathcal{T}'$. If $M \in \mathcal{T}$, then $M = MM$, $M \in \mathcal{T} \subset BR(\mathcal{T})$, so $M \in \mathcal{T}'$ -- that is, $\mathcal{T}' \supset \mathcal{T}$. It appears immediately that $\mathcal{T}'$ fulfills (α_2), (β_2), (γ_2) (because $BR(\mathcal{T})$ fulfills (α_1), (β_1), (γ_1)); thus it is a restricted Borel-ring. Therefore $\mathcal{T}' \supset BR(\mathcal{T})$. Consider now those sets N, for which $M \in \mathcal{T}$ implies $MN \in BR'(\mathcal{T})$. Denote their set by $\mathcal{T}''$. Clearly $\mathcal{T}'' \supset \mathcal{T}$, and $\mathcal{T}''$ is a Borel-ring, as it fulfills (β_1), (γ_1) because $BR'(\mathcal{T})$ fulfills (β_2), (γ_2), and (α_1) owing to the remark after Definition 10.1.2. Thus $\mathcal{T}'' \supset BR(\mathcal{T})$; that is, if $M \in \mathcal{T}$, $N \in BR(\mathcal{T})$, then $MN \in BR'(\mathcal{T})$. (α_2) extends this to all $(M_1 + \dots + M_n) \cdot N$ with $M_1, \dots, M_n \in \mathcal{T}$. Thus $\mathcal{T}' \subset BR'(\mathcal{T})$, completing the proof.

Corollary 1. *If* $\mathcal{T}$ *is a half-ring, we may assume the* $M_1, \dots, M_n$ *disjunct; if* $\mathcal{T}$ *is a ring, we may assume* $n = 1$.

Corollary 2. *If* $M \in BR'(\mathcal{T})$, $N \in BR(\mathcal{T})$, *then* $MN \in BR'(\mathcal{T})$.

Proof: If $M_1, \dots, M_n \in \mathcal{T}$, then $M_1 + \dots + M_n \in R(\mathcal{T})$, thus the first statement of Corollary 1 follows from Theorem 10.1.2, while the second one is obvious. Corollary 2 follows immediately from Theorem 10.1.4.

THEOREM 10.1.5. *If* $\mathcal{T} \subset \mathcal{P}(S)$, *then* $BR(\mathcal{T})$ *is the set of all* $\sum_i M_i$, *where* $\{M_i\}$ *is any sequence of sets* $M_i \in BR'(\mathcal{T})$.

Proof: Denote the set in question by $\mathcal{T}'$. Since $BR'(\mathcal{T}) \subset BR(\mathcal{T})$, (α_1) implies $\mathcal{T}' \subset BR(\mathcal{T})$. On the other hand we have $\mathcal{T}' \supset BR'(\mathcal{T}) \supset \mathcal{T}$. $\mathcal{T}'$ clearly fulfills (α_1). If $\{N_i\}$ is a sequence of sets, $N_i \in \mathcal{T}'$, then write $N_1 = \sum M_j$, $M_j \in BR(\mathcal{T}')$. Now $N_i \in BR(\mathcal{T})$, $\prod_i N_i \in BR(\mathcal{T})$,

$M_j \cdot \prod_i N_i \in BR(\mathcal{Y})$, and thus $\sum_j M_j \prod_i N_i \in \mathcal{Y}'$, but it is $= N_1 \prod_i N_i = \prod_i N_i$, proving (β_1). Similarly (γ_1) is proved by decomposing M. So $\mathcal{Y}'$ is a Borel-ring and therefore $\mathcal{Y}' \supset BR(\mathcal{Y})$, completing the proof.

Corollary 1. The M_i may be assumed to be disjunct, or again they may be assumed to satisfy $M_1 \subset M_2 \subset \dots$.

Proof: Putting $M_i' = \sum_{j=1}^{i} M_j$ ((α_2)!) leads to $M_1' \subset M_2' \subset \dots$; putting $M_1'' = M_1'$, $M_{i+1}'' = M_{i+1}' - M_i'$, gives disjunct M_i'''s.

Corollary 2. If $\mathcal{R}$ is a half-ring, then every $M \in BR(\mathcal{R})$ is $M \subset \sum_i M_i$, where the M_i are all disjunct and $M_i \in \mathcal{R}$.

Proof: By Theorem 10.1.4, Corollary 1 and Theorem 10.1.5, $M \subset \sum_i M_i$, where $M_i \in R(\mathcal{R})$. Replacing the M_i by $M_1' = M_1$, $M_2' = M_2 - M_2 M_1$, $M_3' = M_3 - M_3 (M_1 + M_2)$, ..., have $M \subset \sum_i M_i'$, the M_i''s disjunct and $M_i' \in R(\mathcal{R})$. Then Theorem 10.1.2 proves the statement.

As was indicated in the introduction to this section, the definitions of the four kinds of rings were chosen with a view to defining a measure function over these collections of sets. We now proceed to say what is meant by a measure function and we deduce a number of its properties.

Definition 10.1.7. Let $\mathcal{R}$ be a half-ring. A numerically valued function $\mu(M)$ defined for each $M \in \mathcal{R}$ is called a measure function on $\mathcal{R}$ if

(a) for any two disjunct sets M, N of $\mathcal{R}$, such that $M + N \in \mathcal{R}$

$$\mu(M) + \mu(N) = \mu(M + N)$$

(b) a sequence $\{M_i\}$ of sets $M_i \in \mathcal{R}$ exists such that $\sum_i M_i = S$.

Definition 10.1.8. A measure function $\mu(M)$ is called a complex valued measure function when its values are finite complex numbers.

Definition 10.1.9. A measure function $\mu(M)$ is called a real measure function when its values are finite real numbers.

Definition 10.1.10. *A measure function $\mu(M)$ is called non-negative when each of its values is a non-negative real number or is $+\infty$.*

Definition 10.1.11. *A non-negative measure function $\mu(M)$ is called finite when the value $+\infty$ is disallowed.*

Definition 10.1.12. *A measure function $\mu(M)$ is called totally additive provided that if $\{M_i\}$ is a denumerable sequence of disjunct sets from $\mathcal{R}$ such that $\sum M_i \in \mathcal{R}$ then $\mu(\sum_i M_i) = \sum_i \mu(M_i)$.*

The assumption in Definition 10.1.12 concerning the numerical series $\sum_i \mu(M_i)$ is that it should be convergent, except when the left side is infinite (which can only occur in the non-negative case, Definition 10.1.10), when it should be properly divergent to $+\infty$. But if the M_i are disjoint and $\sum_i M_i = M$, then any permutation of the M_i does the same. So the series $\sum_i \mu(M_i)$ is convergent in any order, which, as is well-known, implies that it is absolutely convergent.

The apparent prolixity of Definitions 10.1.7 to 10.1.10 is deliberate. The measure functions with which we shall concern ourselves will without exception be required to fall under at least one of these definitions. For example, we shall never speak of a measure function which is sometimes negative and sometimes infinite, nor one which takes on both the values $+\infty$ and $-\infty$.

Since every ring (restricted Borel-ring, Borel-ring) is a half ring, it has a sense to speak of a measure function defined on a ring (restricted Borel-ring, Borel-ring).

In the course of the ensuing discussion we shall develop a connection between a measure on a half-ring $\mathcal{R}$ and a measure on $R(\mathcal{R})$ which measures agree on $\mathcal{R}$. Later, a similar connection between measure on a half-ring and its restricted Borel-ring and Borel-ring will be discussed.

THEOREM 10.1.6. *Every non-empty half-ring* $\mathcal{R}$ *contains the set* 0. *If* $\mu(M)$ *is a measure function defined on* $\mathcal{R}$, *and if it is not true that* $\mu(M) = +\infty$, *then* $\mu(0) = 0$.

Proof: If $M \in \mathcal{R}$ then put in (γ_4) $N = M$, the $N_0, N_1, \ldots, N_k$ must all be $= M$, thus $M \subseteq M$, and $0 = M - M \in \mathcal{R}$.

If some $\mu(M)$ is finite, then $\mu(M) + \mu(0) = \mu(M)$ necessitates $\mu(0) = 0$.

If a measure function $\mu(M)$ is defined on the half-ring $\mathcal{R}$, then $\mathcal{R}$ cannot be empty, owing to Definition 10.1.7(b). Thus $0 \in \mathcal{R}$. Since the case in which $\mu(M) = +\infty$ is devoid of interest, we exclude it. Thus we shall always have $0 \in \mathcal{R}$, $\mu(0) = 0$.

Definition 10.1.13. *Let* $\mathcal{R}$ *be a half-ring. If a set* $M \in \mathcal{R}$ *is the sum of a finite number of disjunct sets* $M_1, M_2, \ldots, M_n$ *of* $\mathcal{R}$, *the collection* $\{M_i\}$ *is called a partition of* M.

Definition 10.1.14. *If* $\mu(M)$ *is a measure function defined on* $\mathcal{R}$, *then a partition* $\{M_i\}$ *of* $M \in \mathcal{R}$ *is called an m-partition of* M *(with respect to* μ*) provided that for any* $N \in \mathcal{R}$ *we have* $\mu(MN) = \sum_i \mu(M_i N)$.

Definition 10.1.15. *If* $\{M_i\}$ *and* $\{N_j\}$ *are partitions of* M, *the partition* $\{N_j\}$ *is called a sub-partition of* $\{M_i\}$ *provided that each* N_j *is included in one of the sets* M_i.

Evidently the relation whereby one partition is a sub-partition of another is a transitive one.

THEOREM 10.1.7. *If* $\{M_i\}$, $\{N_j\}$ *are partitions of* M *then the totality* $\{M_i N_j\}$ *of sets* $M_i N_j$ *forms a common subpartition of* $\{M_i\}$ *and* $\{N_j\}$.

Proof: $M = \sum_i M_i = \sum_i M_i M = \sum_i \sum_j M_i N_j$. The sets $M_i N_j$ all belong to $\mathcal{R}$, and if $M_i N_j$ amd $M_{i_1} N_{j_1}$ have a point in common, then $i = i_1$, $j = j_1$. Furthermore $M_i N_j \subset M_i$ and $M_i N_j \subset N_j$.

We shall speak of this special kind of common sub-partition as the <u>intersection</u> of the two partitions.

<u>THEOREM 10.1.8. If a sub-partition $\{N_j\}$ of a partition $\{M_i\}$ is an m-partition, then $\{M_i\}$ is an m-partition.</u>

Proof: For $N \in \mathcal{R}$, $\mu(M_iN) = \mu(M_iNM) = \sum_j \mu(M_iNN_j)$. Since the sets M_i are exclusive, M_iNN_j is always void unless $N_j \subset M_i$, and in this case $M_iNN_j = NN_j$. Therefore $\mu(M_iN) = \sum_{N_j \subset M_i} \mu(NN_j)$. Observing that each N_j occurs in one and only one M_i we have

$$\sum_i \mu(M_iN) = \sum_i \sum_{N_j \subset M_i} \mu(N_jN) = \sum_j \mu(N_jN) = \mu(MN)$$

This proves the theorem.

<u>THEOREM 10.1.9. If $\{M_i\}$ and $\{N_j\}$ are m-partitions of M, then the intersection of these two partitions is an m-partition.</u>

Proof: For any $N \in \mathcal{R}$ we have $\mu(MN) = \sum_i \mu(M_iN) = \sum_i \mu(M_iMN) =$
$= \sum_i \sum_j \mu(M_iN_jN)$.

<u>THEOREM 10.1.10. Let $M \in \mathcal{R}$, $\overline{M} \in \mathcal{R}$, $M \supset \overline{M}$. Let $\overline{M} = N_1 \Subset N_2 \Subset \dots \Subset N_k = M$ where $N_i \in \mathcal{R}$. Then if $L_1 \equiv N_1$, $L_i \equiv N_i - N_{i-1}$ $(i = 2,3,\dots,k)$ the partition $\{L_i\}$ is an m-partition of M.</u>

Proof: For any $N \in \mathcal{R}$ we have $L_iN + N_{i-1}N = N_iN$ $(i = 2,3,\dots,k)$. Since μ is a measure function,

(1) $\mu(L_1N) = \mu(N_1N)$,

(i) $\mu(L_iN) + \mu(N_{i-1}N) = \mu(N_iN)$ $(i = 2,3,\dots,k)$.

Substituting (1) into (2), the result into (3), the result into (4), ..., until we reach (k), we finally obtain

$$\sum_{i=1}^{k} \mu(L_iN) = \mu(N_kN) = \mu(MN)$$

THEOREM 10.1.11. Every partition of M is an m-partition.

Proof: Let $M_1, M_2, \ldots, M_n$ be a partition of M. From Theorem 10.1.10 with $\overline{M} = M_r$ there is for each r (r = 1, 2, ..., n) an m-partition $\{L_i^{(r)}\}$ in which $L_1^{(r)} = M_r$. Now define the partitions $\{M_i^{(r)}\}$ as follows:

Let $\{M_i^{(1)}\}$ be defined by $M_i^{(1)} = L_i^{(1)}$.

Let $\{M_i^{(2)}\}$ be the intersection of the partitions $\{M_i^{(1)}\}$ and $\{L_i^{(2)}\}$.

Let $\{M_i^{(3)}\}$ be the intersection of the partitions $\{M_i^{(2)}\}$ and $\{L_i^{(3)}\}$.

etc., etc.

We observe that by Theorem 10.1.9 $\{M_i^{(r)}\}$ are all m-partitions. Further $\{M_i^{(n)}\}$ is a sub-partition of $\{L_i^{(r)}\}$ for r = 1, 2, ..., n. But $\{M_i^{(n)}\}$ is a sub-partition of $\{M_i\}$, for if any set $M_{io}^{(n)}$ has a point, that point must lie in some M_r, hence in some $L_1^{(r)}$; since $\{M_i^{(n)}\}$ is a sub-partition of $\{L_i^r\}$ the set $M_{io}^{(n)}$ must lie wholly in $L_1^{(r)}$; that is, wholly in M_r. Hence by Theorem 10.1.8 $\{M_i\}$ is an m-partition.

THEOREM 10.1.12. Let $\mu(M)$ be a measure function for a half-ring. There is a unique measure function $\mu_1(M)$ defined on $R(\mathcal{R})$ such that for $M \in \mathcal{R}$ $\mu_1(M) = \mu(M)$.

Proof: By Theorem 10.1.2 every $M \in R(\mathcal{R})$ can be represented as the sum of a finite number of disjunct sets from $\mathcal{R}$. For each $M \in R(\mathcal{R})$ choose a particular such representation: $M = \sum_i M_i$ and define $\mu_1(M) = \sum_i \mu(M_i)$. We shall show that the value of $\mu_1(M)$ is independent of the particular choice of the representation of M as the sum of disjunct sets from $\mathcal{R}$. Let $M = \sum_j N_j$ be another representation. We have to show that $\sum_j \mu(N_j) = \sum_i \mu(M_i)$. Now since $N_j = MN_j = \sum_i M_iN_j$, the sets M_iN_j form for a fixed j a partition of N_j, therefore by Theorem 10.1.11 an m-partition. Thus $\mu(N_j) = \sum_i \mu(M_iN_j)$, from which $\sum_j \mu(N_j) = \sum_j \sum_i \mu(M_iN_j)$. By a similar argument we have $\sum_i \mu(M_i) = = \sum_i \sum_j \mu(M_iN_j)$ and we thus prove $\sum_j \mu(N_j) = \sum_i \mu(M_i)$. That $\mu_1(M)$ is additive, and that it agrees with $\mu(M)$ on $\mathcal{R}$ is now evident from its definition. That there cannot be another function $\mu_2(M)$ satisfying the requirements of the theorem is also clear, for if $\mu_2(M)$ is additive it must in particular be additive when the addends are elements of $\mathcal{R}$, i.e., we must have $\mu_2(M) = \sum_i \mu(M_i)$ where the M_i are disjunct sets from $\mathcal{R}$.

We shall in the future refer to the function $\mu_1(M)$ defined in the preceding theorem as the extension of $\mu(M)$ from the half-ring $\mathcal{R}$ to the ring $R(\mathcal{R})$.

THEOREM 10.1.13. *If $\mu(M)$ is a real (resp. non-negative, finite) measure function, then so is its extension $\mu_1(M)$.*

Proof: We omit the proof.

THEOREM 10.1.14. *If $\mu(M)$ is totally additive, then so is its extension $\mu_1(M)$.*

Proof: Let $M \in R(\mathcal{R})$; let $\{M_i\}$ be a sequence of disjunct sets from $R(\mathcal{R})$ such that $M = \sum_i M_i$. We must prove that $\mu_1(M) = \sum_i \mu_1(M_i)$. Suppose first that $M \in \mathcal{R}$. As an element of $R(\mathcal{R})$ each M_i is the sum of finitely many disjunct sets from $\mathcal{R}$, let us say $M_i = \sum_k N_{ik}$. Furthermore $\mu_1(M_i) = = \sum_k \mu(N_{ik})$. The sets N_{ik} are disjunct and at most denumerably many; their sum is M. Owing to the total additivity of $\mu(M)$, $\mu_1(M) = \mu(M) = \sum_{i=1}^{\infty} \sum_k \mu(N_{ik}) = = \sum_{i=1}^{\infty} \mu_1(M_i)$. If M does not belong to $\mathcal{R}$, it is the sum of finitely many disjunct sets from $\mathcal{R}$ say $N_1, N_2, \ldots, N_n$. Using the result just obtained, we have $\mu_1(N_j) = \sum_{i=1}^{\infty} \mu_1(N_jM_i)$. Thus $\mu_1(M) = \sum_{j=1}^{n} \mu_1(N_j) =$
$\sum_{j=1}^{n} \sum_{i=1}^{\infty} \mu_1(N_jM_i) = \sum_{i=1}^{\infty} \sum_{j=1}^{n} \mu(N_jM_i) = \sum_{i=1}^{\infty} \mu_1(M_i)$.

THEOREM 10.1.15. *Let $\mu(M)$ be a finite measure function defined on a ring $\mathcal{R}$. A necessary and sufficient condition for the total additivity of $\mu(M)$ is this: If $\{M_i\}$ is a sequence of sets $M_i \in \mathcal{R}$ with $M_1 \supset M_2 \supset \dots \supset 0 = \lim_{i\to\infty} M_i$, then $\lim_{i\to\infty} \mu(M_i) = 0$.*

Proof: The original definition was: if $N_1, N_2, \dots$ are disjunct sets, $N_i \in \mathcal{R}$, $\sum_i N_i = N \in \mathcal{R}$, then $\sum_i \mu(N_i) = \mu(N)$. Now with $M_i = N - \sum_{j=1}^{i} N_j$, we have $M_1 \supset M_2 \supset \dots \supset 0 = \lim_{i\to\infty} M_i$, and any sequence of such M_i's can be obtained in this way (put $N = M_1$, $N_1 = 0$, $N_i = M_i - M_{i-1}$). Further $\sum_{i=1}^{n} \mu(N_i) = \mu(\sum_{i=1}^{n} N_i) = \mu(N - M_n) = \mu(N) - \mu(M_n)$, and thus $\sum_i \mu(N_i) = \mu(N)$ is equivalent to $\lim_{i\to\infty} \mu(M_i) = 0$.

We shall close this section by developing a sufficient criterium for the total additivity of a non-negative finite measure function. It will be used in an application in section 5, as was pointed out in the introduction of this section.

Definition 10.1.16. *By a topological space we mean simply a space S in which a notion of limit has been defined in such a way that for each given infinite sequence $\{x_n\}$ of points of S, either there is associated exactly one point x of S called the limit of $\{x_n\}$ or else there is associated no such point, and further such that if x is the limit if a sequence $\{x_n\}$ then x is the limit of any subsequence $\{x_{n_i}\}$ of $\{x_n\}$.*

The customary definitions of topological spaces, which are used in general topological theory (Frechet, Hausdorff), are narrower than the above definition - but this one will be suitable for our immediate purposes.

Definition 10.1.17. *A set M is called closed if every sequence of points from M which has a limit has its limit in M.*

Definition 10.1.18. A set M is called open if its complement is closed.

Definition 10.1.19. A set M is called compact provided that it is closed and provided that every infinite sequence from M has a subsequence possessing a limit.

THEOREM 10.1.16. *A product of closed sets is closed; a sum of open sets, open.*

Proof: We omit the proof.

THEOREM 10.1.17. *A closed subset of a compact set is compact.*

Proof: We omit the proof.

THEOREM 10.1.18. (Cantor): *If $\{M_n\}$ is a sequence of closed, non-empty sets such that M_1 is compact and such that $M_1 \supset M_2 \supset M_3 \supset \dots \supset M = \lim M_n$ then M is non-empty.*

Proof: Select a sequence $\{x_n\}$ such that for each n, $x_n \in M_n$. Since for each n, $x_n \in M_1$, the compactness of M_1 shows that there is a subsequence $\{x_{n_i}\}$ of $\{x_n\}$ which has a limit say x. Now for every i the subsequence $\{x_{n_i}, x_{n_{i+1}}, \dots\}$ has the same limit x and is contained in the closed set M_{n_i}. Thus x belongs to infinitely many M_n. Since the sequence $\{M_n\}$ is non-increasing, x belongs to every M_n, hence to M.

THEOREM 10.1.19. (Borel): *If a compact set M is contained in an infinite sum $\sum_i \phi_i$ of a sequence of open sets ϕ_i, it is contained in a finite subsum.*

Proof: Assume the contrary. Then for no n is the set $M_n = M - M(\sum_{i=1}^{n} \phi_i)$ empty (if it were for $n = n_0$ then we should have $M \subset \sum_{i=1}^{n_0} \phi_i$). The sets M, M_1, M_2, ... form a decreasing sequence, by Theorem 10.1.13 they are closed, and M is by hypothesis compact. By the preceding theorem their limit is not empty. But this is absurd for since $\sum_{i=1}^{\infty} \phi_i \supset M$ we have $\lim M_n = M - M \sum_{i=1}^{\infty} \phi_i = 0$.

THEOREM 10.1.20. Let S be a topological space. In it let a half-ring $\mathcal{R}$ be given and let $\mu(M)$ be a finite non-negative measure function in $\mathcal{R}$. Assume further that for every positive ε and every $M \in \mathcal{R}$ there exist two sets P, Q from $\mathcal{R}$, a compact set C and an open set $\emptyset$ such that $P \subset C \subset M \subset \emptyset \subset Q$ and such that $\mu(P) \geqq \mu(M) - \varepsilon$ and $\mu(Q) \leqq \mu(M) + \varepsilon$. Under these assumptions $\mu(M)$ is totally additive.

Proof: Assume $M, N_1, N_2, \ldots \in \mathcal{R}$, $M \subset \sum_i N_i$, we will prove $\mu(M) \leqq \sum_i \mu(N_i)$. If the N_i are disjunct and $M = \sum_i N_i$ we have besides $\sum_{i=1}^{n} \mu(N_i) = \mu(\sum_{i=1}^{n} N_i) \leqq \mu(M)$ for all n, and therefore $\sum_i \mu(N_i) \leqq \mu(M)$. Thus $\sum_i \mu(N_i) = \mu(M)$, proving the total additivity. Select $\varepsilon > 0$. Since $M \in \mathcal{R}$ we can find a set $P \in \mathcal{R}$ and a compact set C such that $P \subset C \subset M$ and $\mu(P) \geqq \mu(M) - \frac{\varepsilon}{2}$. Also since $N_i \in \mathcal{R}$ we can find a $Q_i \in \mathcal{R}$ and an open $\emptyset_i$ such that $N_i \subset \emptyset_i \subset Q_i$ and $\mu(Q_i) \leqq \mu(N_i) + \frac{\varepsilon}{2^{i+1}}$. Now $C \subset M \subset \sum_{i=1}^{\infty} N_i \subset \sum_{i=1}^{\infty} \emptyset_i$. By the Borel theorem (10.1.19) there is an n such that $C \subset \sum_{i=1}^{n} \emptyset_i$. We therefore have $P \subset C \subset \sum_{i=1}^{n} \emptyset_i \subset \sum_{i=1}^{n} Q_i$ and thus $\mu(M) \leqq \mu(P) + \frac{\varepsilon}{2} \leqq \frac{\varepsilon}{2} + \sum_{i=1}^{n} \mu(Q_i) \leqq$
$\leqq \frac{\varepsilon}{2} + \sum_{i=1}^{\infty}(\mu(N_i) + \frac{\varepsilon}{2^{i+1}}) \leqq \sum_{i=1}^{\infty} \mu(N_i) + \varepsilon$. The number ε being arbitrary this proves what we wanted. Theorem 10.1.20 contains the desired criterium.

§2. Outer measure and inner measure.

In the previous section we concerned ourselves with additive measure functions defined for a restricted class of sets, the half-rings. In this section we shall deal with functions of sets which are generalizations respectively of Lebesgue outer and inner measure. These functions, which we shall call outer (resp. inner) measure functions, are defined over the whole of $\mathcal{P}(S)$ but are not necessarily additive.

Definition 10.2.1. Let S be any set. A function $\mu^*(M)$ defined for all $M \in \mathcal{P}(S)$ is called an outer (exterior) measure provided that

I. $0 \leqq \mu^*(M) \leqq +\infty$ (all $M \in \mathcal{P}(S)$).

II. If $M \supset N$ then $\mu^*(M) \geqq \mu^*(N)$.

III. If $M = \sum_{i=1}^{\infty} M_i$ then $\mu^*(M) \leqq \sum_{i=1}^{\infty} \mu^*(M_i)$.

Definition 10.2.2. Let $\mu^*(M)$ be an outer measure function. A set M is called measurable (with respect to μ^*) if for every $P \in \mathcal{P}(S)$

$$\mu^*(P) = \mu^*(MP) + \mu^*(P - MP). \tag{1}$$

Measurable sets need not exist. For example, the function $\mu^*(M) \equiv 1$ is an outer measure. No set is measurable with respect to it since the equation (1) can never be satisfied, If on the other hand at least one measurable set M_1 exists, then equation (1) with $M = M_1$, $P = 0$, the empty set, shows that either $\mu^*(0) = 0$ or $\mu^*(0) = +\infty$. In the latter event since 0 is contained in every set, postulate II shows that $\mu^*(M) = +\infty$. This case is without interest. Henceforth we assume that measurable sets do exist and that there is not $\mu^*(M) = +\infty$ (cf. the similar restriction made for the measure functions after Theorem 10.1.6 - that is, we assume $\mu^*(0) = 0$.

In proving the measurability of a given set M it is sufficient merely to prove that for all P of finite outer measure

$$\mu^*(P) \geqq \mu^*(MP) + \mu^*(P - MP) \tag{2}$$

This is true because (2) combined with postulate III implies (1). Furthermore (2) is automatically satisfied if $\mu^*(P) = +\infty$.

If we define $-M = S - M$ the condition (1) for measurability takes the form

$$\mu^*(P) = \mu^*(MP) + \mu^*((-M)P) \tag{3}$$

THEOREM 10.2.1. A set M for which $\mu^*(M) = 0$ is measurable.

Proof: Since for any P we have $MP \subset M$ and $P - MP \subset P$, postulate II shows that $\mu^*(P) = \mu^*(M) + \mu^*(P) \geqq \mu^*(MP) + \mu^*(P - MP)$.

THEOREM 10.2.2. If M is measurable, so is -M.

Proof: Since $-(-M) = M$ the equation (3) remains unaltered if we replace M by -M.

THEOREM 10.2.3. If M and N are measurable, so is M + N.

Proof: Since M is measurable we have for P of finite measure, (3) above:

$$\mu^*(P) = \mu^*(MP) + \mu^*((-M)P)$$

Since N is also measurable we have $\mu^*(MP) = \mu^*((MP)N) + \mu^*((MP)(-N))$ and $\mu^*((-M)P) = \mu^*((-M)PN) + \mu^*((-M)P(-N))$. Therefore

$$\mu^*(P) = \mu^*(MNP) + \mu^*(M(-N)P) + \mu^*((-M)NP) + \mu^*((-M)(-N)P) \tag{4}$$

If in (4) we replace P by P(M+N) the first three terms of the right-hand side remain unaltered and the last drops out; we get

$$\mu^*(M + N)P) = \mu^*(MNP) + \mu^*(M(-N)P) + \mu^*((-M)NP) \tag{5}$$

Then since $(-M(-N)) = -(M + N)$, (4) and (5) give

$$\mu^*(P) = \mu^*((M + N)P) + \mu^*(-(M + N)P)$$

Corollary 1. The sum of any finite number of measurable sets is measurable. (Induction.)

Corollary 2. If M and N are measurable sets, so is M - MN. (By Theorems 10.2.2, 10.2.3, and the fact that M - MN = -((-M) + N)).

THEOREM 10.2.4. If $M_1, M_2, \ldots, M_n$ are measurable sets without common points and if P is any set then

$$\mu^*\left(\sum_{i=1}^{n} M_i P\right) = \sum_{i=1}^{n} \mu^*(M_i P) \tag{6}$$

Proof: Since M_m is measurable

$$\mu^*(P) = \mu^*(M_m P) + \mu^*((-M_m)P) \qquad (m = 1, 2, \ldots, n)$$

If in the m-th of these equations we replace P by $\sum_{i=1}^{m} M_i P$ we get

$$\mu^*\left(\sum_1^m M_i P\right) = \mu^*(M_m P) + \mu^*\left(\sum_1^{m-1} M_i P\right) \quad (m = 2, 3, \ldots, n) \tag{7_m}$$

Substituting (7_2) into (7_3), the result into (7_4), ..., until we reach (7_n), we finally obtain (6).

Corollary. If M and N are disjunct sets and if either of them is measurable, then $\mu^*(M + N) = \mu^*(M) + \mu^*(N)$.

Proof: Assume that M is measurable. Use Theorem 10.2.4 with n = 2, P = M + N, $M_1 = M$, $M_2 = -M$.

THEOREM 10.2.5. If $\{M_i\}$ is a denumerable sequence of disjunct measurable sets and P is any set, then $\mu^*\left(\sum_{i=1}^{\infty} M_i P\right) = \sum_{i=1}^{\infty} \mu^*(M_i P)$.

Proof: For any m, $\sum_1^{\infty} M_i P \supset \sum_1^m M_i P$. Therefore by postulate II and Theorem 10.2.4.

$$\mu^*\left(\sum_1^{\infty} M_i P\right) \geqq \mu^*\left(\sum_1^m M_i P\right) = \sum_1^m \mu^*(M_i P)$$

Hence

$$\mu^*\left(\sum_1^{\infty} M_i P\right) \geqq \lim_{m\to\infty} \sum_1^m \mu^*(M_i P) = \sum_1^{\infty} \mu^*(M_i P) \geqq \mu^*\left(\sum_1^{\infty} M_i P\right),$$

the last step in this last inequality being a consequence of postulate III.

THEOREM 10.2.6. *If $\{M_i\}$ is a denumerable sequence of measurable sets then $M = \sum_{i=1}^{\infty} M_i$ is also measurable.*

Proof: It is sufficient to prove the theorem when the sets M_i are without common points, for if they have common points then the sets $N_i = M_i - M_i(\sum_{j=1}^{i-1} M_j)$ are measurable by Corollaries 1 and 2 to Theorem 10.2.3, they have no common points, and their sum is M.

Let P be any set. For any m, $\sum_{i=1}^{m} M_i$ is measurable; therefore $\mu^*(P) = \mu^*(\sum_1^m M_iP) + \mu^*((-\sum_1^m M_i)P)$. If we now use the fact that $(-\sum_1^m M_i) \supset (-M)$, use postulate II, and Theorem 10.2.4, we get

$$\mu^*(P) \geqq \sum_1^m \mu^*(M_iP) + \mu^*((-M)P)$$

A passage to the limit with m and an application of Theorem 10.2.5 gives

$$\mu^*(P) \geqq \sum_1^{\infty} \mu^*(M_iP) + \mu^*((-M)P) = \mu^*(MP) + \mu^*((-M)P)$$

THEOREM 10.2.7. *The sets measurable with respect to a fixed outer measure $\mu^*(M)$ form a Borel-ring, while the measurable sets with a finite $\mu^*(M)$ form a restricted Borel-ring.*

Proof: Postulate (α_1) is Theorem 10.2.6; postulate (γ_1) is corollary 2 to Theorem 10.2.3, thus proving the first statement. Owing to postulates II, III, $\mu^*(L)$ is finite if $L \subset M, \mu^*(M)$ finite, and $\mu^*(M + N)$ is finite if $\mu^*(M)$, $\mu^*(N)$ are finite. Now (α_2) refers directly to M + N, while (β_2), (γ_2) refer to $\prod_i M_i \subset M$, and $M - M \cdot N \subset M$ respectively, so that the second statement follows from the first.

Definition 10.2.3. *The sets occurring in Theorem 10.2.7 will be denoted as follows: the Borel-ring of all sets measurable with respect to $\mu^*(M)$*

by MS_{μ^*}, *the restricted Borel-ring of all measurable sets with a finite measure by* MS'_{μ^*}.

THEOREM 10.2.8. *If* $\{M_i\}$ *is a sequence of measurable sets such that* $M_1 \subset M_2 \subset \dots \subset M = \lim_{n\to\infty} M_n$, *then* M *is measurable and if* P *is any set* $\lim_{n\to\infty} \mu^*(M_nP) = \mu^*(MP)$.

Proof: M is measurable since $M = \sum_n M_n$. Define $N_1 = M_1$, $N_n = M_n - M_{n-1}$ $(n > 1)$. The sets N_n are measurable and disjunct; furthermore $M_n = \sum_{i=1}^{n} N_i$, $M = \sum_{i=1}^{\infty} N_i$. By Theorems 10.2.5 and 10.2.4

$$\mu^*(MP) = \sum_{i=1}^{\infty} \mu^*(N_iP) = \lim_{n\to\infty} \sum_{i=1}^{n} \mu^*(N_iP) = \lim_{n\to\infty} \mu^*(\sum_{i=1}^{n} N_iP) = \lim_{n\to\infty} \mu^*(M_nP).$$

THEOREM 10.2.9. *If* $\{M_i\}$ *is a sequence of measurable sets such that* $M_1 \supset M_2 \supset \dots \supset M = \lim_{n\to\infty} M_n$, *then* M *is measurable. If further* P *is a set such that for some* n_o $\mu^*(M_{n_o}P)$ *is finite, then* $\lim_{n\to\infty} \mu^*(M_nP) = \mu^*(MP)$.

Proof: M is measurable since $M = \prod_n M_n$. For the second part of the theorem we may by throwing away a finite number of terms assume that $\mu^*(M_1P)$ is finite. All the outer measures occurring are then finite. The sequence $M_1 - M_n$ is a non-decreasing sequence of measurable sets and has for limit the set $M_1 - M$. Since $\mu^*(M_nP) = \mu^*(M_1P) - \mu^*((M_1 - M_n)P)$, Theorem 10.2.9 gives

$$\lim_{n\to\infty} \mu^*(M_nP) = \mu^*(M_1P) - \lim_{n\to\infty} \mu^*((M_1 - M_n)P) = \mu^*(M_1P) - \mu^*((M_1 - M)P) = \mu^*(MP).$$

Definition 10.2.4. *The outer measure* $\mu^*(M)$ *of a measurable set* M *is called its measure and is denoted by* $\mu(M)$. *We shall assume in this part of the discussion that if we write* $\mu(M)$ *we mean to assert that* M *is measurable.*

Definition 10.2.5. *An outer measure function* $\mu^*(M)$ *is called regular (Caratheodory) if*

(a) for any M

$$\mu^*(M) = \underset{N \supset M}{\text{g.l.b.}}\ \mu(N)$$

(b) a sequence of sets $\{M_i\}$ with finite $\mu^*(M_i)$'s and $\sum_i M_i = S$ exists.

In the sequel the outer measure functions will, unless otherwise indicated be assumed to be regular. What amounts to the same thing, we add a fourth postulate to the definition of outer measure:

IV. $\mu^*(M)$ is regular.

Definition 10.2.6. Let $\mu^*(M)$ be an outer measure function. (Regularity is not assumed.) A half-ring $\mathcal{R} \subset \mathcal{P}(S)$ will be called a determining set for $\mu^*(M)$ provided that

(a) Every $M \in \mathcal{R}$ is measurable and of finite measure.

(b) A sequence $\{M_i\}$ of sets $M_i \in \mathcal{R}$ with $\sum_i M_i = S$ exists.

(c) For any $M \in \mathcal{P}(S)$, $\mu^*(M) = \underset{\substack{N \supset M \\ N \in BR(\mathcal{R})}}{\text{g.l.b.}}\ \mu(N)$

We note that condition (a) implies that every $M \in BR(\mathcal{R})$ is measurable, for the Borel-ring of measurable sets contains $\mathcal{R}$, hence $BR(\mathcal{R})$; condition (b) implies $S \in BR(\mathcal{R})$ and thus the set of N's to which the g.l.b. in (c) refers is not empty.

Corollary: The sets M_i in Definition 10.2.6, (b), may be assumed disjunct.

Proof: Put $N_1 = M_1$, $N_2 = M_1 - M_2M_1$, $N_3 = M_3 - M_3(M_1 + M_2)$, ... Then $\sum_i N_i = S, N_i \in R(\mathcal{R})$ and the N_i are disjunct. Write $N_i = \sum_j M_{ij}$, $M_{ij} \in \mathcal{R}$, the M_{ij} disjunct by Theorem 10.1.2. If the M_{ij} are written as a simple sequence, they do what we desired.

THEOREM 10.2.10. The statement that an outer measure $\mu^*(M)$ is regular is equivalent to the statement that there is a determining set for $\mu^*(M)$.

Proof: Suppose $\mu^*(M)$ is regular. The set MS'_{μ^*} from Definition 10.2.3 is a restricted Borel-ring, therefore we may put $\mathcal{R} = MS'_{\mu^*}$; (a) is clearly fulfilled, (b) coincides with Definition 10.2.5, (b). As for (c), note that the g.l.b. of a set of numbers is not changed if the number $+\infty$ is removed from it (if it belonged to it at all; the g.l.b. of any empty set is $+\infty$!). Therefore

$$\mu^*(M) = \underset{N \supset M}{\text{g.l.b}}\ \mu(N) = \underset{\substack{N \supset M \\ \mu(N) \text{ finite}}}{\text{g.l.b}}\ \mu(N) = \underset{N \in MS'_{\mu^*}}{\text{g.l.b.}}\ \mu(N) \geqq$$

$$\geqq \underset{N \in BR(MS'_{\mu^*})}{\text{g.l.b.}}\ \mu(N) \geqq \mu^*(M).$$

On the other hand suppose that there is a determining set

Then

$$\mu^*(M) = \underset{\substack{N \supset M \\ N \in BR(\mathcal{R})}}{\text{g.l.b.}}\ \mu(N) \geqq \underset{N \supset M}{\text{g.l.b.}}\ \mu(N) \geqq \mu^*(M),$$

proving Definition 10.2.5, (a), while Definition 10.2.5, (b) follows from Definition 10.2.6, (a) and (b). Thus $\mu^*(M)$ is regular.

THEOREM 10.2.11. If a half-ring $\mathcal{R}$ determines $\mu^*(M)$, then for every set M there exists a set $N \in BR(\mathcal{R})$ such that $M \subset N$ and $\mu^*(M) = \mu(N)$.

Proof: If $\mu^*(M) = +\infty$, then $\mu^*(S) \geqq \mu^*(M) = +\infty$, $\mu^*(S) = +\infty$, and so $N = S$ meets the requirements. Let therefore $\mu^*(M)$ be finite. Since $\mu^*(M) = \underset{\substack{N \supset M \\ N \in BR(\mathcal{R})}}{\text{g.l.b}}\ \mu(N)$, there is for each $\varepsilon > 0$ a set N_ε such that $N_\varepsilon \supset M$, $N_\varepsilon \in BR(\mathcal{R})$ and $\mu(N_\varepsilon) \leqq \mu^*(M) + \varepsilon$. Now let $N = \prod_{n=1}^{\infty} N_{\frac{1}{n}}$. We have $N \in BR(\mathcal{R})$ and for each n, $M \subset N \subset N_{\frac{1}{n}}$. Therefore $\mu^*(M) \leqq \mu(N) \leqq \mu(N_{\frac{1}{n}}) \leqq \mu^*(M) + \frac{1}{n}$. Since n is arbitrary, $\mu^*(M) = \mu(N)$.

THEOREM 10.2.12. *If* $\mathcal{R}$ *determines* $\mu^*(M)$, *then in order that a set* M *be measurable it is necessary and sufficient that there be two sets* N', N" *from* $BR(\mathcal{R})$ *such that* $N' \subset M \subset N''$ *and* $\mu(N'' - N') = 0$.

Proof: Suppose that M is measurable. If M is of finite measure, then by Theorem 10.2.11 there is a set $N'' \in BR(\mathcal{R})$ such that $M \subset N''$ and $\mu(N'') = \mu(M)$. Now $N'' - M$ is measurable and $\mu(N'' - M) = \mu(N'') - \mu(M) = 0$. Again, by Theorem 10.2.11 there is a set $P \in BR(\mathcal{R})$ such that $N'' - M \subset P$ and $\mu(P) = \mu(N'' - M) = 0$. Now put $N' = N'' - N''P$. Then $N' \subset M \subset N''$, N'_1, $N'' \in BR(\mathcal{R})$ and $\mu(N'' - N') = \mu(N''P) = 0$. If M is not of finite measure, use the sequence $\{M_i\}$ from the Corollary to Definition 10.2.6. $M = \sum MM_i$, and each MM_i is measurable and of finite measure. By what we have just proved there are sets N'_i and N''_i of $BR(\mathcal{R})$ such that $N'_i \subset MM_i \subset N''_i$ and $\mu(N''_i - N'_i)$. Hence $N' = \sum_i N'_i \subset M \subset \sum_i N''_i = N''$. But $N'' - N' = \sum_i N''_i - \sum_i N'_i \subset \sum_i (N''_i - N'_i)$, so that $0 \leqq \mu(N'' - N') \leqq \sum_i \mu(N''_i - N'_i) = 0$.

Conversely the condition is sufficient, for if M is any set and N', N" are two sets of $BR(\mathcal{R})$ for which $N' \subset M \subset N''$ and $\mu(N'' - N') = 0$, then since $N'' - M \subset N'' - N'$ we have $0 \leqq \mu^*(N'' - M) \leqq \mu(N'' - N') = 0$. Theorem 10.2.1 shows that $N'' - M$ is measurable; therefore so is $M = N'' - (N'' - M)$.

Corollary: If M *is measurable, then it is the sum of a set from* $BR(\mathcal{R})$ *and a set of measure* 0. *Also conversely.*

Proof: We omit the proof.

We remark in passing that the argument used to prove sufficiency in Theorem 10.2.12 uses essentially only the measurability of the sets N', N". Hence in order to prove that a set M is measurable, it is sufficient to show that there are two *measurable* sets N', N" such that $N' \subset M \subset N''$ and $\mu(N'' - N') = 0$.

The last theorem and its corollary establish a criterium for measurability directly in terms of sets from $\mathcal{R}$ and BR($\mathcal{R}$) and the sets of measure zero.

THEOREM 10.2.13. *If $\mu^*(M)$ is a regular measure, then it is a totally additive measure function on the Borel-ring MS_{μ^*}* (cf.Definition 10.2.3.)

Proof: The sets M_i from Definition 10.2.5 (b) can be assumed to be measurable by Theorem 10.2.11. This proves the condition (b) in the Definition 10.1.7 of measure function. Definition 10.1.7 (a), as well as the condition in Definition 10.1.12 follow from Theorem 10.2.5 with P = S.

We now proceed to define and discuss a notion of *inner measure*, a notion of which Lebesgue inner measure is an instance.

Definition 10.2.7. *Let $\mu^*(M)$ be a regular outer measure function. The function $\mu_*(M)$ defined by $\mu_*(M) = \underset{N \subset M}{\text{l.u.b.}}\ \mu(N)$ (when we write $\mu(N)$ we mean to assert that N is measurable) is called the inner measure function associated with $\mu^*(M)$.*

In the next theorem we shall demonstrate a number of properties of inner measure which have their close parallels in the postulates I - IV for regular outer measure.

THEOREM 10.2.14. *If $\mu^*(M)$ is a regular outer measure, then the corresponding inner measure $\mu_*(M)$ is defined for all M and satisfies*

I. $0 \leqq \mu_*(M) \leqq +\infty$

II. *If* $M \supset N$ *then* $\mu_*(M) \geqq \mu_*(N)$

III. *If $\{M_i\}$ is a finite or denumerable sequence of disjunct sets whose sum is M, then* $\mu_*(M) \geqq \sum_i \mu_*(M_i)$

IV. *For any set M,* $\mu_*(M) = \underset{N \subset M}{\text{l.u.b.}}\ \mu(N)$.

Proof: Statements I, II are trivial to verify; IV is the definition of $\mu_*(M)$. To show III we shall show first that for each set M there is a measurable set $N \subset M$ such that $\mu(N) = \mu_*(M)$. Let M be any set, then, if $\mu_*(M)$ is finite, by the definition of $\mu_*(M)$ there is for each n = 1, 2, ... a measurable set N_n such that $N_n \subset M$ and $\mu(N_n) > \mu_*(M) - \frac{1}{n}$. Now let $N = \sum_n N_n$; then $N \subset M$ and we have for each n, $\mu_*(M) \geqq \mu(N) \geqq \mu(N_n) > \mu_*(M) - \frac{1}{n}$. As n becomes infinite this gives $\mu_*(M) = \mu(N)$. If $\mu_*(N)$ is infinite, the same result obtains by using n instead of $\mu_*(N) - \frac{1}{n}$. Now let $\{M_i\}$ be a sequence of disjunct sets and let $M = \sum_i M_i$. By what we have just proved there is for each i a set $N_i \subset M_i$ such that $\mu(N_i) = \mu_*(M_i)$; $N \equiv \sum_i N_i$ is by Theorem 10.2.6 measurable and it is obvious that $N \subset M$. Using Theorem 10.2.5 we get $\sum_i \mu_*(M_i) =$

$$= \sum_i \mu(N_i) = \mu(N) \leqq \underset{N \subset M}{\text{l.u.b.}}\ \mu(N) = \mu_*(M).$$

In connection with the parallelism between outer and inner measures it is to be observed that for outer measures the following statement is true: If $\mathcal{R}$ is a determining set for $\mu^*(M)$, then $\mu^*(M) = \underset{N \supset M;\, N \in BR(\mathcal{R})}{\text{g.l.b.}}\ \mu(N)$; and the greatest lower bound is actually attained (Theorem 10.2.11). The corresponding equation $\mu_*(M) = \underset{\substack{N \subset M \\ N \in BR(\mathcal{R})}}{\text{l.u.b.}}\ \mu(N)$ holds in general; and the least upper bound is actually attained. This is clear since in the proof of Theorem 10.2.14 it was shown that there is a measurable set $N \subset M$ such that $\mu(N) = \mu_*(M)$; and Theorem 10.2.12 shows that there is an N' from $BR(\mathcal{R})$ such that $N' \subset N$ and $\mu(N') = \mu(N)$.

If $\mathcal{R}$ is replaced by any system $\mathcal{T}$ of sets, then we have in general only $\mu^*(M) \leqq \underset{\substack{N \supset M \\ N \in BR(\mathcal{T})}}{\text{g.l.b.}}\ \mu^*(N)$ and $\mu_*(M) \geqq \underset{\substack{N \subset M \\ N \in BR(\mathcal{T})}}{\text{l.u.b.}}\ \mu^*(N)$

THEOREM 10.2.15. *If $\mu^*(M)$ is a regular outer measure function, then if M is measurable $\mu^*(M) = \mu_*(M)$, also if M is any set such that $\mu^*(M)$ and*

$\mu_*(M)$ _have a common finite value, then M is measurable._

Proof: If M is measurable then it is clear that the g.l.b. in the condition for regularity and the l.u.b. in the definition of inner measure are both attained by the common value $\mu(M)$. If $\mu^*(M)$ and $\mu_*(M)$ are equal and finite, then there exist measurable sets N', N'' such that $N' \subset M \subset N''$ and such that $\mu(N') = \mu_*(M) = \mu^*(M) = \mu(N'')$. Since these numbers are finite we have $\mu(N'' - N') = 0$. Hence, by the remark following Theorem 10.2.12, M is measurable.

THEOREM 10.2.16. _If_ $\mu^*(M)$ _is a regular outer measure, then for any two disjunct sets M, N we have_

$$\mu_*(M) + \mu_*(N) \leqq \mu_*(M + N) \leqq \begin{Bmatrix} \mu_*(M) + \mu^*(N) \\ \mu^*(M) + \mu_*(N) \end{Bmatrix} \leqq \mu^*(M + N) \leqq \mu^*(M) + \mu^*(N) .$$

Proof: Because of postulate III for outer measure and because of property III for inner measure (Theorem 10.2.14), and also because of the symmetry in M and N of the above inequality, it is sufficient to prove that $\mu_*(M + N) \leqq \mu^*(M) + \mu_*(N) \leqq \mu^*(M + N)$.

Let P be a measurable set contained in M + N and such that $\mu_*(M + N) = \mu(P)$. Let Q be a measurable set containing M such that $\mu(Q) = \mu^*(M)$. Then $N \supset P - PQ$ and we have $\mu_*(M + N) = \mu(P) = \mu(PQ) + \mu(P - PQ) \leqq \mu(Q) + \mu_*(N) = \mu^*(M) + \mu_*(N)$.

Now let P be a measurable set such that $P \supset M + N$ and $\mu^*(M + N) = \mu(P)$. Let Q be a measurable set such that $Q \subset N$ and $\mu(Q) = \mu_*(N)$. Then $P - QP \supset M$ and we have $\mu^*(M + N) = \mu(P) = \mu(P - PQ) + \mu(PQ) = \mu(P - PQ) + \mu(Q) \geqq \mu^*(M) + \mu_*(Q)$.

THEOREM 10.2.17. _Theorems 10.2.4, 10.2.5, 10.2.8, and 10.2.9 remain true if_ $\mu^*(N)$ _is replaced in them everywhere by_ $\mu_*(N)$.

Proof: As Theorems 10.2.8 and 10.2.9 are direct consequences of 10.2.4 and 10.2.5, and as 10.2.4 is a special case of 10.2.5, we need only to consider Theorem 10.2.5. Because of property III for inner measure (Theorem 10.2.14) we must only prove $\mu_*(\sum_i M_i P) \leqq \sum_i \mu_*(M_i P)$.

If N is any measurable set $\subset \sum_i M_i P$, then $M_i N$ is measurable and $\subset M_i P$. Thus

$$\mu(N) = \mu(\sum_i M_i N) = \sum_i \mu(M_i N) \leqq \sum_i \mu_*(M_i P).$$

The $\mu(N)$ on the left side can be replaced by $\underset{N \subset \sum_i M_i P}{\text{g.l.b.}} \mu(N) = \mu_*(\sum_i M_i P)$, and we obtain $\mu_*(\sum_i M_i P) \leqq \sum_i \mu_*(M_i P)$, as we desired.

<u>THEOREM 10.2.18</u>. <u>If</u> $\mu^*(M)$ <u>is a regular outer measure and if</u> P <u>is a measurable set of finite measure, then for any set</u> $M \subset P$ <u>we have</u> $\mu_*(M) = \mu(P) - \mu^*(P - M)$.

Proof: The preceding theorem with $M = M$, $N = P - M$, $P = M + N$ gives $\mu_*(P) \leqq \mu_*(M) + \mu^*(P - M) \leqq \mu^*(P)$. Since P is measurable we have $\mu_*(P) = \mu^*(P) = \mu(P)$. Therefore $\mu(P) = \mu_*(M) + \mu^*(P - M)$.

This last theorem suggests an alternative method of defining inner measure. This method is in fact precisely that used by Lebesgue. Lebesgue first defines outer measure in terms of interval coverings, defines the inner measure of a bounded set M by means of the equation $\mu_*(M) = \mu^*(P) - \mu^*(P - M)$ where P is some interval containing M, shows that this definition is independent of the particular choice of P; he then defines measurability for bounded M in terms of the coincidence of the values of $\mu^*(M)$ and $\mu_*(M)$, and finally defines a measurable set as a set whose intersection with any interval is measurable. The procedure could be carried over <u>mutatis mutandis</u> to the present discussion replacing the intervals by sets of a determining ring. Of course it would be desirable to reformulate the definition of determining ring in such a way that the concept of measurability is there avoided. This could be done by requiring that the outer measure $\mu^*(M)$ be totally additive and finite over the ring (these properties being described in Definitions 10.1.6, 10.1.10, 10.1.11. Note that condition (b) from Definition 10.1.6 is required) and that it satisfy the condition $\mu^*(M) = \underset{\substack{\sum N_i \supset M \\ N_i \in \mathcal{R}}}{\text{g.l.b.}} \sum_i \mu^*(N_i)$. That this definition is equivalent will be seen later in the discussion. (See remark after Theorem 10.3.2.)

THEOREM 10.2.19. *If $\mathcal{R}$ is a determining set for $\mu^*(M)$, then if P is any set of $\mathcal{R}$ and if $M \supset P$ then in order that M be measurable it is sufficient that $\mu^*(P) = \mu^*(M) + \mu^*(P - M)$. If M is an arbitrary set then in order that it be measurable it is sufficient that for all $P \in \mathcal{R}$, $\mu^*(P) =$*
$= \mu^*(PM) + \mu^*(P - MP)$.

Proof: As for the first part of the theorem, Theorem 10.2.19 shows that $\mu_*(M) = \mu^*(P) - \mu^*(P - M)$ and by hypothesis $\mu^*(M) = \mu^*(P) - \mu^*(P - M)$, hence by Theorem 10.2.17 M is measurable. The second part of the theorem is obtained by considering the sets $M_i \in \mathcal{R}$ with $\sum_i M_i = S$ from Definition 10.2.6, (b), and applying the above result to the sets $MM_i \subset M_i$ with $P = M_i$. Thus the MM_i are all measurable, and as $M = \sum_i MM_i$, M is too.

§3. General extension theorems.

In the first part of this section we apply the properties of outer and inner measure to the problems of extending a non-negative, finite and totally additive measure function defined on a half-ring $\mathcal{R}$ to measure functions defined on $BR'(\mathcal{R})$ and on $BR(\mathcal{R})$. As a preliminary we prove

THEOREM 10.3.1. *Let $\mu_1^*(M)$ and $\mu_2^*(M)$ be two regular outer measure functions. Let $\mathcal{R}$ be a half-ring such that it determines $\mu_1^*(M)$, and such that every set of $\mathcal{R}$ is measurable with respect to $\mu_2^*(M)$, and such that for $M \in \mathcal{R}, \mu_1(M) = \mu_2(M)$. Under these assumptions we have*

1) *For every $M \in \mathcal{P}(S)$, $\mu_1^*(M) \geqq \mu_2^*(M)$*
2) *For every $M \in \mathcal{P}(S)$, $\mu_{1_*}(M) \leqq \mu_{2_*}(M)$*
3) *If M is a set measurable with respect to $\mu_1^*(M)$, then it is measurable with respect to $\mu_2^*(M)$ and $\mu_1(M) = \mu_2(M)$.*

Proof: We lose nothing by assuming that $\mathcal{R}$ is a ring, for if $\mathcal{R}$ is a half-ring satisfying the hypotheses of the theorem, then since the sets of $R(\mathcal{R})$ are also measurable with respect to both outer measures, Theorems 10.1.12 and 10.1.13 show that the hypotheses of the theorem with $\mathcal{R}$ replaced everywhere by $R(\mathcal{R})$ are satisfied. Since $BR(\mathcal{R}) = BR(R(\mathcal{R}))$, the conclusions of the theorem are the same in either case.

Denote by $\mathcal{T}$ the totality of sets M which are measurable with respect to both outer measures and which are such that if $N \in \mathcal{R}$ then $\mu_1(MN) = \mu_2(MN)$. The collection $\mathcal{T}$ has the properties:

(α) If $\{M_i\}$ is a sequence of disjunct sets from $\mathcal{T}$, then $M \equiv \sum_i M_i \in \mathcal{T}$.

(γ) If $N \in \mathcal{R}$ and $M \in \mathcal{T}$, then $N - NM \in \mathcal{T}$.

To prove (α) let N be any set of $\mathcal{R}$; then by Theorem 10.2.5, $\mu_1(MN) = \sum_i \mu_1(M_iN) = \sum_i \mu_2(M_iN) = \mu_2(MN)$. To prove ($\gamma$) let $N\ M \in \mathcal{T}$, $P \in \mathcal{R}$; then $\mu_1((N-NM)P) = \mu_1(NP) - \mu_1(NPM) = \mu_2(NP) - \mu_2(NPM) = \mu_2((N-NM)P)$.

By Theorem 10.1.3, $\mathcal{T} \supset BR(\mathcal{R})$. Hence for every $M \in BR(\mathcal{R})$ and every $N \in \mathcal{R}$, $\mu_1(MN) = \mu_2(MN)$, that is, by Theorem 10.1.4. and its Corollary 1, for every $M \in BR'(\mathcal{R})$, $\mu_1(M) = \mu_2(M)$. Theorem 10.1.5 and its Corollary 1 extend this at once to all $M \in BR(\mathcal{R})$.

To prove conclusion 1) of the theorem we now observe that

$$\mu_1^*(M) = \underset{\substack{N \supset M \\ N \in BR(\mathcal{R})}}{\text{g.l.b.}} \mu_1(N) = \underset{\substack{N \supset M \\ N \in BR(\mathcal{R})}}{\text{g.l.b.}} \mu_2(N) \geqq \mu_2^*(M)$$

As for conclusion 2) it was shown in the remarks immediately following Theorem 10.2.14 that for any regular outer measure function $\mu^*(M)$ determined by a set $\mathcal{R}$ and for any M there is an $N' \in BR(\mathcal{R})$ such that $N' \subset M$ and $\mu(N') = \mu_*(M)$. Applying this remark with $\mu^* = \mu_1^*$ we have

$$\mu_{1_*}(M) = \mu_1(N') = \mu_2(N') \leqq \underset{N \subset M}{\text{l.u.b.}}\ \mu_2(N) = \mu_{2_*}(M).$$

As for 3), if M is measurable with respect to $\mu_1^*(M)$, then by Theorem 10.2.14 there are two sets N', N" from $BR(\mathcal{R})$ such that $N' \subset M \subset N''$ and $\mu_1^*(N'' - N') = 0$. Now N', N" are also measurable with respect to $\mu_2^*(M)$; hence by the remark following Theorem 10.2.12, M is measurable with respect to $\mu_2^*(M)$.

Corollary: *If to the hypotheses of the theorem we add that* $\mathcal{R}$ *determines* $\mu_2^*(M)$, *then conclusions* 1), 2), 3) *may be replaced by*

1') *For every* $M \in \mathcal{P}(S)$, $\mu_1^*(M) = \mu_2^*(M)$

2') *For every* $M \in \mathcal{P}(S)$, $\mu_{1_*}(M) = \mu_{2_*}(M)$

3') *A set is measurable with respect to* $\mu_1^*(M)$ *if and only if it is measurable with respect to* $\mu_2^*(M)$.

Proof: Conclusions 1') - 3') follow from the theorem by interchanging the roles of $\mu_1^*(M)$ and $\mu_2^*(M)$.

THEOREM 10.3.2. *Let* $\mathcal{R}$ *be a half-ring; let* $\mu(M)$ *be a finite non-negative, totally additive measure function on* $\mathcal{R}$. *There is exactly one outer measure function* $\mu_1^*(M)$ *such that* (1) $\mu_1^*(M) = \mu(M)$ *for* $M \in \mathcal{R}$, *and* (2) $\mathcal{R}$ *determines* $\mu_1^*(M)$.

Remark: The reason for the use of the notation μ_1^* instead of the simpler one μ^* is this: $\mu_1(M)$ is $\mu_1^*(M)$ restricted to the sets M of $MS_{\mu_1^*}$. $\mu_1(M)$ has the domain $MS_{\mu_1^*}$, while the original $\mu(M)$ has the domain $\mathcal{R}$. Now $MS_{\mu_1^*} \supset \mathcal{R}$ and in general $MS_{\mu_1^*} \neq \mathcal{R}$. Hence $\mu_1(M)$ and $\mu(M)$ have in general different domains and therefore they should not be designated by the same symbol. This precludes dropping the index 1 of μ_1^* and of μ_1.

Proof: If such a function $\mu_1^*(M)$ does exist it is unique by the corollary to Theorem 10.3.1.

No generality is lost in the theorem if we assume that $\mathcal{R}$ is a ring. For, if $\mathcal{R}$ is a half-ring satisfying the hypotheses of the theorems, then using Theorems 10.1.12 and 10.1.13 we can extend the measure function $\mu(M)$ to the ring $R(\mathcal{R})$. The hypotheses of the theorem are satisfied if $\mathcal{R}$ is replaced by $R(\mathcal{R})$ and the conclusion remains the same.

Define for all sets $M \in \mathcal{P}(S)$ the function $\mu_1^*(M) = \underset{\substack{\sum N_i \supset M \\ N_i \in \mathcal{R}}}{\text{g.l.b.}} \sum_i \mu(N_i)$

We shall show that $\mu_1^*(M)$ is an outer measure function with the desired properties.

In the first place, if $M \in \mathcal{R}$ then $\mu_1^*(M) = \mu(M)$. Since M is a covering for itself, the sum consisting of the single term $\mu(M)$ is one of the sums over which the g.l.b. defining $\mu_1^*(M)$ is taken. Therefore $\mu_1^*(M) \leqq \mu(M)$. If $\mu_1^*(M)$ were actually less than $\mu(M)$ there would be a covering $M \subset \sum_i N_i$, $N_i \in \mathcal{R}$, such that $\sum_i \mu(N_i) < \mu(M)$. If we define $L_i = M(N_i - N_i \sum_{j=1}^{i-1} N_j)$ then the L_i are disjunct, $L_i \in \mathcal{R}$, $L_i \subset N_i$ and $\sum_i L_i = M$. Since $\mu(L_i) \leqq \mu(N_i)$ we should have

$$(7) \qquad \sum_i \mu(L_i) \leqq \sum_i \mu(N_i) < \mu(M),$$

contradicting the total additivity of $\mu(M)$.

In the second place, $\mu_1^*(M)$ is an outer measure function. To show that $\mu_1^*(M)$ satisfies postulates I and II for outer measure is trivial. That postulate III is satisfied is shown as follows: Let M be any set; let $\{M_i\}$ be a sequence of sets whose sum is M. Let ε be a positive number. For each M_i choose a covering from $\mathcal{R}$: $M_i \subset \sum_j N_{ij}$, $N_{ij} \in \mathcal{R}$ in such a way that

$$\mu_1^*(M_i) \geqq \sum_j \mu(N_{ij}) - \frac{\varepsilon}{2^i} .$$

Now since $\sum_i \sum_j N_{ij}$ is a covering for M, we have

$$\sum_i \mu_1^*(M_i) \geqq \sum_i \sum_j \mu(N_{ij}) - \varepsilon \geqq \mu_1^*(M) - \varepsilon .$$

Allowing ε to approach 0 we have $\mu_1^*(M) \leqq \sum_i \mu_1^*(M_i)$. Thus postulate III is satisfied, and $\mu_1^*(M)$ is an outer measure.

Finally $\mathcal{R}$ determines $\mu_1^*(M)$. We show that conditions (a) - (c) from Definition 10.2.6 are fulfilled. As for (a) let M be a set of $\mathcal{R}$, and let P be any set of finite outer measure. Choose a positive ε and let $\sum_i N_i \supset P$ be a covering of P with sets N_i of $\mathcal{R}$ such that $\mu_1^*(P) + \varepsilon \geqq \sum_i \mu(N_i)$. Since μ is additive over $\mathcal{R}$ we have $\mu(N_i) = \mu(M_iM) + \mu(N_i - N_iM)$. Also since $\sum_i N_iM \supset PM$ and $\sum_i (N_i - N_iM) \supset P - PM$ we have $\sum_i \mu(N_iM) \geqq \mu_1^*(PM)$ and $\sum_i \mu(N_i - N_iM) \geqq$ $\geqq \mu_1^*(P - PM)$. Therefore

$$\mu_1^*(P) + \varepsilon \geqq \sum_i \mu(N_i) = \sum_i \mu(N_iM) + \sum_i \mu(N_i - N_iM) \geqq \mu_1^*(PM) + \mu_1^*(P - PM).$$

If ε approaches 0, the last inequality shows that M is measurable. That the measure of M is finite is clear since $\mu_1(M) = \mu(M)$. (b) is an immediate consequence of Definition 10.2.5 (b), if we remember that the sets M_i occurring in it can be chosen measurable by Theorem 10.2.11. To show (c) we observe that

$$\mu_1^*(M) = \underset{\substack{\sum N_i \supset M \\ N_i \in \mathcal{R}}}{\text{g.l.b.}} \sum_i \mu(N_i) = \underset{\substack{\sum N_i \supset M \\ N_i \in \mathcal{R}}}{\text{g.l.b.}} \sum_i \mu_1(N_i) \geqq \underset{\substack{\sum N_i \supset M \\ N_i \in \mathcal{R}}}{\text{g.l.b.}} \mu_1(\sum_i N_i) \geqq$$

$$\geqq \underset{\substack{N \supset M \\ N \in BR(\mathcal{R})}}{\text{g.l.b.}} \mu_1(N) \geqq \mu_1^*(M).$$

This completes the proof.

In the discussion following Theorem 10.2.19 it was asserted that a definition equivalent to Definition 10.2.6 for a determining set is the following: A half-ring $\mathcal{R}$ is called a determining set for an outer measure $\mu^*(M)$ provided that (a') $\mu^*(M)$ is finite and totally additive over $\mathcal{R}$ (this is meant to include condition (b) from Definition 10.1.6) and (b') for any M,

$$\mu^*(M) = \underset{\substack{\sum N_i \supset M \\ N_i \in \mathcal{R}}}{\text{g.l.b.}} \sum_i \mu^*(N_i).$$

This equivalence becomes clear in the light of Theorem 10.3.2. Let $\mu_2^*(M)$ be an outer measure admitting a half-ring $\mathcal{R}$ as a determining set in the sense of Definition 10.2.6. By conditions (a) and (b) of Definition 10.2.6 and Theorem 10.2.13, $\mu_2^*(M)$ is a finite and totally additive measure function over $\mathcal{R}$. Thus $\mu_2^*(M)$ satisfies condition (a') above. From the proof of Theorem 10.3.2 (defining the $\mu(M)$ of that theorem as $\mu_2^*(M)$ for $M \in \mathcal{R}$) it follows that $\mu_1^*(M) \equiv \underset{\substack{\sum N_i \supset M \\ N_i \in \mathcal{R}}}{\text{g.l.b.}} \sum_i \mu_2^*(N_i)$ is the unique outer measure function agreeing with $\mu(M) = \mu_2^*(M)$ on $\mathcal{R}$ and admitting $\mathcal{R}$ for a determining set. Since $\mu_2^*(M)$ is also such an outer measure, we have for all M, $\mu_1^*(M) = \mu_2^*(M)$. Thus $\mu_2^*(M) = \underset{\substack{\sum N_i \supset M \\ N_i \in \mathcal{R}}}{\text{g.l.b.}} \sum_i \mu_2^*(N_i)$ and condition (b') above is satisfied. We have therefore proved that if an outer measure $\mu^*(M)$ admits a determining set $\mathcal{R}$ in the sense of Definition 10.2.6, it admits it in the above sense. Conversely, if $\mu^*(M)$ admits $\mathcal{R}$ for a determing set in the above sense, then an exact transcription of the argument applied to $\mu_1^*(M)$ in the proof of Theorem 10.3.2 shows that $\mathcal{R}$ is a determining set in the sense of Definition 10.2.6.

<u>THEOREM 10.3.3.</u> <u>Let $\mathcal{R}$ be a half-ring; let $\mu(M)$ be a finite, non-negative, totally additive measure function defined on $\mathcal{R}$. Form the outer</u>

<u>measure function</u> $\mu_1^*(M)$ <u>as defined by Theorem</u> 10.3.2, <u>and the corresponding sets</u> $MS'_{\mu_1^*}$, $MS_{\mu_1^*}$ (cf. Definition 10.2.3).

<u>For every half-ring</u> $\mathcal{T}$ <u>with</u> $\mathcal{R} \subset \mathcal{T} \subset MS_{\mu_1^*}$ <u>there exists uniquely a non-negative, totally additive measure function</u> $\tilde{\mu}(M)$ <u>such that for</u> $M \in \mathcal{R}$ $\tilde{\mu}(M) = \mu(M)$.

<u>For a given</u> M, <u>all</u> $\tilde{\mu}(M)$ <u>belonging to any</u> $\mathcal{T}$ <u>with</u> $M \in \mathcal{T}$ <u>have the same value</u>.

<u>The measure function</u> $\tilde{\mu}(M)$ <u>is finite if and only if</u> $\mathcal{T} \subset MS'_{\mu_1^*}$.

Proof: We prove the statements of this theorem in successive steps:

(i) By Theorem 10.2.13 $\tilde{\mu}(M) = \mu_1^*(M) = \mu_1(M)$ is a totally additive measure function on $\mathcal{T} = MS_{\mu_1^*}$, and by the construction of $\mu_1^*(M)$ we have for $M \in \mathcal{R}$ $\tilde{\mu}(M) = \mu_1^*(M) = \mu(M)$. All these characteristics are conserved if we pass to a subset $\mathcal{T} \subset MS_{\mu_1^*}$, except Definition 10.1.7 (b). But this holds for $\mathcal{R}$ and therefore for every $\mathcal{T} \supset \mathcal{R}$. Thus the existence of $\tilde{\mu}(M)$ is established for every half-ring $\mathcal{T}$ with $\mathcal{R} \subset \mathcal{T} \subset MS_{\mu_1^*}$.

(ii) If another totally additive measure function $\tilde{\mu}_2(M)$ was defined on a half-ring $\mathcal{T}$ with $\mathcal{R} \subset \mathcal{T} \subset MS_{\mu_1^*}$, such that for $M \in \mathcal{R}$ $\tilde{\mu}_2(M) = \mu(M)$, then we argue as follows: Let $\mathcal{T}'$ be the set of all $M \in \mathcal{T}$ with a finite $\tilde{\mu}_2(M)$; clearly $\mathcal{R} \subset \mathcal{T}' \subset \mathcal{T} \subset MS_{\mu_1^*}$. Form the outer measure which extends $\tilde{\mu}_2(M)$ by Theorem 10.3.2: $\mu_2^*(M)$. Then we have two outer measures $\mu_1^*(M)$, $\mu_2^*(M)$, such that $\mathcal{R}$ determines $\mu_1^*(M)$, and for $M \in \mathcal{R}$, $\mu_1^*(M) = \mu_2^*(M)$ (both are $= \mu(M)$). Thus Theorem 10.3.1 applies: if $M \in MS_{\mu_1^*}$, then $\mu_1^*(M) = \mu_2^*(M)$ (use statement 3)). If $M \in \mathcal{T}'$, then we have $M \in MS_{\mu_1^*}$, and so $\mu_1(M) = \mu_1^*(M) = \mu_2^*(M) = \tilde{\mu}_2(M)$.

If $M \in \mathcal{T}$, then form the $M_i \in \mathcal{R}$ of the corollary to Definition 10.2.5. $MM_i \in \mathcal{T}$ (because $\mathcal{T}$ is a half-ring), $\mu_2(MM_i) = \mu_2^*(MM_i) \leqq \mu_2^*(M_i) =$

$= \mu(m_i)$, so $\tilde{\mu}_2(MM_i)$ is finite, $MM_i \in \mathcal{T}'$. Thus $\mu_1(MM_i) = \mu_2(MM_i)$, and as $M = \sum_i MM_i$, the total additivity implies $\mu_1(M) = \mu_2(M)$.

Thus the uniqueness of $\tilde{\mu}_1(M)$ in $\mathcal{T}$ is established.

(iii) As we saw in (i) the value of $\tilde{\mu}_1(M)$ is for every $\mathcal{T}$ with $M \in \mathcal{T}$ equal to $\mu_1^*(M) = \mu_1(M)$; thus it is independent of $\mathcal{T}$.

(iv) Owing to $\tilde{\mu}_1(M) = \mu_1^*(M)$, $\tilde{\mu}_1(M)$ is finite if and only if $\mu_1^*(M)$ is finite for all $M \in \mathcal{T}$, that is if $\mathcal{T} \subset MS'_{\mu_1^*}$.

<u>Corollary. The $\mathcal{T}$ of the preceding theorem may be chosen equal to $BR'(\mathcal{R})$ or to $BR(\mathcal{R})$, for the former choice $\tilde{\mu}(M)$ is finite.</u>

Proof: $MS'_{\mu_1^*}$ is a restricted Borel-ring, $MS_{\mu_1^*}$ a Borel-ring, both $\supset \mathcal{R}$. Thus they are $\supset BR'(\mathcal{R})$, $BR(\mathcal{R})$ respectively, and therefore $\mathcal{R} \subset BR'(\mathcal{R}) \subset MS'_{\mu_1^*}$, $\mathcal{R} \subset BR(\mathcal{R}) \subset MS_{\mu_1^*}$.

<u>THEOREM 10.3.4. For all $\mathcal{T}$ of Theorem 10.3.3 to which Theorem 10.3.2. can be applied (that is, for those with a finite $\tilde{\mu}(M)$, which means $\mathcal{T} \subset MS'_{\mu_1^*}$), the outer measure $\tilde{\mu}_1^*(M)$ formed according to Theorem 10.3.2 coincides with $\mu_1^*(M)$.</u>

<u>Therefore the derived notions (measurability with respect to $\tilde{\mu}_1^*(M)$ or $\mu_1^*(M)$, the inner measures $\tilde{\mu}_{1*}(M)$ and $\mu_{1*}(M)$) coincide, too.</u>

Proof: $\mathcal{R}$ determines $\mu_1^*(M)$, and for $M \in \mathcal{R}$, $\mu_1^*(M) = \tilde{\mu}_1^*(M)$ (both are $= \mu_1(M)$). Therefore Theorem 10.3.1 applies for all M, $\mu_1^*(M) \geqq \tilde{\mu}_1^*(M)$, and if $M \in MS_{\mu_1^*}$ then $\mu_1^*(M) = \tilde{\mu}_1^*(M)$. If M is arbitrary, apply Theorem 10.2.11 to $\tilde{\mu}_1^*(M)$, which is determined by $\mathcal{T}$: An $N \in BR(\mathcal{T})$ with $N \supset M$, $\tilde{\mu}_1^*(N) = \tilde{\mu}_1^*(M)$ exists. As $\mathcal{T} \subset MS_{\mu_1^*}$, and $MS_{\mu_1^*}$ is a Borel ring, $BR(\mathcal{T}) \subset MS_{\mu_1^*}$, and thus $N \in MS_{\mu_1^*}$, $\mu_1^*(N) = \tilde{\mu}_1^*(N)$. Therefore $\tilde{\mu}_1^*(M) = \mu_1^*(N) \geqq \mu_1^*(M)$, and, together with what we proved above, $\mu_1^*(M) = \tilde{\mu}_1^*(M)$.

The rest of this section will be devoted to a general discussion of the various types of extensions of measure functions.

THEOREM 10.1.2 shows that an unrestricted measure function - unrestricted, that is, for the blanket requirement made after Definition 10.1.12 - defined on a half-ring $\mathcal{R}$ can be extended in a unique way to the ring $R(\mathcal{R})$. The character of the measure function with respect to the range of admissible values and with respect to total additivity is the same after the extension as before (cf. Theorems 10.1.13 and 10.1.14. The extensions to the restricted Borel-ring $BR'(\mathcal{R})$ and to the Borel-ring $BR(\mathcal{R})$ (cf. Theorem 10.3.3 and its corollary) are of a different type. There would be no reason why the extension (to $BR'(\mathcal{R})$ or $BR(\mathcal{R})$) should be unique, if its total additivity (on $BR'(\mathcal{R})$ or $BR(\mathcal{R})$ respectively) were not required, but in order that such an extension exists, we must require finiteness, non-negativity and total additivity of the given measure function already on $\mathcal{R}$. The necessity of the last requirement is obvious, the reasons for the two first are of a technical character, and are connected with the method of extension we used (based on the outer measures). In the next chapter we shall discuss the corresponding extension-processes for measure functions the non-negativity of which is not assumed.

Theorem 10.3.3 shows, however, that this type of extension can generally be carried essentially farther than to $BR'(\mathcal{R})$ and $BR(\mathcal{R})$: the maximal ranges on which we found the same conditions being (according to whether or not we required the finiteness of the extended measure function), $MS'_{\mu_1^*}$ and $MS_{\mu_1^*}$ respectively. The reason that $BR'(\mathcal{R})$ and $BR(\mathcal{R})$ still play an essential role in considerations concerning extensions of a given measure function $\mu(M)$, is that they depend only upon the range $\mathcal{R}$ of $\mu(M)$, and not, like $MS'_{\mu_1^*}$ and $MS_{\mu_1^*}$, upon $\mu(M)$ itself. They are defined by purely set-theoretical operations (cf. Definitions 10.1.1 and 10.1.2), and not by arithmetical means like the maximal ranges (cf. Definitions 10.2.2, 10.2.3). As one frequently considers many $\mu(M)$'s defined on the same range $\mathcal{R}$, these common ranges of extension are particularly useful.

Without wishing to enter more deeply on this particular subject, we remark that $BR'(\mathcal{R})$ and $BR(\mathcal{R})$ are not even maximal ranges of (finite, non-negative, totally additive) measure function extensions which depend on $\mathcal{R}$ alone. The so-called "analytic sets with respect to $\mathcal{R}$" form a range which is in general wider, and which is not maximal either. The subject, connected with the general theory of "analytic" and "projective" sets is far from being exhausted at present.

Returning to the ranges of extension which depend upon $\mu(M)$ itself, the greatest importance must be attached to the maximal ones from Theorem 10.3.3: $MS'_{\mu_1^*}$ and $MS_{\mu_1^*}$. Their relation to other measure functions will be clarified further by the following considerations.

Definition 10.3.1. Two finite, non-negative, totally additive measure functions $\mu(M)$, $\nu(N)$ are equivalent, if the outer measures $\mu_1^*(M)$, $\nu_1^*(N)$ associated with them by Theorem 10.3.2, are identical.

In what follows we restrict ourselves to finite, non-negative, totally additive measure functions.

THEOREM 10.3.5. If a measure function $\mu(M)$ is given, its extension $\tilde{\mu}_1(M)$ on $MS'_{\mu_1^*}$ in the sense of Theorem 10.3.3, is the maximal measure function equivalent to $\mu(M)$. That is

(i) $\tilde{\mu}_1(M)$ is equivalent to $\mu(M)$.

(ii) If $\nu(M)$ is equivalent to $\mu(M)$, then its definition domain is contained in that one of $\tilde{\mu}_1(M)$, and in it $\nu(M) \equiv \tilde{\mu}_1(M)$ holds identically.

Proof: (i) follows from Theorem 10.3.4. As to (ii), form $\nu_1^*(M)$ (by Theorem 10.3.2), if $\nu(M)$ is defined, M is measurable with respect to $\nu_1^*(M)$, and $\nu(M) = \nu_1^*(M)$ -- thus the same is true for $\mu_1^*(M)$. So $M \in MS'_{\mu_1^*}$, $\nu(M) =$ $= \mu_1^*(M) = \tilde{\mu}_1(M)$.

<u>Definition 10.3.2. The measure function $\tilde{\mu}_1(M)$ with the definition domain $MS'_{\mu_1^*}$ (in the sense of Theorems 10.3.3 and 10.3.5) will be called the maximal equivalent extension of $\mu(M)$.</u>

<u>THEOREM 10.3.6. $\mu(M)$ and $\nu(M)$ are equivalent if and only if their respective maximal equivalent extensions $\tilde{\mu}_1(M)$ and $\tilde{\nu}_1(M)$ are identical.</u>

Proof: If $\mu(M)$ and $\nu(M)$ are equivalent, then the coincidence of $\tilde{\mu}_1(M)$ and $\tilde{\nu}_1(M)$ follows from Theorem 10.3.5, or from the fact that $\mu_1^*(M) \equiv \nu_1^*(M)$ and that $\tilde{\mu}_1(M)$ and $\tilde{\nu}_1(M)$ have been defined with their help. If, conversely, $\tilde{\mu}_1(M)$ and $\tilde{\nu}_1(M)$ coincide, then as $\mu(M)$ and $\nu(M)$ are equivalent to them respectively by Theorem 10.3.5, $\mu(M)$ and $\nu(M)$ must be equivalent.

<u>THEOREM 10.3.7. If $\nu(M)$ is an extension of $\mu(M)$, then $\tilde{\nu}_1(M)$ is an extension of $\tilde{\mu}_1(M)$.</u>

Proof: The definition domain $\mathcal{R}$ of $\mu(M)$ determines $\mu_1^*(M)$. Every $M \in \mathcal{R}$ belongs to the definition domain of $\nu(M)$ too, and therefore it is measurable with respect to $\nu_1^*(M)$. Besides $\mu_1^*(M) = \nu_1^*(M)$ (because both are $= \mu(M) =$ $= \nu(M)$). Thus Theorem 10.3.1 applies, and its statement 3 contains our theorem.

<u>Corollary. If $\mu(M)$ is an extension of $\lambda(M)$ and $\nu(M)$ an extension of $\mu(M)$, and if $\lambda(M)$ and $\nu(M)$ are equivalent, then $\mu(M)$ is equivalent to them.</u>

Proof: Results by application of Theorem 10.3.7 to $\lambda(M)$, $\mu(M)$ and to $\mu(M)$, $\nu(M)$, and then using Theorem 10.3.6.

Theorem 10.3.6 shows that we cannot be sure whether an arbitrary extension $\nu(M)$ of $\mu(M)$ is equivalent to it. An actual example of an inequivalent extension will be given in the next Chapter, Section 5. The definition domain $\mathcal{V}$ of an equivalent extension $\nu(M)$ must be $\supset \mathcal{R}$ and $\subset MS'_{\nu_I^*} = MS'_{\mu_I^*}$. Thus the $\tilde{\mu}_1(M)$ of Theorem 10.3.3 with $\mathcal{R} \subset \mathcal{V} \subset MS'_{\mu_I^*}$ are the only equivalent extensions of $\mu(M)$ (remember Theorem 10.3.5). The extensions to $R(\mathcal{R})$ and $BR'(\mathcal{R})$ are therefore equivalent ones.

§4. Measure in product spaces.

This section is concerned with the construction in a product space of a measure function defined in terms of measure functions in the spaces over which the product is taken. One has in mind, of course, the familiar situation in which the area of a plane interval (rectangle) is defined as the product of the lengths of two adjacent sides. Here, assuming that a measure is given for intervals of a line (the measure being the length of the interval) there is defined on the direct product of the line with itself (the plane) a measure for plane intervals. This is merely an illustrative instance; the general procedure will carry us beyond. Further examples of the process will be found in Section 5.

Definition 10.4.1. Let $\mathcal{E}$ be a set of elements which will be used as indices. If for every $\nu \in \mathcal{E}$ a set M_ν is defined, then we shall define the direct product $\prod\prod_{\nu \in \mathcal{E}} M_\nu$ of the M_ν's as the set of all functions $x(\nu)$ such that $x(\nu)$ is defined for every $\nu \in \mathcal{E}$ and such that $x(\nu) \in M_\nu$.

If $\mathcal{E}$ is finite or countably infinite, $\mathcal{E} = \{\nu_1, \nu_2, \ldots\}$ (this sequence may end with a ν_k or not at all), then we shall sometimes write $M_{\nu_1} \times M_{\nu_2} \times \ldots$ instead of $\prod\prod_{\nu \in \mathcal{E}} M_\nu$ and $[x(\nu_1), x(\nu_2), \ldots]$ instead of $x(\nu)$, $\nu \in \mathcal{E}$.

If for every $\nu \in \mathcal{E}$ a space S_ν is given, we can form a new space $\mathcal{S} = \prod\prod_{\nu \in \mathcal{E}} S_\nu$ called the direct product space of the S_ν's. For every choice of $M_\nu \subset S_\nu$ (for each $\nu \in \mathscr{E}$) the direct product set $\mathscr{M} = \prod_{\nu \in \mathscr{E}} M_\nu$ is a subset of $\mathscr{S}$.

We first discuss some properties of these sets $\mathscr{M}$ of $\mathscr{S}$ which are direct product sets.

THEOREM 10.4.1. $\prod_{\nu \in \mathscr{E}} M_\nu = 0$ if and only if $M_\nu = 0$ for some $\nu \in \mathscr{E}$.

Proof: If for some ν_1, $M_{\nu_1} = 0$, then there exists no function $x(\nu)$ defined for all $\nu \in \mathscr{E}$ such that $x(\nu) \in M_\nu$ (since the statement $x(\nu) \in M_\nu$ is false for $\nu = \nu_1$). On the other hand, if all M_ν's are non-empty, the assertion of the existence of a function $x(\nu)$ of the type described is simply a statement of the axiom of choice.

THEOREM 10.4.2. Let $\mathscr{M} = \prod_{\nu \in \mathscr{E}} M_\nu$ and $\mathscr{N} = \prod_{\nu \in \mathscr{E}} N_\nu$. If for all $\nu \in \mathscr{E}$ $M_\nu \subset N_\nu$, then $\mathscr{M} \subset \mathscr{N}$; if $\mathscr{M} \neq 0$ and $\mathscr{M} \subset \mathscr{N}$ then $M_\nu \subset N_\nu$ for all $\nu \in \mathscr{E}$. Thus if $\mathscr{M} \neq 0$, the conditions $\mathscr{M} \subset \mathscr{N}$ and $M_\nu \subset N_\nu (\nu \in \mathscr{E})$ are equivalent.

Proof: The first part is clear since if a function $x(\nu)$ satisfies $x(\nu) \in M_\nu$ for all $\nu \in \mathscr{E}$ it also satisfies $x(\nu) \in N_\nu$ for all $\nu \in \mathscr{E}$.

For the second part, suppose that for $\nu = \nu_0$, $M_{\nu_0} - M_{\nu_0} N_{\nu_0}$ contains a point x_0. Then since $\mathscr{M} \neq 0$, $\mathscr{M}$ has an element $[x(\nu)]$. Now the function

$$\bar{x}(\nu) = \begin{cases} x_0, & \nu = \nu_0 \\ x(\nu), & \nu \neq \nu_0 \end{cases}$$

belongs to $\mathscr{M}$ but not to $\mathscr{N}$. Thus if $\mathscr{M} \subset \mathscr{N}$, the set $M_\nu - M_\nu N_\nu$ must be empty for all $\nu \in \mathscr{E}$. That is $M_\nu \subset N_\nu$.

Corollary: If $\mathscr{M} \neq 0$ is of the form $\prod_{\nu \in \mathscr{E}} M_\nu$, then the M_ν are uniquely determined.

Proof: If $\mathscr{M} = \prod_{\nu \in \mathscr{E}} M_\nu = \prod_{\nu \in \mathscr{E}} N_\nu$, then $M_\nu \subset N_\nu$ and $M_\nu \supset N_\nu$.

THEOREM 10.4.3. If $\mathscr{M} = \prod_{\nu \in \mathscr{E}} M_\nu$, $\mathscr{N} = \prod_{\nu \in \mathscr{E}} N_\nu$, $\mathscr{P} = \prod_{\nu \in \mathscr{E}} P_\nu$, and if $\mathscr{N} \neq 0$, $\mathscr{P} \neq 0$, then $\mathscr{M} = \mathscr{N} + \mathscr{P}$, $\mathscr{N}\mathscr{P} = 0$ is equivalent to the existence of a (unique) $\nu_0 \in \mathscr{E}$ such that for $\nu \neq \nu_0$, $M_\nu = N_\nu = P_\nu$ while for

$\nu = \nu_o$, $M_\nu = N_\nu + P_\nu$, $N_\nu P_\nu = 0$.

Proof: Assume first that $\mathcal{M} = \mathcal{N} + \mathcal{P}$, $\mathcal{N}\mathcal{P} = 0$. Since $\mathcal{M} \supset \mathcal{N}$ and $\mathcal{M} \supset \mathcal{P}$ we have (Theorem 10.4.2) $M_\nu \supset N_\nu$ and $M_\nu \supset P_\nu$, thus $M_\nu \supset N_\nu + P_\nu$. Also we have again, by Theorem 10.4.2, $\mathcal{N} = \prod\!\!\prod_{\nu \in \mathcal{E}} N_\nu \subset \prod\!\!\prod_{\nu \in \mathcal{E}} (N_\nu + P_\nu)$ and similarly $\mathcal{P} \subset \prod\!\!\prod_{\nu \in \mathcal{E}} (N_\nu + P_\nu)$, from which $\mathcal{M} = \mathcal{N} + \mathcal{P} \subset \prod\!\!\prod_{\nu \in \mathcal{E}} (N_\nu + P_\nu)$ and hence $M_\nu \subset N_\nu + P_\nu$. Therefore $M_\nu = N_\nu + P_\nu$. Again, $\prod\!\!\prod_{\nu \in \mathcal{E}} (N_\nu P_\nu) = 0$; for if not, Theorem 10.4.2 shows that it is contained in $\mathcal{N}$ and in $\mathcal{P}$ hence in $\mathcal{N}\mathcal{P} = 0$. Theorem 10.4.1 then shows that for some ν, say ν_o, we have $N_{\nu_o} P_{\nu_o} = 0$. Now for every $\nu \neq \nu_o$ the sets $M_\nu - N_\nu$ and $M_\nu - P_\nu$ are empty. Suppose not; for example, let x_1 be a point of $M_{\nu_1} - N_{\nu_1}$ with $\nu_1 \neq \nu_o$; let x_o be a point of N_{ν_o}; let $x(\nu)$, $\nu \in \mathcal{E}$, be a point of the product set $\mathcal{M}$.

Define $\overline{x}(\nu) = \begin{cases} x_o, & \nu = \nu_o \\ x_1, & \nu = \nu_1 \\ x(\nu), & \text{otherwise} \end{cases}$ Now $[\overline{x}(\nu)]$ is clearly in $\mathcal{M}$ but cannot belong either to $\mathcal{P}$ (on account of $\overline{x}(\nu_o) = x_o \in N_{\nu_o}$, $N_{\nu_o} P_{\nu_o} = 0$) or to $\mathcal{N}$ (on account of $\overline{x}(\nu_1) = x_1 \in M_{\nu_1} - N_{\nu_1}$). Thus we have for $\nu \neq \nu_o$, $M_\nu = N_\nu = P_\nu$.

Assume conversely that for $\nu \in \mathcal{E}$, $\nu \neq \nu_o$, $M_\nu = N_\nu = P_\nu$ and for $\nu = \nu_o$, $M_\nu = N_\nu + P_\nu$, $N_\nu P_\nu = 0$. Then for all $\nu \in \mathcal{E}$, $M_\nu \supset N_\nu$ $M_\nu \supset P_\nu$, hence $\mathcal{M} \supset \mathcal{N} + \mathcal{P}$. Any function $[x(\nu)]$ of $\mathcal{M}$ belongs to $\mathcal{N}$ or $\mathcal{P}$ respectively, according as $x(\nu_o) \in N_{\nu_o}$ or $x(\nu_o) \in P_{\nu_o}$. From this we have $\mathcal{M} \subset \mathcal{N} + \mathcal{P}$, $\mathcal{N}\mathcal{P} = 0$.

Definition 10.4.2. *An admissible system is defined as follows: Let $\mathcal{E}$ be a set (of indices); for each $\nu \in \mathcal{E}$ let a space S_ν and a half-ring $\mathcal{R}_\nu \subset \mathcal{P}(S_\nu)$ be given in such a way that, except possibly for at most a finite number of ν's, $S_\nu \in \mathcal{R}_\nu$. There may be no exceptions at all. If $\mathcal{E}$ is a finite set the condition that there be at most a finite number of exceptions to $S_\nu \in \mathcal{R}_\nu$ is vacuous. Similar remarks apply to analogous exceptions in this* definition and in Definition 10.4.3. We shall upon occasion use the term "almost all" to mean "with a finite number and perhaps no exceptions".

In the direct product space $\mathcal{S} = \prod\prod_{\nu \in \mathcal{E}} S_\nu$ let there be defined the system of sets $\mathcal{R} \subset \mathcal{P}(\mathcal{S})$ consisting of all sets $\prod\prod_{\nu \in \mathcal{E}} M_\nu$ in which for all $\nu \in \mathcal{E}$, $M_\nu \in \mathcal{R}_\nu$ and for almost all $\nu \in \mathcal{E}$, $M_\nu = S_\nu$. A system $(\mathcal{E}, S_\nu, \mathcal{R}_\nu)$ having these properties, we shall call an admissible system.

We remark that every exceptional ν for which S_ν does not belong to $\mathcal{R}_\nu$ is necessarily a ν for which $M_\nu \neq S_\nu$; on the other hand, a ν for which $M_\nu \neq S_\nu$ may very well be one for which $S_\nu \in \mathcal{R}_\nu$.

THEOREM 10.4.4. The system of sets $\mathcal{R} \subset \mathcal{P}(\mathcal{S})$ defined in Definition 10.4.2 is a half-ring in $\mathcal{S}$.

Proof: Let $\mathcal{M} = \prod\prod_{\nu \in \mathcal{E}} M_\nu$ and $\mathcal{N} = \prod\prod_{\nu \in \mathcal{E}} N_\nu$. We observe first that $\mathcal{M} \cdot \mathcal{N} = \prod\prod_{\nu \in \mathcal{E}} M_\nu N_\nu$, for if $x(\nu)$ is a point of $\mathcal{M} \cdot \mathcal{N}$ then for each $\nu \in \mathcal{E}$, $x(\nu) \in M_\nu$ and $x(\nu) \in N_\nu$ hence $x(\nu) \in M_\nu N_\nu$ and we have $\mathcal{M} \cdot \mathcal{N} \subset \prod\prod_{\nu \in \mathcal{E}} M_\nu N_\nu$. Theorem 10.4.2 shows that the latter set is contained in both $\mathcal{M}$ and $\mathcal{N}$, therefore in $\mathcal{M} \cdot \mathcal{N}$. Hence $\mathcal{M} \cdot \mathcal{N} = \prod\prod_{\nu \in \mathcal{E}} M_\nu N_\nu$.

If now $\mathcal{M} \in \mathcal{R}$ and $\mathcal{N} \in \mathcal{R}$ then for each $\nu \in \mathcal{E}$, $M_\nu \in \mathcal{R}_\nu$, $N_\nu \in \mathcal{R}_\nu$ and by condition (β_4) for half-rings $M_\nu N_\nu \in \mathcal{R}_\nu$. Furthermore with at most a finite number of exceptions $M_\nu N_\nu = M_\nu = N_\nu = S_\nu$. Therefore $\mathcal{M} \cdot \mathcal{N} = \prod\prod_{\nu \in \mathcal{E}} M_\nu N_\nu \in \mathcal{R}$. Thus $\mathcal{R}$ satisfies (β_4) (Definition 10.1.5).

To show that $\mathcal{R}$ satisfies (γ_4) suppose that $\mathcal{M} \in \mathcal{R}$, $\mathcal{N} \in \mathcal{R}$ and $\mathcal{M} \supset \mathcal{N}$. If $\mathcal{N} = 0$ then clearly $\mathcal{N} \Subset \mathcal{M}$. If $\mathcal{N} \neq 0$, Theorem 10.4.2 shows that $M_\nu \supset N_\nu$ for all $\nu \in \mathcal{E}$. Owing to the requirements of Definition 10.4.2 there is an at most finite, possibly void, set of ν's, say $\nu_1, \nu_2, \ldots, \nu_k$, for which $N_\nu \neq S_\nu$, and for all other ν's we have $S_\nu = N_\nu \subset M_\nu \subset S_\nu$. For each ν_i $(i = 1, 2, \ldots, k)$ form a chain $N_{\nu_i} = N^0_{\nu_i} \Subset N^1_{\nu_i} \Subset \ldots \Subset N^{n_i}_{\nu_i} = M_{\nu_i}$.

Now define recursively the sets $\mathcal{M}_{i,j}$ $(1 \leqq i \leqq k\ ;\ 0 \leqq j \leqq n_i)$ in the following way: Let $\mathcal{M}_{1,0} = \mathcal{N}$. Having defined $\mathcal{M}_{i-1,j}$ for all $j = 0, 1, \ldots, n_{i-1}$ define $\mathcal{M}_{i,o} = \mathcal{M}_{i-1,\ n_{i-1}}$. Having defined $\mathcal{M}_{i,j-1}$ and shown that it belongs to $\mathcal{R}$, define $\mathcal{M}_{i,j}$ as the direct product obtained from $\mathcal{M}_{i,j-1}$ by simply replacing by $N^j_{\nu_i}$ the ν_i'th direct factor in the product $\mathcal{M}_{i,j-1}$. It is clear that $\mathcal{M}_{i,j} \in \mathcal{R}$ and that $\mathcal{M}_{k,n_k} = \mathcal{M}$. Now for $j > 0$ define $\mathcal{L}_{ij}$ as the direct product obtained by replacing by $N^j_{\nu_i} - N^{j-1}_{\nu_i}$ the ν_i'th direct factor in $\mathcal{M}_{i,j}$. Evidently $\mathcal{L}_{i,j} \in \mathcal{R}$. It follows from Theorem 10.4.3 that $\mathcal{M}_{i,j} = \mathcal{L}_{i,j} + \mathcal{M}_{i,j-1}$ and $\mathcal{L}_{i,j} \circ \mathcal{M}_{i,j-1} = 0$. Hence $\mathcal{M}_{i,j-1} \Subset \mathcal{M}_{i,j}$. Thus we have $\mathcal{N} = \mathcal{M}_{1,0} \Subset \mathcal{M}_{1,1} \Subset \ldots \Subset \mathcal{M}_{1,n_1} = \mathcal{M}_{2,0} \Subset \ldots \Subset \mathcal{M}_{k,n_k} = \mathcal{M}$.

Definition 10.4.3. Let $(\mathcal{E}, S_\nu, \mathcal{R}_\nu)$ be an admissible system. Suppose that for each $\nu \in \mathcal{E}$ a measure function $\mu_\nu(M)$ is defined on $\mathcal{R}_\nu$ in such a manner that for almost all $\nu \in \mathcal{E}$ we have $\mu_\nu(S_\nu) = 1$. Such a system of measure functions we call admissible. (Every exceptional ν of the first half of Definition 10.4.2, i.e., a ν for which S_ν is not in $\mathcal{R}_\nu$ is an exceptional ν here, but not conversely.)

Associated with each admissible system of measure functions we define a measure function $\mu(\mathcal{M})$ on $\mathcal{R}$ as follows: write $\mathcal{M} = \prod\prod_{\nu \in \mathcal{E}} M_\nu$ and define $\mu(\mathcal{M}) = \prod_{\nu \in \mathcal{E}} \mu_\nu(M_\nu)$.

Concerning the Definition 10.4.3 we make the following observations:

(i) The product $\prod_{\nu \in \mathcal{E}} \mu_\nu(M_\nu)$ is actually only a finite product, because except for a finite number of ν's, $S_\nu \in \mathcal{R}_\nu$, $\mu_\nu(S_\nu) = 1$ and $M_\nu = S_\nu$ -- and thus $\mu_\nu(M_\nu) = 1$.

(ii) The numerical value of $\mu(\mathcal{M}) = \prod_{\nu \in \mathcal{E}} \mu_\nu(M_\nu)$ is uniquely determined by $\mathcal{M}$ itself. If $\mathcal{M} = 0$, Theorem 10.4.1 shows that at least one $M_\nu = 0$ and hence $\mu(\mathcal{M}) = 0$. If $\mathcal{M} \neq 0$, the corollary to Theorem 10.4.2 shows that the M_ν are uniquely determined by $\mathcal{M}$.

(iii) In case the factors 0 and $+\infty$ occur in a product we shall employ the convention $+\infty \cdot 0 = 0$.

THEOREM 10.4.5. The function $\mu(\mathcal{M})$ constructed in Definition 10.4.3 has the following properties:

(i) If all $\mu_\nu(M)$'s are non-negative, then $\mu(\mathcal{M})$ is a non-negative measure function on $\mathcal{R}$.

(ii) If all $\mu_\nu(M)$'s are finite, then $\mu(\mathcal{M})$ is a finite measure function on $\mathcal{R}$.

Proof: We first show that $\mu(\mathcal{M})$ is a measure function. We must prove (a) If $\mathcal{M} \in \mathcal{R}$, $\mathcal{N} \in \mathcal{R}$, $\mathcal{P} \in \mathcal{R}$ and $\mathcal{M} = \mathcal{N} + \mathcal{P}$, $\mathcal{N} \cdot \mathcal{P} = 0$, then $\mu(\mathcal{M}) = \mu(\mathcal{N}) + \mu(\mathcal{P})$; and (b) A countable set from $\mathcal{R}$ covers $\mathcal{S}$.

If either $\mathcal{N}$ or $\mathcal{P}$ is empty, (a) is trivial (remark (ii) following Definition 10.4.3). If not, Theorem 10.4.3 shows, writing $\mathcal{M} = \prod\prod_{\nu \in \mathcal{E}} M_\nu$, $\mathcal{N} = \prod\prod_{\nu \in \mathcal{E}} N_\nu$, $\mathcal{P} = \prod\prod_{\nu \in \mathcal{E}} P_\nu$, that for all except one ν, say ν_0, we have $M_\nu = N_\nu = P_\nu$ and for $\nu = \nu_0$, $M_{\nu_0} = N_{\nu_0} + P_{\nu_0}$, $N_{\nu_0} P_{\nu_0} = 0$. Hence

$$\mu(\mathcal{M}) = \prod_{\nu \in \mathcal{E}} \mu_\nu(M_\nu) = \mu_{\nu_0}(M_{\nu_0}) \prod_{\substack{\nu \in \mathcal{E} \\ \nu \neq \nu_0}} \mu_\nu(M_\nu) =$$

$$= \mu_{\nu_0}(N_{\nu_0}) \prod_{\substack{\nu \in \mathcal{E} \\ \nu \neq \nu_0}} \mu_\nu(M_\nu) + \mu_{\nu_0}(P_{\nu_0}) \prod_{\substack{\nu \in \mathcal{E} \\ \nu \neq \nu_0}} \mu_{\nu_0}(M_{\nu_0}) =$$

$$= \prod_{\nu \in \mathcal{E}} \mu_\nu(N_\nu) + \prod_{\nu \in \mathcal{E}} \mu_\nu(P_\nu) = \mu(\mathcal{N}) + \mu(\mathcal{P}).$$

As for (b) the condition that for each ν a measure function $\mu_\nu(M)$ be defined over $\mathcal{R}_\nu$ imposes on $\mathcal{R}_\nu$ the requirement that a countable set from $\mathcal{R}_\nu$ cover S_ν, cf. Definition 10.1.7. Since we are dealing with an admissible system we have for almost all ν, $S_\nu \in \mathcal{R}_\nu$. Denote the exceptional ν's by $\nu_1, \nu_2, \ldots, \nu_k$. For each $i = 1, 2, \ldots, k$, there is a sequence $\{M_{\nu_i}^{(n)}\}$ of sets $M_{\nu_i}^{(n)}$ from $\mathcal{R}_{\nu_i}$ which covers S_{ν_i}. Consider now those $\mathcal{M} \equiv \prod\prod_{\nu \in \mathcal{E}} M_\nu$ from $\mathcal{R}$ which are such that if ν is distinct from $\nu_1, \nu_2, \ldots, \nu_k$ then $M_\nu = S_\nu$, and if $\nu = \nu_i$ then M_ν is a set from the sequence $\{M^{(n)}_{\nu_i}\}$. The totality of such $\mathcal{M}$'s is obviously countable and covers $\mathcal{S}$.

The finiteness and non-negativity are evident when the original $\mu_\nu(M)$'s have the respective properties.

We now prove two lemmas which will be needed in the proofs of Theorems 10.4.6 and 10.4.7. For Theorem 10.4.6 a somewhat weaker formulation than that given below would suffice: The outer measures could be replaced by measure functions on rings, and the discussion carried out on the basis of the results of Section 1 alone. Besides, in Lemma 2 it would be sufficient to consider the case where only a finite number of ν's with $M_\nu \neq S_\nu$ occur.

Lemma 1. *Let $\nu^*(M)$ be an outer measure in a space T; Q a set in T having a finite outer measure; $P_1, P_2, \ldots$ measurable sets in T having finite measures; $\beta, \alpha_1, \alpha_2, \ldots$ non-negative real numbers. Assume that for each $x \in Q$*

(1) $$\sum_{x \in P_i} \alpha_i \geqq \beta \qquad \text{(the summation is over } i\text{);}$$

then

(2) $$\sum_i \alpha_i \nu(P_i) \geqq \beta \nu^*(Q).$$

Proof: Let δ be an arbitrary positive number. Now for each $x \in Q$, $\sum_{x \in P_i} \alpha_i > \beta - \delta$; hence the inequality also holds when the (possibly) infinite sum $\sum_{x \in P_i} \alpha_i$ is replaced by a suitably chosen finite subsum. In other words, there is for each x a finite subset J(x) of the integers such that for each $i \in J(x)$, $x \in P_i$ and such that $\sum_{i \in J(x)} \alpha_i > \beta - \delta$. Regarded as finite sets of integers the totality of distinct J(x)'s is countable (a given J may correspond to many x's). Let $J_1, J_2, \ldots$ be an enumeration. Now define $F_k \equiv \prod_{i \in J_k} P_i$. The sets F_k are measurable. Furthermore since for every $x \in Q$ there is an integer $k(x)$ such that $J(x) = J_{k(x)}$, it follows that $x \in F_{k(x)}$. Hence $\sum_k F_k \supset Q$. Define $G_k = F_k - F_k(\sum_1^{k-1} F_m)$. The sets G_k are measurable and disjunct and $\sum_k G_k = \sum_k F_k$. Let it be noted that for $i_1 \in J_k$, $G_k \subset F_k = \prod_{i \in J_k} P_i \subset P_{i_1}$. Now

$$\sum_i \alpha_i \nu(P_i) \geqq \sum_i \alpha_i \nu^*(P_i Q) = \sum_i \alpha_i \nu^*(P_i(\sum_k G_k)Q) = \sum_i \sum_k \alpha_i \nu^*(P_i G_k Q) =$$

$$= \sum_k \sum_i \alpha_i \nu^*(P_i G_k Q) \geqq \sum_k \sum_{i \in J_k} \alpha_i \nu^*(P_i G_k Q) = \sum_k \sum_{i \in J_k} \alpha_i \nu^*(G_k Q) =$$

$$= \sum_k \nu^*(G_k Q)(\sum_{i \in J_k} \alpha_i) > (\beta - \delta) \sum_k \nu^*(G_k Q) = (\beta - \delta) \nu^*(Q).$$

Since δ is arbitrary this inequality gives (2).

Lemma 2. Consider a (finite or infinite) sequence of spaces $\{S_i\}$ $i = 1, 2, \ldots$ and in each S_i an outer measure $\mu_i^*(M)$. Except for a finite number of i's let $\mu_i(S_i) = \mu_i^*(S_i) = 1$. Assume that in each S_i an arbitrary set $M_i \subset S_i$ and an (infinite) sequence of measurable sets $N_i^h \subset S_i$, $h = 1, 2, \ldots$ (measurable with respect to $\mu_i^*(M)$) are given in such a manner that all $\mu_i^*(M_i)$, $\mu_i(N_i^h) = \mu_i^*(N_i^h)$ are finite and that for fixed h the number of i's for which $N_i^h \neq S_i$ is finite.

Then

$$\prod\!\!\!\prod_i M_i \subset \sum_h \prod\!\!\!\prod_i N_i^h$$

implies

$$\prod_i \mu_i^*(M_i) \leqq \sum_h \prod_i \mu_i(N_i^h).$$

(All the products here are convergent or divergent to 0, because for all except a finite number of i's the value of $\mu_i^*(M)$ lies always between 0 and 1, inclusive.)

Proof: Assume the contrary, that is

$$\prod_i \mu_i^*(M_i) > \sum_h \prod_i \mu_i(N_i^h) . \tag{3}$$

Denote by $\bar{n}$ the number of i's, so that $\bar{n} = 1, 2, \ldots$ or $+\infty$. A system $[x_1, \ldots, x_n]$, $n = 0, 1, 2, \ldots$, will be called an X-system, if $x_j \in M_j$ for $j = 1, 2, \ldots, n$ and

$$\prod_{i=n+1}^{\bar{n}} \mu_i^*(M_i) > \sum_{h=1}^{\infty}{}_{[x_j \in N_j^h \text{ for } j=1,2,\ldots,n]} \prod_{i=n+1}^{\bar{n}} \mu_i^*(N_i^h) \tag{4}$$

where the $[\ldots]$ after $\sum_{h=1}^{\infty}$ means that the sum is extended only over those h's which satisfy the condition in $[\ldots]$. If $n = 0$ there is only one possible choice of $[x_1, \ldots, x_n]$, the empty one: $[\quad]$. As (4) then coincides with (3) this is an X-system.

Assume $n < \bar{n}$, and let $[x_1, \ldots, x_n]$ be an X-system. Apply Lemma 1, putting $T = S_{n+1}$, $Q = M_{n+1}$, and the $P_1, P_2, \ldots$ equal to those N_{n+1}^h, $h=1,2,\ldots,$ for which $x_j \in N_j^h$ $(j = 1, 2, \ldots, n)$; putting β equal to $\prod_{i=n+2}^{\bar{n}} \mu_i^*(M_i)$, and $\alpha_1, \alpha_2, \ldots$ to the $\prod_{i=n+2}^{\bar{n}} \mu_i(N_i^h)$, $h = 1, 2, \ldots,$ for which $x_j \in N_j^h$, $j = 1,2,\ldots,n$. (If $n+2 > \bar{n}$ --- which occurs only if $\bar{n}$ is finite and $n = \bar{n}-1$ --- every $\prod_{i=n+2}^{\bar{n}} = 1$.) Then (4) states that (2) in Lemma 1 is not fulfilled. Therefore an $x \in Q$ must exist for which (1) is not true. Put this $x = x_{n+1}$; then $x_{n+1} \in M_{n+1}$ and the opposite of (1) is just (4) with n+1 instead of n. Thus $[x_1, \ldots, x_{n+1}]$ is an X-system.

We have thereby shown that it is possible to construct a sequence of X-systems beginning with the empty one in such a way that each contains its predecessor: $[\quad]$, $[x_1]$, $[x_1, x_2]$, $\ldots$. If $\bar{n}$ is finite all $n \leqq \bar{n}$ occur;

if $\bar{n}$ is infinite all n occur. Thus we obtain a sequence $[x_1, x_2, \ldots]$ with $x_i \in M_i$ for every i. Now by assumption $\prod_i M_i \subset \sum_h \prod_i N_i^h$. Since $[x_1, x_2, \ldots] \subset \prod_i M_i$, there must be a fixed $\bar{h}$ such that $[x_1, x_2, \ldots] \subset \prod_i N_i^{\bar{h}}$, that is $x_i \in N_i^{\bar{h}}$ for a fixed $\bar{h}$ and all i's.

If $\bar{n}$ is finite choose $n = \bar{n}$; if $\bar{n}$ is infinite choose n so large that for all $i > n$, $\mu_i(S_i) = 1$, $N_i^{\bar{h}} = S_i$ ($\bar{h}$ as defined above). The latter implies that for $i > n$, $\mu_i^*(M_i) \leqq 1$, $\mu_i(N_i^{\bar{h}}) = 1$. Thus we have

$$\prod_{i=n+1}^{\bar{n}} \mu_i^*(M_i) \leqq 1 , \quad \prod_{i=n+1}^{\bar{n}} \mu_i(N_i^{\bar{h}}) = 1 .$$

Now return to (4) with this n. The $\sum_{h=1}^{\infty} [\ldots]$ will certainly contain the term for $h = \bar{h}$ (because $x_j \in N_j^{\bar{h}}$ for all j's). Thus (4) implies $1 > 1$, which is impossible.

THEOREM 10.4.6. If all $\mu_\nu(M)$'s occurring in Definition 10.4.3 are finite, non-negative, and totally additive, then the measure function $\mu(\mathcal{M})$ constructed there is also finite, non-negative and totally additive.

Proof: Theorem 10.4.5 shows that $\mu(\mathcal{M})$ is finite and non-negative. To show the total additivity it is sufficient, as in the proof of Theorem 10.1.20, to show that if $\mathcal{M} \in \mathcal{R}$ is contained in the sum of a sequence $\{\mathcal{N}_h\}$ where $\mathcal{N}_h \in \mathcal{R}$, then $\mu(\mathcal{M}) \leqq \sum_h \mu(\mathcal{N}_h)$.

Extend the measure function $\mu(\mathcal{M})$ to its outer measure function $\mu_1^*(\mathcal{M})$ and the outer measure functions $\mu_\nu(M)$ to their outer measure functions $\mu_{\nu 1}^*(M)$ as in Theorem 10.3.2. In order to avoid confusions in the notation we shall throughout this section write $\mu_I^*(\mathcal{M})$ instead of $\mu_1^*(\mathcal{M})$, and apply a consistent notation for all other extended outer measures that may occur. (For the need of such an index cf. the Remark after Theorem 10.3.2)

Let $\mathcal{M} = \prod_{\nu\in\mathcal{E}} M_\nu$, $\mathcal{N}_h = \prod_{\nu\in\mathcal{E}} N_\nu^h$. The restrictions of Definition 10.4.3 require that for each $\mathcal{N}_h$ with the exception of a finite number of ν's, $N_\nu^h = S_\nu$, $\mu_\nu(S_\nu) = 1$. Similarly for $\mathcal{M}$. The totality of all ν's which are exceptional for $\mathcal{M}$ or some $\mathcal{N}_h$ is countable. Let these exceptional ν's be written as a sequence: $\{\nu_1, \nu_2, \ldots\}$. Then for all ν not belonging to $\{\nu_i\}$ and all h we have $M_\nu = N_\nu^h = S_\nu$, $\mu_\nu(S_\nu) = 1$. Hence $\mu(\mathcal{M}) =$

$$= \prod_{\nu\in\mathcal{E}} \mu_{\nu I}(M_\nu) = \prod_{i=1}^{\infty} \mu_{\nu_i I}(M_{\nu_i}), \text{ and } \mu(\mathcal{N}_h) = \prod_{\nu\in\mathcal{E}} \mu_{\nu I}(N_\nu^h) = \prod_{i=1}^{\infty} \mu_{\nu_i I}(N_{\nu_i}^h).$$

We prove that $\prod_i M_{\nu_i} \subset \sum_h \prod_i N_{\nu_i}^h$. Let $x(\nu_i)$ be a function such that for each i, $x(\nu_i) \in M_{\nu_i}$, that is, a point of $\prod_i M_{\nu_i}$. Define for $\nu\in\mathcal{E}$ the function $\overline{x}(\nu)$ as $x(\nu_i)$ for $\nu = \nu_i$ $(i = 1, 2, \ldots)$ and as any point of S_ν when ν does not belong to $\{\nu_1, \nu_2, \ldots\}$. Then $[\overline{x}(\nu)]$ is a point of $\mathcal{M}$, hence it belongs to some $\mathcal{N}_h$. For this particular h we have for all $\nu\in\mathcal{E}$, $\overline{x}(\nu) \in N_\nu^h$, hence in particular $x(\nu_i) \in N_{\nu_i}^h$ for all i. Thus $[x(\nu_i)]$ is a point in $\prod_i N_{\nu_i}^h$. This proves the statement at the outset of this paragraph.

We are now in a position to apply Lemma 2 with $S_i = S_{\nu_i}$, $M_i = M_{\nu_i}$, $N_i^h = N_{\nu_i}^h$, $\mu_i^*(M) = \mu_{\nu_i I}^*(M)$, and obtain $\mu(\mathcal{M}) \leq \sum_h \mu(\mathcal{N}_h)$.

This extension of measure functions from the spaces S_ν to their direct product $\mathcal{S} = \prod_{\nu\in\mathcal{E}} S_\nu$ admits of many important applications.

First, it enables us in section 5 to introduce the Lebesgue-measure, in one dimension as well as in several dimensions, without the help of any but elementary (untopological) concepts. Second, it gives an easy approach to the theory of integration; for the integral in a space S can be considered as a measure in the direct product, $S \times R_1$ of S with the one-dimensional Euclidean space R_1 (the set of real numbers). (Cf. the theory of Lebesgue integration in Chapter VIII, and our general theory of integration in Chapter XI.) Third, it provides exactly the mathematical tool needed for modern theory of probability in the form given to it by Fréchet, Kolmogoroff, Steinhaus, and others. Likewise, it is useful in some problems which are closely connected to probability theory, like those which group around the "Ergodic Theorem".

We now continue the investigation of the properties of the measure function $\mu(\mathcal{M})$ in the direct product space $\mathcal{S} = \prod\prod_{\nu \in \mathcal{E}} S_\nu$. We are chiefly interested in questions concerning outer measures, extensions, and equivalence.

First, it is convenient to introduce the notion of numerical products with arbitrarily many factors.

Definition 10.4.4. Let there be given for every $\nu \in \mathcal{E}$ a finite or infinite non-negative real number m_ν; and, except for a finite number of ν's, let $m_\nu \leq 1$. Then define the product $\prod_{\nu \in \mathcal{E}} m_\nu$ as follows: Denote the $m_\nu > 1$ by $m_{\rho_1}, \dots, m_{\rho_s}$, $(s = 0, 1, 2, \dots)$, then

$$\prod_{\nu \in \mathcal{E}} m_\nu \equiv m_{\rho_1} \cdots m_{\rho_s} \cdot \underset{\{\sigma_1, \dots, \sigma_n\} \subset \mathcal{E} - \{\rho_1, \dots, \rho_s\}}{\text{g.l.b.}} (m_{\sigma_1} \cdots m_{\sigma_n}).$$

In this g.l.b. n runs over all numbers 0, 1, 2, ..., and $\sigma_1, \dots, \sigma_n$ are any n different elements of $\mathcal{E}$, all $\neq \rho_1, \dots, \rho_s$. If it happens that factors $+\infty$ and 0 occur in the product on the right side, we use the convention $+\infty \cdot 0 = 0$.

The following remarks serve to clarify the nature of this definition:

(i) If $m_\nu < 1$ occurs an uncountably infinite number of times then $\prod_{\nu \in \mathcal{E}} m_\nu = 0$ and this even in spite of the fact that some m_ν's may be infinite. This is clear for if there are non-denumerably many m_ν's less than 1, there must be an integer p such that infinitely many m_ν's satisfy $m_\nu < 1 - \frac{1}{p}$. Thus

$\underset{\{\sigma_1,\ldots,\sigma_n\}\subset \mathcal{E} - \{\rho_1,\ldots,\rho_s\}}{\text{g.l.b.}} (m_{\sigma_1} \ldots m_{\sigma_n}) = 0$ and therefore $\prod_{\nu \in \mathcal{E}} m_\nu = 0.$

(ii) Assume that $m_\nu < 1$ occurs only a finite or countably infinite number of times. Then the ν's with $m_\nu < 1$ can be written as a (finite or infinite) sequence $\nu_1, \nu_2, \ldots$. As $0 \leqq m_{\nu_i} < 1$ for all i's, the product $\prod_i m_{\nu_i}$, if infinite at all, is convergent or it is divergent to the limit 0. Since for $\nu \neq \rho_1, \ldots, \rho_s, \nu_1, \nu_2, \ldots, m_\nu = 1$, we have

$$\underset{\{\sigma_1,\ldots,\sigma_n\} \subset \mathcal{E} - \{\rho_1,\ldots,\rho_s\}}{\text{g.l.b.}} (m_{\sigma_1} \ldots m_{\sigma_n}) = \prod_i m_{\nu_i}$$

Therefore:

(iii) If under the assumptions of (ii) $\prod_i m_{\nu_i} = 0$ (either because some $m_{\nu_i} = 0$, or because the product is infinite and divergent to the limit 0), then, even if some m_{ρ_i} are infinite, we have $\prod_{\nu \in \mathcal{E}} m_\nu = 0$.

(iv) If under the assumptions of (ii) $\prod_i m_{\nu_i} \neq 0$ (because all $m_{\nu_i} \neq 0$, and the product is either finite, or infinite and convergent), then $\prod_{\nu \in \mathcal{E}} m_\nu = m_{\rho_1} \ldots m_{\rho_s} \cdot \prod_i m_{\nu_i}$. (Now $\prod_{\nu \in \mathcal{E}} m_\nu = +\infty$ if and only if some m_{ρ_i} is infinite.)

(v) The product $\prod_{\nu \in \mathcal{E}} m_\nu$ has in common with the ordinary infinite product the property that if $\nu_1,\ldots,\nu_n$ is any finite set from $\mathcal{E}$, then $\prod_{\nu \in \mathcal{E}} m_\nu = m_{\nu_1} m_{\nu_2} \ldots m_{\nu_n} \prod_{\nu \in \mathcal{E}} m_\nu$. This is trivial if there are uncountably many ν's for which $m_\nu < 1$, for the removal of finitely many cannot change this. It is a consequence of the representation as an ordinary finite or infinite product in case the assumptions of (ii) hold.

THEOREM 10.4.7. *Let there be given an admissible system* $(\mathcal{E}, S_\nu, \mathcal{R}_\nu)$ *and in it an admissible system of finite, non-negative, and totally additive* measure functions $\mu_\nu(M)$ ($\mu_\nu(M)$ is defined in $\mathcal{R}_\nu$); construct in the sense of Definitions 10.4.1, 10.4.2 and 10.4.3 the product space $\mathcal{S} = \prod\prod_{\nu \in \mathcal{E}} S_\nu$, the half-ring $\mathcal{R}$, and in $\mathcal{R}$ construct the finite non-negative and totally additive measure function $\mu(\mathcal{M})$.

Form the outer measures $\mu^*_{\nu I}(M)$ in S_ν and $\mu^*_I(\mathcal{M})$ in $\mathcal{S}$ determined in the sense of Theorem 10.3.2 by $\mu_\nu(M)$ and $\mu(\mathcal{M})$ respectively.

If now for an arbitrary system of sets $M_\nu \subset S_\nu$ we write $\mathcal{M} = \prod\prod_{\nu \in \mathcal{E}} M_\nu$, we have

$$\mu^*_I(\mathcal{M}) = \prod\prod_{\nu \in \mathcal{E}} \mu^*_{\nu I}(M_\nu).$$

Proof: We first remark that the requirement that $\mu_\nu(M)$ be an admissible system of measure functions necessitates that for almost all ν, $\mu^*_{\nu I}(S_\nu) = \mu_\nu(S_\nu) = 1$, thus for almost all ν, $\mu^*_{\nu I}(M) \leqq 1$. Since also the $\mu^*_{\nu I}(M)$ are non-negative, the Definition 10.4.4 applies to the product $\prod\prod_{\nu \in \mathcal{E}}$ on the right-hand side of the above equation.

Denote the ν's for which $\mu^*_{\nu I}(M_\nu) > 1$ --- their number is, by the remark just made, finite --- by $\rho_1, \dots, \rho_s$. Denote the ν's which do not occur among $\rho_1, \dots, \rho_s$ and for which $S_\nu \in \mathcal{R}_\nu$, $\mu_\nu(S_\nu) = 1$ does not hold --- their number is by the admissibility of the system also finite --- by $\rho_{s+1}, \dots, \rho_t$.

We now proceed to carry out the proof of the theorem in several successive steps.

(i) It is sufficient to prove the theorem for the case where all $\mu^*_{\nu I}(M_\nu)$ are finite. Assume that the theorem has been proved for this case. The ν's for which $\mu^*_{\nu I}(M_\nu)$ is infinite, if such occur, occur among $\rho_1, \dots, \rho_s$, hence among $\rho_1, \dots, \rho_t$. For each $i = 1, 2, \dots, t$ cover the space S_{ρ_i} with a sequence $\{P^h_{\rho_i}\}$ of disjunct sets $P^h_{\rho_i}$ from $\mathcal{R}_{\rho_i}$. The existence of such a covering is implied by the existence of the measure functions $\mu_{\rho_i}(M)$.

Now for each set of t integers $h_1, h_2, \ldots, h_t$ define $\mathcal{P}^{h_1,\ldots,h_t} \equiv$ $\equiv \prod_{\nu \in \mathcal{E}} P_\nu^{h_1,\ldots,h_t}$ where $P_\nu^{h_1,\ldots,h_t} = S_\nu$ for $\nu \neq \rho_1, \ldots, \rho_t$ and $P_\nu^{h_1,\ldots,h_t} = P_{\rho_i}^{h_i}$ for $\nu = \rho_i$. It is evident that the sets $\mathcal{P}^{h_1,\ldots,h_t}$ are disjunct, belong to $\mathcal{R}$, and that their sum is $\mathcal{S}$. In $\mathcal{M}\mathcal{P}^{h_1,\ldots,h_t}$ the direct factor corresponding to S_{ρ_i} is $M_{\rho_i} P_{\rho_i}^{h_i}$ and $\mu^*_{\rho_i I}(M_{\rho_i} P_{\rho_i}^{h_i}) \leq \mu_{\rho_i}(P_{\rho_i}^{h_i}) < +\infty$. Since, by what we have assumed the theorem holds for $\mathcal{M}\mathcal{P}^{h_1,\ldots,h_t}$ we have with the help of Theorem 10.2.5

$$\mu_I^*(\mathcal{M}) = \sum_{h_1,\ldots,h_t} \mu_I^*(\mathcal{M}\mathcal{P}^{h_1,\ldots,h_t}) = \sum_{h_1,\ldots,h_t} \left(\prod_{i=1}^{t} \mu^*_{\rho_i I}(M_{\rho_i} P_{\rho_i}^{h_i}) \cdot \prod_{\substack{\nu \in \mathcal{E} \\ \nu \neq \rho_1,\ldots,\rho_t}} \mu^*_{\nu I}(M_\nu)\right) =$$

$$= \prod_{i=1}^{t} \left(\sum_{h_i=1}^{\infty} \mu^*_{\rho_i I}(M_{\rho_i} P_{\rho_i}^{h_i})\right) \cdot \prod_{\substack{\nu \in \mathcal{E} \\ \nu \neq \rho_1,\ldots,\rho_t}} \mu^*_{\nu I}(M_\nu) = \prod_{i=1}^{t} \mu^*_{\rho_i I}(M_{\rho_i}) \cdot \prod_{\nu \in \mathcal{E}} \mu^*_{\nu I}(M_\nu) =$$

$$= \prod_{\nu \in \mathcal{E}} \mu^*_{\nu I}(M_\nu) .$$

From now on we assume that $u^*_{\nu I}(M_\nu)$ are all finite.

(ii) We prove that $\mu_I^*(\mathcal{M}) \leq \prod_{\nu \in \mathcal{E}} \mu^*_{\nu I}(M_\nu)$. Choose an arbitrary positive ε. Select an integer $u \geq t$ depending on ε and u-t elements $\rho_{t+1}, \rho_{t+2}, \ldots, \rho_u$ from $\mathcal{E} - \{\rho_1, \ldots, \rho_t\}$ in such a way that $\prod_{i=t+1}^{u} \mu^*_{\rho_i I}(M_{\rho_i}) < \prod_{\nu \in \mathcal{E}} \mu^*_{\nu I}(M_\nu) + \varepsilon$. Now for each $i = 1, 2, \ldots, u$ select a sequence of sets $\{P_{\rho_i}^h\}$ from $\mathcal{R}_{\rho_i}$ in such a way that $M_{\rho_i} \subset \sum_h P_{\rho_i}^h$ and $\sum_h \mu_{\rho_i}(P_{\rho_i}^h) < \mu^*_{\rho_i I}(M_{\rho_i}) + \varepsilon$. For each set of u integers $h_1, \ldots, h_u$ form the set $\mathcal{P}^{h_1,\ldots,h_u}$ in a manner exactly analogous to that used in (i) above.

Again we have $\mathcal{P}^{h_1,\ldots,h_u}$ contained in $\mathcal{R}$ and we have $\sum_{h_1,\ldots,h_u} \mathcal{P}^{h_1,\ldots,h_u} \supset \mathcal{M}$. Therefore

$$\mu_I^*(\mathcal{M}) \leqq \sum_{h_1,\ldots,h_u} \mu(\mathcal{P}^{h_1,\ldots,h_u}) = \sum_{h_1,\ldots,h_u} \prod_{i=1}^{u} \mu_{\rho_i}(P_{\rho_i}^{h_i}) = \prod_{i=1}^{u}\left(\sum_{h_i=1}^{\infty} \mu_{\rho_i}(P_{\rho_i}^{h_i})\right) \leqq$$

$$\leqq \prod_{i=1}^{u}(\mu^*_{\rho_i I}(M_{\rho_i}) + \varepsilon) \leqq \prod_{i=1}^{t}(\mu^*_{\rho_i I}(M_{\rho_i}) + \varepsilon) \circ \left(\prod_{\substack{\nu \in \mathcal{E} \\ \nu \neq \rho_1,\ldots,\rho_t}} \mu^*_{\nu I}(M_\nu) + \varepsilon\right) .$$

and allowing ε to become zero, we get $\mu_I^*(\mathcal{M}) \leqq \prod_{\nu \in \mathcal{E}} \mu^*_{\nu I}(M_\nu)$.

(iii) To complete the proof it is sufficient to establish the reverse of this last inequality. We proceed to do this. Select a positive ε and a sequence of sets $\{\mathcal{N}^h\}$ from $\mathcal{R}$ in such a way that $\mathcal{M} \subset \sum_h \mathcal{N}^h$ and $\sum_h \mu(\mathcal{N}^h) < \mu_I^*(\mathcal{M}) + \varepsilon$. Let $\mathcal{N}^h = \prod\prod_{\nu \in \mathcal{E}} N_\nu^h$. By the definition of $\mathcal{R}$ we have for fixed h that for almost all ν, $N_\nu^h = S_\nu$. The ν's for which $\mu^*_{\nu I}(M_\nu) > 1$ or for which it is false that for all h, $N_\nu^h = S_\nu$, $\mu_\nu(S_\nu) = 1$ are therefore finite or countably infinite in number. Write them as a sequence $\nu_1, \nu_2, \ldots$. Since $\prod\prod_{\nu \in \mathcal{E}} M_\nu \subset \sum_h \prod\prod_{\nu \in \mathcal{E}} N_\nu^h$ we have evidently $\prod\prod_i M_{\nu_i} \subset \sum_h \prod\prod_i N_{\nu_i}^h$. We may therefore apply Lemma 2 and obtain

$$\prod_i \mu^*_{\nu_i I}(M_{\nu_i}) \leqq \sum_h \prod_i \mu_{\nu_i}(N_{\nu_i}^h)$$

Since for $\nu \neq \nu_1, \nu_2, \ldots$ we have $\mu^*_{\nu I}(M_\nu) \leqq 1$ and $\mu_\nu(N_\nu^h) = 1$, it follows that

$$\prod_{\nu \in \mathcal{E}} \mu^*_{\nu I}(M_\nu) \leqq \prod_i \mu^*_{\nu_i I}(M_{\nu_i}) \leqq \sum_h \prod_i \mu_{\nu_i}(N_{\nu_i}^h) = \sum_h \prod_{\nu \in \mathcal{E}} \mu_\nu(N_\nu^h) =$$

$$= \sum_h \mu(\mathcal{N}^h) < \mu_I^*(\mathcal{M}) + \varepsilon .$$

As ε makes the limiting transition to zero the last inequality together with (i) and (ii) gives the theorem.

Corollary: Let $(\mathcal{E}, S_\nu, \mathcal{R}_\nu)$, $\mu_\nu(M)$, $\mu^*_{\nu I}(M)$ and $\mu(\mathcal{M})$ and $\mu^*_I(\mathcal{M})$ be as in the preceding theorem. If $\mathcal{M}$ is an arbitrary subset of $\mathcal{S}$ then

$$\mu^*_I(\mathcal{M}) = \underset{\substack{N^h_\nu \subset S_\nu \\ \mathcal{M} \subset \sum_h \prod_{\nu \in \mathcal{E}} N^h_\nu}}{\text{g.l.b.}} \mu^*_{\nu I}(N^h_\nu)$$

Proof: The g.l.b. in question cannot exceed the g.l.b. of the same expression where the N^h_ν are restricted so that $N^h_\nu \in \mathcal{R}_\nu$ and that for fixed h for almost all ν, $N^h_\nu = S_\nu$. The latter g.l.b. is by definition $\mu^*_I(\mathcal{M})$. On the other hand the theorem shows that if $\mathcal{M} \subset \sum_h \prod_{\nu \in \mathcal{E}} N^h_\nu$, then

$$\mu^*_I(\mathcal{M}) \leqq \mu^*_I(\sum_h \prod_{\nu \in \mathcal{E}} N^h_\nu) \leqq \sum_h \mu^*_I(\prod_{\nu \in \mathcal{E}} N^h_\nu) = \sum_h \prod_{\nu \in \mathcal{E}} \mu^*_{\nu I}(N^h_\nu).$$

THEOREM 10.4.8. Let $\mathcal{E}$, S_ν, $\mathcal{S}$ be as in Theorem 10.4.7, similarly the half-rings and measure functions $\mathcal{R}_\nu$, $\mu_\nu(M)$, $\mathcal{R}$, $\mu(\mathcal{M})$, and let there be given another set of half-rings and measure functions with the same properties, $\mathcal{T}_\nu$, $\lambda_\nu(M)$, $\mathcal{T}$, $\lambda(\mathcal{M})$. If, for each $\nu \in \mathcal{E}$, $\lambda_\nu(M)$ is an extension of $\mu_\nu(M)$, then $\lambda(\mathcal{M})$ is an extension of $\mu(\mathcal{M})$. If for each $\nu \in \mathcal{E}$, $\mu_\nu(M)$ and $\lambda_\nu(M)$ are equivalent, then $\mu(\mathcal{M})$ and $\lambda(\mathcal{M})$ are equivalent.

Proof: The first statement is obvious. The second statement means that $\mu^*_{\nu I}(M) \equiv \lambda^*_{\nu I}(M)$ implies $\mu^*_I(\mathcal{M}) \equiv \lambda^*_I(\mathcal{M})$. This follows from the preceding corollary, which shows that $\mu^*_I(\mathcal{M})$ and $\lambda^*_I(\mathcal{M})$ can be defined respectively by means of the $\mu^*_{\nu I}(M)$, $\lambda^*_{\nu I}(M)$ alone.

We finally turn our attention to the notions of inner measure and of measurability. Theorems 10.4.9 and 10.4.10 will give exhaustive criteria concerning these. The Lemmas 3,4 with which we begin are special cases of these theorems, but necessary for their proof.

<u>Lemma 3. Let $\mathcal{E}$, S_ν, $\mathcal{R}_\nu$, $\mu_\nu(M)$, $\mathcal{S}$, $\mathcal{R}$, $\mu(\mathcal{M})$ be as in Theorem 10.4.7, similarly the outer measures $\mu^*_{\nu I}(M)$, $\mu^*_I(\mathcal{M})$. Form the corresponding inner measures $\mu_{\nu I*}(M)$, $\mu_{I*}(\mathcal{M})$. (Cf. Definition 10.2.7. All these outer measures are based on Theorem 10.3.2, thus regular by Theorem 10.2.10.)</u>

<u>Assume that $\mathcal{E}$ is finite. If for every $\nu \in \mathcal{E}$ an $M_\nu \subset P_\nu$, $P_\nu \in \mathcal{R}_\nu$, is given, and $\mathcal{M} = \prod\prod_{\nu \in \mathcal{E}} M_\nu$, then</u>

$$\mu_{I*}(\mathcal{M}) \leqq \prod_{\nu \in \mathcal{E}} \mu_{\nu I*}(M_\nu).$$

Proof: Let $\mathcal{E} = \{1, 2, \ldots, \bar{n}\}$. For $\bar{n} = 1$, the lemma is trivial. If it is established for $\bar{n} = 2$ and for $\bar{n} = n$ it follows at once for $\bar{n} = n+1$ as follows: Replace S_n, S_{n+1} by $S_n \times S_{n+1}$; M_n, M_{n+1} by $M_n \times M_{n+1}$; $\mathcal{R}_n$, $\mathcal{R}_{n+1}$ by $\mathcal{R}'$ (the $\mathcal{R}$ of Definition 10.4.2) and $\mu_n(M)$, $\mu_{n+1}(M)$ by $\mu'(M)$ (the $\mu(\mathcal{M})$ of Definition 10.4.3). An application of the lemma for the two factors M_n, M_{n+1} and then for the n factors $M_1, \ldots, M_{n-1}, M_n \times M_{n+1}$ gives it for the n+1 factors $M_1, M_2, \ldots, M_{n+1}$.

We now consider the case $\bar{n} = 2$. Let $\mathcal{S} = S_1 \times S_2$, $\mathcal{P} = P_1 \times P_2$, $\mathcal{M} = M_1 \times M_2$. By Theorem 10.2.18 we have $\mu_{I*}(\mathcal{M}) = \mu_I(\mathcal{P}) - \mu^*_I(\mathcal{P} - \mathcal{M})$ therefore we have only to prove that $\mu^*_I(P_1 \times P_2 - M_1 \times M_2) \geqq \mu_1(P_1)\mu_2(P_2) - \mu_{1I*}(M_1)\mu_{2I*}(M_2)$ or considering that $\mu_{1I*}(M_1) = \mu_1(P_1) - \mu^*_{1I}(P_1 - M_1)$, $\mu_{2I*}(M_2) = \mu_2(P_2) - \mu^*_{2I}(P_2 - M_2)$, we have to prove that $\mu^*_I(P_1 \times P_2 - M_1 \times M_2) \geqq \mu_1(P_1)\, \mu^*_{2I}(P_2 - M_2) + \mu^*_{1I}(P_1 - M_1)\, \mu_2(P_2) - \mu^*_{1I}(P_1 - M_1)\, \mu^*_{2I}(P_2 - M_2)$.

Now consider two sequences of sets $N_1^i \in \mathcal{R}_1$, $N_2^i \in \mathcal{R}_2$, such that $\sum_i N_1^i \times N_2^i \supset P_1 \times P_2 - M_1 \times M_2$. If $x_1 \in P_1 - M_1$ and $x_2 \in P_2$, then $[x_1, x_2] \in P_1 \times P_2 - M_1 \times M_2$; from which it follows that for some i, $[x_1, x_2] \in N_1^i \times N_2^i$, or $x_1 \in N_1^i$, $x_2 \in N_2^i$. Thus if $x_1 \in P_1 - M_1$, then

$\sum_{x_1 \in N_1^i} N_2^i \supset P_2$. Similarly if $x_1 \in M_1$, then $\sum_{x_1 \in N_1^i} N_2^i \supset P_2 - M_2$. Now choose by Theorem 10.2.14 a set $\bar{M}_1 \subset M_1$ measurable with respect to $\mu_{1I}^*(M)$ and such that $\mu_{1I}(\bar{M}_1) = \mu_{1I*}(M_1) = \mu_1(P_1) - \mu_{1I}^*(P_1 - M_1)$. Since $P_1 - \bar{M}_1 \supset P_1 - M_1$ it follows that, for $x_1 \in P_1 - M_1$, $\sum_{x_1 \in N_1^i(P_1 - \bar{M}_1)} N_2^i \supset P_2$, and therefore that $\sum_{x_1 \in N_1^i(P_1 - \bar{M}_1)} \mu_2(N_2^i) \geqq \mu_2(P_2)$. For $x_1 \in \bar{M}_1$, $\sum_{x_1 \in N_1^i} N_2^i \supset P_2 - M_2$; therefore, $\sum_{x_1 \in N_1^i \bar{M}_1} N_2^i \supset P_2 - M_2$, and hence $\sum_{x_1 \in N_1^i \bar{M}_1} \mu_2(N_2^i) \geqq \mu_{2I}^*(P_2 - M_2)$.

To these two relations Lemma 1 can be applied: replacing Q, P_i, β, α_i by $P_1 - M_1$, $N_1^i(P_1 - \bar{M}_1)$, $\mu_2(P_2)$, $\mu_2(N_2^i)$ in the first case, and by $\bar{M}_1, N_1^i\bar{M}_1$, $\mu_{2I}^*(P_2 - M_2)$, $\mu_2(N_2^i)$ in the second. We obtain:

$$\sum_i \mu_{1I}(N_1^i(P_1 - \bar{M}_1)) \cdot \mu_2(N_2^i) \geqq \mu_{1I}^*(P_1 - M_1)\, \mu_2(P_2)$$

$$\sum_i \mu_{1I}(N_1^i\bar{M}_1) \cdot \mu_2(N_2^i) \geqq \mu_{1I}(\bar{M}_1)\, \mu_{2I}^*(P_2 - M_2)$$

Adding these two inequalities, and noticing that

$$\mu_{1I}(N_1^i(P_1 - \bar{M}_1)) + \mu_{1I}(N_1^i\bar{M}_1) = \mu_{1I}(N_1^iP_1) \leqq \mu_1(N_1^i)\ ,$$

$$\mu_{1I}(\bar{M}_1) = \mu_1(P_1) - \mu_{1I}^*(M_1 - P_1)\ ,$$

we find

$$\sum_i \mu_1(N_1^i)\, \mu_2(N_2^i) \geqq \mu_{1I}^*(P_1 - M_1)\, \mu_2(P_2) +$$

$$+ (\mu_1(P_1) - \mu_{1I}^*(P_1 - M_1))\, \mu_{2I}^*(M_2 - P_2).$$

But the outer measure $\mu_I^*(P_1 \times P_2 - M_1 \times M_2)$ is defined as g.l.b. of terms of the type on the left hand side of the last inequality. Hence the lemma.

Lemma 4. Let $\mathcal{E}$, S_ν, $\mathcal{R}_\nu$, $\mu_\nu(M)$, $\mathcal{S}$, $\mathcal{R}$, $\mu(\mathcal{M})$ be as in Theorem 10.4.7, similarly the outer measures $\mu^*_{\nu I}(M)$, $\mu^*_I(\mathcal{M})$.

If for every $\nu \in \mathcal{E}$ a set M_ν is given which is measurable with respect to $\mu^*_{\nu I}(M)$, and if except for a finite or countably infinite number of ν's, $M_\nu = S_\nu$, then $\mathcal{M} = \prod\prod_{\nu \in \mathcal{E}} M_\nu$ is measurable with respect to $\mu^*_I(\mathcal{M})$.

Proof: We must prove

$$\mu_I(\mathcal{P}) \geqq \mu^*_I(\mathcal{M}\mathcal{P}) + \mu^*_I(\mathcal{P} - \mathcal{M}\mathcal{P})$$

for every $\mathcal{P} \in \mathcal{R}$.

$\mathcal{P} \in \mathcal{R}$ means $\mathcal{P} = \prod\prod_{\nu \in \mathcal{E}} P_\nu$, $P_\nu \in \mathcal{R}_\nu$, and $P_\nu \neq S_\nu$ occurs only for a finite number of ν's. The ν's for which we do not have $S_\nu \in \mathcal{R}_\nu$ $P_\nu = S_\nu$, $\mu_\nu(S_\nu) = 1$ are finite in number; let them be $\nu_1, \ldots, \nu_k$. The ν's with $\nu \neq \nu_1, \ldots, \nu_k$ and $M_\nu \neq S_\nu$ are finite or countable in number, so we can write them as a (finite or infinite) sequence: $\nu_{k+1}, \nu_{k+2}, \ldots$. Thus we have (the notation is self-explanatory)

$$\mathcal{P} = \prod_i\prod P_{\nu_i} \times \prod\prod_{\substack{\nu \in \mathcal{E} \\ \nu \neq \nu_1, \nu_2, \ldots}} S_\nu ; \qquad \mathcal{M} = \prod_i\prod M_{\nu_i} \times \prod\prod_{\substack{\nu \in \mathcal{E} \\ \nu \neq \nu_1, {}_2, \ldots}} S_\nu ,$$

with $P_{\nu_i} = S_{\nu_i}$ for $i > k$.

This implies

$$\mathcal{M}\mathcal{P} = \prod_i\prod P_{\nu_i} M_{\nu_i} \times \prod\prod_{\substack{\nu \in \mathcal{E} \\ \nu \neq \nu_1, \nu_2, \ldots}} S_\nu ,$$

$$\mathcal{P} - \mathcal{M}\mathcal{P} = \sum_j \prod_{i=1}^{j-1}\prod P_{\nu_i} \times (P_{\nu_j} - P_{\nu_j} M_{\nu_j}) \times \prod\prod_{i > j} P_{\nu_i} M_{\nu_i} \times \prod\prod_{\substack{\nu \in \mathcal{E} \\ \nu \neq \nu_1, \nu_2, \ldots}} S_\nu$$

From this we obtain, using Theorem 10.4.7 and the measurability of P_{ν_i} and M_{ν_i}:

$$\mu_I^*(\mathcal{P} - \mathcal{M}\mathcal{P}) \leqq \sum_j \prod_{i=1}^{j-1} P_{\nu_i} \cdot (\mu_{\nu_j}(P_{\nu_j}) - \mu_{\nu_j I}(P_{\nu_j} M_{\nu_j}))\cdot \prod_{i>j} \mu_{\nu_i I}(P_{\nu_i} M_{\nu_i}) =$$

$$= \sum_j [\prod_{i=1}^{j} \mu_{\nu_i}(P_{\nu_j})\cdot \prod_{i>j} \mu_{\nu_i I}(P_{\nu_i} M_{\nu_i}) - \prod_{i=1}^{j-1} \mu_{\nu_i}(P_{\nu_i})\cdot \prod_{i>j-1} \mu_{\nu_i I}(P_{\nu_i} M_{\nu_i})] =$$

$$= \lim_{j \to \infty} \prod_{i=1}^{j} \mu_{\nu_i}(P_{\nu_i})\cdot \prod_{i>j} \mu_{\nu_i I}(P_{\nu_i} M_{\nu_i}) - \prod_i \mu_{\nu_i I}(P_{\nu_i} M_{\nu_i}) =$$

$$= \prod_i \mu_{\nu_i}(P_{\nu_i}) \cdot 1 - \prod_i \mu_{\nu_i I}(P_{\nu_i} M_{\nu_i}) = \mu_I(\mathcal{P}) - \mu_I^*(\mathcal{M}\mathcal{P}),$$

that is

$$\mu_I(\mathcal{P}) \geqq \mu_I^*(\mathcal{M}\mathcal{P}) + \mu_I^*(\mathcal{P} - \mathcal{M}\mathcal{P}),$$

as we desired.

THEOREM 10.4.9. Let $\mathcal{E}$, S_ν, $\mathcal{R}_\nu$, $\mu_\nu(M)$, $\mathcal{S}$, $\mathcal{R}$, $\mu(\mathcal{M})$ be as in Theorem 10.4.7, similarly the outer measures $\mu^*_{\nu I}(M)$, $\mu^*_I(\mathcal{M})$. Form the corresponding inner measures $\mu_{\nu I*}(M)$, $\mu_{I*}(\mathcal{M})$. (Cf. Definition 10.2.7. All these outer measures are based on Theorem 10.3.2, thus regular by Theorem 10.2.10.)

If for every $\nu \in \mathcal{E}$ an $M_\nu \subset S_\nu$ is given, then we have for $\mathcal{M} = \prod\prod_{\nu \in \mathcal{E}} M_\nu$

$$\mu_{I*}(\mathcal{M}) \leqq \prod_{\nu \in \mathcal{E}} \mu_{\nu I*}(M_\nu).$$

If except for a finite or countably infinite number of ν's, $M_\nu = S_\nu$ (if $\mathcal{E}$ is finite or countably infinite, then this condition is void), then the = holds in the above relations.

Remark: The example $\mathcal{M}'$ in part (v) of the proof of Theorem 10.4.10 shows that the restriction in the second half of this theorem is really necessary.

Proof: Denote the ν's for which $\mu_{\nu I*}(M_\nu) > 1$ --- their number is finite --- by $\rho_1, \ldots, \rho_s$; denote the ν's for which $\nu \neq \rho_1, \ldots, \rho_s$ and for which $S_\nu \in \mathcal{R}_\nu$, $\mu_\nu(S_\nu) = 1$ does not hold --- their number too is finite --- by $\rho_{s+1}, \ldots, \rho_t$. ($0 \leq s \leq t$, cf. the beginning of the proof of Theorem 10.4.7.) We now carry the proof in several successive steps.

(i) As in the proof of Theorem 10.4.7, it is sufficient to make the proof of the first part of the theorem for the case where $M_{\rho_i} \subset P_{\rho_i} \in \mathcal{R}_{\rho_i}$. For, as in the proof of Theorem 10.4.7, step (i), we introduce for each S_{ρ_i}, $i = 1, \ldots, t$ a sequence of disjunct sets $P^h_{\rho_i}$, $h = 1, 2, \ldots$, with $P^h_{\rho_i} \in \mathcal{R}_{\rho_i}$ and $\sum_h P^h_{\rho_i} = S_{\rho_i}$. Define the $\mathcal{P}^{h_1,\ldots,h_t}$ as in Theorem 10.4.7. Assuming the theorem true for the sets $\mathcal{P}^{h_1,\ldots,h_t}$, the argument may be repeated literally with the following deviations: We have inner measures instead of outer measures; consequently we use Theorem 10.2.17 instead of Theorem 10.2.4; and finally we add over a sequence of inequalities instead of a sequence of equations. Thus for the first half of our theorem we can assume $M_\nu \subset P_\nu$, $P_\nu \in \mathcal{R}_\nu$ if $\nu = \rho_1, \ldots, \rho_t$. But for all other ν's this is the case too --- trivially with $P_\nu = S_\nu$.

(ii) Choose an arbitrary $u \geq t$ and $u - t$ elements of $\mathcal{E}$, all $\neq \rho_1, \ldots, \rho_t$: $\rho_{t+1}, \ldots, \rho_u$. Then

$$\mu_{I*}\Big(\prod_{i=1}^{u} M_{\rho_i} \times \prod_{\substack{\nu \in \mathcal{E} \\ \nu \neq \rho_1,\ldots,\rho_u}} S_\nu\Big) \leq \prod_{i=1}^{u} \mu_{\rho_i I*}(M_{\rho_i}).$$

If Lemma 3 is applied with u+1 factors, so that $S_1, \ldots, S_u, S_{u+1}$ are replaced by $S_{\rho_1}, \ldots, S_{\rho_u}, \prod_{\substack{\nu \in \mathcal{E} \\ \nu \neq \rho_1,\ldots,\rho_u}} S_\nu$; $M_1, \ldots, M_u, M_{u+1}$ by $M_{\rho_1}, \ldots, M_{\rho_u}, \prod_{\substack{\nu \in \mathcal{E} \\ \nu \neq \rho_1,\ldots,\rho_u}} S_\nu$; $\mu_1(M), \ldots, \mu_u(M), \mu_{u+1}(M)$ by $\mu_{\rho_1}(M), \ldots, \mu_{\rho_u}(M)$ and the measure function resulting from the formation of the direct product $\prod_{\substack{\nu \in \mathcal{E} \\ \nu \neq \rho_1,\ldots,\rho_u}}$. Since

$\mathcal{M} \subset \prod_{i}^{u} M_{\rho_i} \times \prod_{\substack{\nu \in \mathcal{E} \\ \nu \neq \rho_1, \dots, \rho_u}} S_\nu$, this implies

$$\mu_{I*}(\mathcal{M}) \leqq \prod_{i=1}^{u} \mu_{\rho_i I*}(M_{\rho_i}) .$$

But $\{\rho_{s+1}, \dots, \rho_u\}$ is really any finite set $\subset \mathcal{E} - \{\rho_1, \dots, \rho_s\}$ and $\supset \{\rho_{s+1}, \dots, \rho_t\}$, and so we have

$$\mu_{I*}(\mathcal{M}) \leqq \prod_{i=1}^{s} \mu_{\rho_i I*}(M_{\rho_i}) \cdot \underset{\substack{\{\sigma_1, \dots, \sigma_n\} \subset \mathcal{E} - \{\rho_1, \dots, \rho_s\} \\ \supset \{\rho_{s+1}, \dots, \rho_t\}}}{\text{g.l.b.}} \left(\prod_{j=1}^{n} \mu_{\sigma_j I*}(M_{\sigma_j}) \right) =$$

$$= \prod_{i=1}^{s} \mu_{\rho_i I*}(M_{\rho_i}) \cdot \underset{\{\sigma_1, \dots, \sigma_n\} \subset \mathcal{E} - \{\rho_1, \dots, \rho_s\}}{\text{g.l.b.}} \left(\prod_{j=1}^{u} \mu_{\sigma_j I*}(M_{\sigma_j}) \right) =$$

$$= \prod_{\nu \in \mathcal{E}} \mu_{\nu I*}(M_\nu).$$

This completes the proof of the first half of our theorem.

(iii) Let us now consider the second half of our theorem. We conserve the notation $\rho_1, \dots, \rho_s, \rho_{s+1}, \dots, \rho_t$, but forget about the $\rho_{t+1}, \dots, \rho_u$ from (ii). The number of the ν's with $\nu \neq \nu_1, \dots, \nu_t$ and $M_\nu \neq S_\nu$ is finite or countably infinite, so we can write them as a finite or infinite sequence $\rho_{t+1}, \rho_{t+2}, \dots$.

Choose for each $i = 1, 2, \dots$ a set $\overline{M}_{\rho_i}$ which is measurable with respect to $\mu^*_{\rho_i I}(M)$, $\overline{M}_{\rho_i} \subset M_{\rho_i}$, $\mu_{\rho_i I}(\overline{M}_{\rho_i}) = \mu_{\rho_i I*}(M_{\rho_i})$ (by Theorem 10.2.12). Put $\overline{\mathcal{M}} = \prod_{i} \overline{M}_{\rho_i} \times \prod_{\substack{\nu \in \mathcal{E} \\ \nu \neq \rho_1, \rho_2, \dots}} S_\nu$. By Lemma 4, $\overline{\mathcal{M}}$ is measurable with respect to $\mu^*_I(\mathcal{M})$, and we have

$$\overline{\mathcal{M}} = \prod_i \overline{M}_{\rho_i} \times \prod_{\substack{\nu \in \mathcal{E} \\ \nu \neq \rho_1, \rho_2, \ldots}} S_\nu \subset \prod_i M_{\rho_i} \times \prod_{\substack{\nu \in \mathcal{E} \\ \nu \neq \rho_1, \rho_2, \ldots}} S_\nu = \mathcal{M}.$$

So we have (using Theorem 10.4.7),

$$\mu_{I*}(\mathcal{M}) \geq \mu_I(\overline{\mathcal{M}}) = \prod_i \mu_{\rho_i I}(\overline{M}_{\rho_i}) = \prod_i \mu_{\rho_i I*}(M_{\rho_i}) = \prod_{\nu \in \mathcal{E}} \mu_{\nu I*}(M_\nu).$$

This completes the proof of the second half of our theorem.

THEOREM 10.4.10. *Let* $\mathcal{E}$, S_ν, $\mathcal{R}_\nu$, $\mu_\nu(M)$, $\mathcal{S}$, $\mathcal{R}$, $\mu(\mathcal{M})$ *be as in Theorem 10.4.7, similarly the outer measures* $\mu^*_{\nu I}(M)$, $\mu^*_I(\mathcal{M})$.

If for every $\nu \in \mathcal{E}$ *an* $M_\nu \subset S_\nu$ *is given, then* $\mathcal{M} = \prod_{\nu \in \mathcal{E}} M_\nu$ *will be measurable with respect to* $\mu^*_I(\mathcal{M})$ *if and only if one of the two following conditions holds:*

1) *Either* $\mu^*_I(\mathcal{M}) = \prod_{\nu \in \mathcal{E}} \mu^*_{\nu I}(M_\nu) = 0$

2) *Or* $\mu^*_I(\mathcal{M}) = \prod_{\nu \in \mathcal{E}} \mu^*_{\nu I}(M_\nu) \neq 0$, *and besides:*

   a) *For each* $\nu \in \mathcal{E}$, M_ν *is measurable with respect to* $\mu^*_{\nu I}(M)$

   b) *The number of* ν*'s for which* $M_\nu \neq S_\nu$, $\mu_{\nu I}(M_\nu) = \mu_\nu(S_\nu)$ *is finite or countably infinite. (If* $\mathcal{E}$ *is finite or countably infinite, then this condition is void.)*

Proof: We carry the proof in several steps.

(i) As in the proof of Theorems 10.4.7, part (i), introduce the $\rho_1, \ldots, \rho_s, \rho_{s+1}, \ldots, \rho_t$ with $\mu^*_{\nu I}(M_\nu) \leq 1$ for $\nu \neq \rho_1, \ldots, \rho_s$ and $S_\nu \in \mathcal{R}_\nu$, $\mu_\nu(S_\nu) = 1$ for $\nu \neq \rho_1, \ldots, \rho_t$, and for each $i = 1, \ldots, t$ (t, not s!) the disjunct sets $P^h_{\rho_i}$, $h = 1, 2, \ldots$, with $P^h_{\rho_i} \in \mathcal{R}_{\rho_i}$, $\sum_h P^h_{\rho_i} = S_{\rho_i}$. Consider any combination $h_1, \ldots, h_t = 1, 2, \ldots$, and put $\mathcal{P}^{h_1, \ldots, h_t} = \prod_{i=1}^{t} P^{h_i}_{\rho_i} \times \prod_{\substack{\nu \in \mathcal{E} \\ \nu \neq \rho_1, \ldots, \rho_t}} S_\nu$, as in Theorem 10.4.2. Then the $\mathcal{P}^{h_1,\dots,h_t}$ are disjunct, $\mathcal{P}^{h_1,\dots,h_t} \in \mathcal{R}$, $\sum_{h_1,\dots,h_t} \mathcal{P}^{h_1,\dots,h_t} = \mathcal{S}$. The measurability of $\mathcal{M}$ is equivalent to that of all sets $\mathcal{M}\mathcal{P}^{h_1,\dots,h_t}$ which arise if we replace the M_{ρ_i}, $i = 1, \dots, t$, by $M_{\rho_i} P_{\rho_i}^{h_i}$ respectively. The conditions 1), 2) of our theorem go over into themselves by this substitution.

This proves that we can assume $M_\nu \subset P_\nu$, $P_\nu \in \mathcal{R}_\nu$ for $\nu = \rho_1, \dots, \rho_t$. But for all other ν's this is the case, too, trivially with $P_\nu = S_\nu$.

(ii) Condition 1) is sufficient by Theorem 10.2.1, condition 2) is sufficient by Lemma 2, if we remember that the existence of an uncountably infinite number of ν's with $\mu_{\nu I}(M_\nu) = \mu^*_{\nu I}(M_\nu) < 1$ would imply $\prod_{\nu \in \mathcal{E}} \mu^*_{\nu I}(M_\nu) = 0$ by remark (i) after Definition 10.4.4, and thus contradict one of the assumptions of 2). Thus we need only prove that 1) or 2) is necessary, that is: If $\mathcal{M} = \prod\prod_{\nu \in \mathcal{E}} M_\nu$ is measurable with respect to $\mu^*_I(\mathcal{M})$, and $\mu^*_I(\mathcal{M}) = \prod_{\nu \in \mathcal{E}} \mu^*_{\nu I}(M_\nu) \neq 0$, then 2) is fulfilled.

(iii) Decompose $\mathcal{E}$ into two disjunct subsets: $\mathcal{E} = \mathcal{E}' + \mathcal{E}''$. Put $\mathcal{S}' = \prod\prod_{\nu \in \mathcal{E}'} S_\nu$, $\mathcal{S}'' = \prod\prod_{\nu \in \mathcal{E}''} S_\nu$, and similarly $\mathcal{M}' = \prod\prod_{\nu \in \mathcal{E}'} M_\nu$, $\mathcal{M}'' = \prod\prod_{\nu \in \mathcal{E}''} M_\nu$; let $\mu'(\mathcal{N}')$, $\mu''(\mathcal{N}'')$ be the measure functions resulting from forming the direct products $\prod\prod_{\nu \in \mathcal{E}'}$, $\prod\prod_{\nu \in \mathcal{E}''}$, respectively; let $\mu'^*_I(\mathcal{N}')$, $\mu''^*_I(\mathcal{M}'')$ be the corresponding outer measures, $\mu'_{I*}(\mathcal{M}')$, $u''_{I*}(\mathcal{N}'')$ the corresponding inner measures. Then by Theorem 10.4.7 and 10.4.9 applied to two factors

$$\mu^*_I(\mathcal{M}) = \mu'^*_I(\mathcal{M}') \cdot \mu''^*_I(\mathcal{M}''), \quad \mu_{I*}(\mathcal{M}) = \mu'_{I*}(\mathcal{M}') \cdot \mu''_{I*}(\mathcal{M}'') .$$

As $\mathcal{M}$ is measurable with resepct to $\mu^*_I(\mathcal{M})$, we have $\mu^*_I(\mathcal{M}) = \mu_{I*}(\mathcal{M})$, thus $\mu'^*_I(\mathcal{M}') \cdot \mu''^*_I(\mathcal{M}'') = \mu'_{I*}(\mathcal{M}') \cdot \mu''_{I*}(\mathcal{M}'') \leq \mu'_{I*}(\mathcal{M}') \cdot \mu''^*_I(\mathcal{M}'')$. As $\mu^*_I(\mathcal{M}) > 0$, we have $\mu''^*_I(\mathcal{M}'') > 0$, thus we can divide the above inequality by $\mu''^*_I(\mathcal{M}'')$, obtaining $\mu_I'^*(\mathcal{M}') \leqq \mu_{I*}'(\mathcal{M}')$, $\mu_I'^*(\mathcal{M}') = \mu_{I*}'(\mathcal{M}')$. As $\mu_I^*(\mathcal{M})$ is finite and $\mu_I''^*(\mathcal{M}'') > 0$, $\mu_I'^*(\mathcal{M}')$ must be finite, too, and so we can apply Theorem 10.2.15 and obtain the fact that $\mathcal{M}'$ is measurable with respect to $\mu_I'^*(\mathcal{N}')$.

(iv) Apply (iii) to $\mathcal{E}' = \{\nu\}$. It shows that M_ν is measurable with respect to $\mu_{\nu I}^*(M)$, that is the necessity of 2) a). It remains to prove the necessity of 2) b), that is this: Under our present assumptions (including the measurability of all M_ν's) the number of the ν's with $M_\nu \neq S_\nu$, $\mu_{\nu I}(M_\nu) = \mu_\nu(S_\nu)$ cannot be uncountably infinite. It is, of course, sufficient to prove that the number of the ν's with $M_\nu \neq S_\nu$, $\mu_{\nu I}(M_\nu) = u_\nu(S_\nu) = 1$ cannot be uncountably infinite. Assume that it is, and denote the set of these ν's by $\mathcal{E}'$. Apply now (iii): $\mathcal{M}' = \prod\prod_{\nu \in \mathcal{E}'} M_\nu$ must be measurable with respect to $\mu_I'^*(\mathcal{N}')$. So if we can prove that it is not, the contradiction is obtained.

(v) Thus we have an uncountably infinite $\mathcal{E}'$; for every $\nu \in \mathcal{E}'$, M_ν is measurable with respect to $\mu_{\nu I}^*(M)$, $M_\nu \neq S_\nu$, and $\mu_{\nu I}(M_\nu) = \mu_\nu(S_\nu) = 1$. We wish to prove that $\mathcal{M}' = \prod\prod_{\nu \in \mathcal{E}'} M_\nu$ is not measurable. Now $\mu_I'^*(\mathcal{M}') = \prod\prod_{\nu \in \mathcal{E}} \mu_{\nu I}^*(M_\nu) = 1$ (Theorem 10.4.7); $\mu_{I*}'(\mathcal{M}') = \mu'(\mathcal{S}') - \mu_I'^*(\mathcal{S}' - \mathcal{M}') = 1 - \mu_I'^*(\mathcal{S}' - \mathcal{M}')$. We will prove $\mu_{I*}'(\mathcal{M}') = 0$, thus establishing the non-measurability of $\mathcal{M}'$ or, what is the same thing, the equation $\mu_I'^*(\mathcal{S}' - \mathcal{M}') = 1$. (Note that $\prod\prod_{\nu \in \mathcal{E}'} \mu_{\nu I*}(M\ ) = 1$, so that it will turn out to be $> \mu_{I*}'(\mathcal{M}')$, in accordance with the Remark after Theorem 10.4.9.)

Remembering the definition of outer measure in Theorem 10.3.2, we see that this means: If for each $\nu \in \mathcal{E}'$ as a sequence of sets $N_\nu^h \in \mathcal{R}_\nu$ is given, such that for every fixed h all $N_\nu^h = S_\nu$, except for a finite number of ν's (depending on h), and if $\sum_h \prod\prod_{\nu \in \mathcal{E}'} N_\nu^h \supset \mathcal{S}' - \mathcal{M}' = \prod\prod_{\nu \in \mathcal{E}'} S_\nu - \prod\prod_{\nu \in \mathcal{E}'} M_\nu$, then $\sum_h \prod\prod_{\nu \in \mathcal{E}'} \mu_\nu(N_\nu^h) \geqq 1$.

The ν's for which $N_\nu^h \neq S_\nu$ holds for any $h = 1, 2, \ldots$, form a (finite or infinite) sequence: $\nu_1, \nu_2, \ldots$. Thus $\mathcal{E}'$ contains an element $\bar{\nu} \neq \nu_1, \nu_2, \ldots$. Choose a $\xi \in S_{\bar{\nu}} - M_{\bar{\nu}}$. If any sequence $x_{\nu_i} \in S_{\nu_i}$, $i = 1, 2, \ldots$, is given, we can define a function $\bar{x}(\nu) \in S_\nu$ with $\bar{x}(\nu_i) = x_{\nu_i}$ for $i = 1, 2, \ldots$, $\bar{x}(\bar{\nu}) = \xi$. Then $[\bar{x}(\nu)] \in \prod\prod_{\nu \in \mathcal{E}'} S_\nu - \prod\prod_{\nu \in \mathcal{E}'} M_\nu \subset \sum_h \prod\prod_{\nu \in \mathcal{E}'} N_\nu^h$, and thus for some $h = 1, 2, \ldots$, $[\bar{x}(\nu)] \in \prod\prod_{\nu \in \mathcal{E}'} N_\nu^h$, $\bar{x}(\nu) \in N_\nu^h$, and so in particular $x_{\nu_i} \in N_{\nu_i}^h$ for all $i = 1, 2, \ldots$. This proves that $\sum_h \prod\prod_i N_{\nu_i}^h = \prod\prod_i S_{\nu_i}$. Therefore

$$\sum_h \prod_i \mu_{\nu_i}(N_{\nu_i}^h) \geqq \prod_i \mu_{\nu_i}(S_{\nu_i}) = 1,$$

but $\sum_h \prod_{\nu \in \mathcal{E}'} \mu_\nu(N_\nu^h) = \sum_h \prod_i \mu_{\nu_i}(N_{\nu_i}^h)$, thus proving our statement.

§5. Examples of measure functions

In what follows we shall develop a number of non-trivial examples of the foregoing theory. These examples will not only be illustrative but one of them, the Stieltjes measure (Example 4), will have important applications to operator theory. In the course of examples 2 and 3 we shall give an entirely new and untopological derivation of the properties of the Lebesgue measure, which has already been discussed in Chapters 2-5.

Example 1: The discrete space

Let $S = \{x_1, x_2, \ldots\}$ be a finite or countably infinite set. Assign to each x_n a finite, non-negative "weight" α_n.

Definition 10.5.1. *Denote by $\mathcal{R}_\alpha$ the totality of subsets of S which are either empty or contain exactly one element. For each $M \in \mathcal{R}_\alpha$ define $\mu_\alpha(M) = \alpha_n$ if $M = \{x_n\}$, and $\mu_\alpha(M) = 0$ if $M = 0$.*

THEOREM 10.5.1. *$\mathcal{R}_\alpha$ is a half-ring; $\mu_\alpha(M)$ is a finite, non-negative, and totally-additive measure function on $\mathcal{R}_\alpha$.*

Proof: We omit a formal proof. It is trivial to verify that $\mathcal{R}_\alpha$ is a half-ring. The additivity and total additivity of $\mu_\alpha(M)$ are also trivial since none but one-element and empty sets occur in $\mathcal{R}_\alpha$. The special character of S assures that a countable set from $\mathcal{R}_\alpha$ covers S .

THEOREM 10.5.2. Concerning the space S, the system $\mathcal{R}_\alpha$, and the measure function $\mu_\alpha(M)$ we may make the following assertions:

1) $BR(\mathcal{R}_\alpha) = \mathcal{P}(S)$.

2) With respect to the outer measure $\mu^*_{\alpha I}(M)$ associated with $\mu_\alpha(M)$, every set $M \in \mathcal{P}(S)$ is measurable; i.e., $MS_{\mu^*_{\alpha I}} = \mathcal{P}(S)$.

3) $\tilde{\mu}_\alpha(M) = \mu_{\alpha I}(M) = \mu^*_{\alpha I}(M) = \sum_{x_n \in M} \alpha_n$

4) A necessary and sufficient condition that $M \in \mathcal{P}(S)$ belong to $MS'_{\mu^*_{\alpha I}}$ is that $\sum_{x_n \in M} \alpha_n$ be finite.

Cf. Theorem 10.3.2. As in §4, we use the notation $\mu^*_I(M)$ instead of $\mu^*_1(M)$.

Proof: 1) follows from the fact that every $M \in \mathcal{P}(S)$ is the sum of finitely or countably many sets from $\mathcal{R}_\alpha$, hence the fact that $BR(\mathcal{R}_\alpha) \supset \mathcal{P}(S)$ 2) follows from 1) because $\mathcal{P}(S) = BR(\mathcal{R}_\alpha) \subset MS_{\mu^*_{\alpha I}} \subset \mathcal{P}(S)$. 3) is a consequence of 3) and the definition of $MS'_{\mu^*_{\alpha I}}$.

Definition 10.5.2. We shall call $\tilde{\mu}_\alpha(S) = \sum_n \alpha_n$ the total weight of S.

It is evident that the finiteness of the total weight of S is equivalent to the finiteness of the measure function $\tilde{\mu}_\alpha(M)$ --- that is, equivalent to $MS'_{\mu^*_{\alpha I}} = MS_{\mu^*_{\alpha I}}$, or $\tilde{\mu}'_\alpha(M) \equiv \tilde{\mu}_\alpha(M)$. The cases where the total weight of S is 1 are of special interest.

We proceed to give some examples.

Example 1a. Let S be a finite set: $S = \{1, 2, \ldots, m\}$; let $\alpha_1 = \ldots = \alpha_m = \frac{1}{m}$. $\tilde{\mu}_\alpha(M)$ is the number of elements of M, divided by m. The total weight is 1.

Example 1b. Let S be the set $\{1, 2, \ldots\}$; $\alpha_1 = \alpha_2 = \ldots = 1$. Then $\tilde{\mu}_\alpha(M)$ is the number of elements of M; $MS'_{\mu^*_{\alpha I}}$ is the system of all finite sets; the total weight is $+\infty$.

Example 1c. Again, let S be the set $\{1, 2, \ldots\}$; let $\alpha_1 = \frac{1}{2}$, $\alpha_2 = \frac{1}{4}$, ... Then $\tilde{\mu}_\alpha(M) = \sum_{n \in M} \frac{1}{2^n}$; the total weight is 1.

Example 2. Lebesgue measure in the interval I: $0 < x \leqq 1$.

In terms of the preceding examples we proceed to give non-topological definitions of Lebesgue measure. This can be done in two different ways.

We shall denote by $\sum_2$, and $\mu_{\Delta_2}(M)$ respectively, the S and $\tilde{\mu}_\alpha(M)$ of Example 1a with $m = 2$, and by $\sum_\omega$, $\mu_{\Delta_\omega}(M)$ respectively, the S and $\tilde{\mu}_\alpha(M)$ of Example 1c.

First Definition. Consider the direct product space $\overline{S}' = \sum_2 \times \sum_2 \times \ldots$ where there are denumerably many factors $\sum_2$. In each factor space $\sum_2$ consider the half-ring $\mathcal{R}(\sum_2)$ and on it the measure function $\mu_{\Delta_2}(M)$. Since $\sum_2 \in \mathcal{R}(\sum_2)$ and $\mu_{\Delta_2}(\sum_2) = 1$, we have an admissible system in the sense of Definition 10.4.3. We can therefore form in $\overline{S}'$ the half-ring $\overline{\mathcal{R}}'$ and on it the measure function $\overline{\mu}'(M)$. ($\overline{S}'$, $\overline{\mathcal{R}}'$, $\overline{\mu}'(M)$ are respectively the $\mathcal{S}$, $\mathcal{R}$, $\mu(\mathcal{M})$ of Definition 10.4.3.)

Denote by ξ the general point of $\overline{S}'$. Thus ξ is a sequence $\xi = [x_1, x_2, \ldots]$ where each $x_n \in \{1, 2\}$. Define $x(\xi) = \sum_{n=1}^{\infty} \frac{x_n - 1}{2^n}$. We have

THEOREM 10.5.3. Let $S°$ be the set of all points $\xi = [x_1, x_2, \ldots]$ of $\overline{S}'$ such that for almost all n, $x_n = 1$. Then

1) $S°$ is a countable set.

2) S° <u>is measurable with respect to</u> $\bar{\mu}_I'^{*}(M)$ <u>and we have</u>

$$\tilde{\bar{\mu}}'(S^\circ) = \bar{\mu}_I'(S^\circ) = \bar{\mu}_I'^{*}(S^\circ) = 0.$$

3) $\overline{S}' - S^\circ$ <u>is mapped by</u> $x = x(\xi)$ <u>in a one-to-one way upon the interval</u> I: $0 < x \leq 1$.

Proof: Conclusion 1) is evident from the definition of S°. If $\{\xi\}$ is a set consisting of the single point $\xi = [x_1, x_2, \ldots]$, then $\{\xi\} = \{x_1\} \times \{x_2\} \times \ldots$, and we have

$$\bar{\mu}'(\{\xi\}) = \mu_{\Delta_2}(\{x_1\}) \cdot \mu_{\Delta_2}(\{x_2\}) \ldots = \tfrac{1}{2} \cdot \tfrac{1}{2} \ldots = 0.$$

Thus for every countable M, $\bar{\mu}'(M) = 0$. This establishes conclusion 2). 3) is a well known property of "dyadic expansions".

In view of the last theorem, it is possible to interpret in the interval I certain of the results obtained in $\overline{S}'$. As S° is a zero set in $\overline{S}'$, the measurability properties of a set $M \subset \overline{S}'$ are not affected if M is replaced by $M(\overline{S}' - S^\circ) = M - MS^\circ$. If M is any set of $\overline{S}'$ we shall call the $x = x(\xi)$ image of $M - MS^\circ$ the image of M in I. This correspondence between sets is of course not one-to-one, but it has the property that any two sets of $\overline{S}'$ which have the same image in I differ at most on S°. In general, if $\mathcal{T}$ is a collection of sets in $\mathcal{P}(\overline{S}')$ we shall speak of the totality of sets of $\mathcal{P}(I)$ which are images of some set of $\mathcal{T}$ as the image of $\mathcal{T}$ in $\mathcal{P}(I)$. Since the mapping $x = x(\xi)$ is a one-to-one between $\overline{S}' - S^\circ$ and I, there exists for each subset of M of I a set of $\overline{S}'$, indeed of $\overline{S}' - S^\circ$, whose image is M. The nature of this mapping makes it possible to speak of the $\tilde{\bar{\mu}}'$ - measure of a set $M \subset I$. By this we mean the $\tilde{\bar{\mu}}'$-measure of (any) set of $\overline{S}'$ whose image is M.

<u>THEOREM 10.5.4.</u> <u>The image of</u> $BR(\mathcal{R}')$ <u>in</u> $\mathcal{P}(I)$ <u>is the set of all Borel sets.</u>

Proof: (i) The image of every set of $BR(\overline{\mathcal{R}}')$ is a Borel set in I. We shall have proved this statement if we can show that the image of every set of $\overline{\mathcal{R}}'$ is a Borel set. For, let $\mathcal{Y}$ be the totality of sets of $\overline{S}'$ whose image is a Borel set. Then it is easily verified that $\mathcal{Y}$ is a Borel -ring. The temporary hypothesis that the image of every set of $\overline{\mathcal{R}}'$ is a Borel set says that $\mathcal{Y} \supset \overline{\mathcal{R}}'$. We should therefore have $\mathcal{Y} \supset BR(\overline{\mathcal{R}}')$. We proceed to establish that $\mathcal{Y} \supset \overline{\mathcal{R}}'$. A set $M \in \overline{\mathcal{R}}'$ has the form

$M = M_1 \times M_2 \times \ldots \times M_p \times \Sigma_2 \times \Sigma_2 \times \ldots$, where $M_1, \ldots, M_p \subset \Sigma_2 = \{(1,2)\}$.

It is sufficient to consider those M's where each M_m has exactly one element, as all other M's are finite sums of such. Thus we have $M_m = \{x_m\}$ where x_m is either 1 or 2. Then $M = M_1 \times \ldots \times M_p \times \Sigma_2 \times \Sigma_2 \ldots$ clearly corresponds in I to the half-open interval $\sum_{n=1}^{p} \frac{x_n - 1}{2^n} < x \leqq \sum_{n=1}^{p} \frac{x_n - 1}{2^n} + \frac{1}{2^p}$ which is a Borel set.

(ii) Every Borel set in I is the image of a set from $BR(\overline{\mathcal{R}}')$. Denote by $\mathcal{T}$ those sets of $\mathcal{P}(I)$ which are images of some set of $BR(\overline{\mathcal{R}}')$. It is again easily verified that $\mathcal{T}$ is a Borel-ring in $\mathcal{P}(I)$. The Borel sets of I are the sets of the Borel-ring generated from the totality of the intersections of I with open sets. Thus we shall have established the statement above if we can show that the intersection of I with any open set belongs to $\mathcal{T}$. Now the intersection of I with an open set is expressible as the sum of countably many (they may be assumed disjunct) half-open intervals of the type $a < x \leqq b$ where $0 < a < b \leqq 1$ and a and b are "dyadically rational". Each such interval in turn may be written as the sum of finitely many disjunct half-open intervals of the form $\frac{N}{2^p} < x \leqq \frac{N+1}{2^p}$ where N and p are integers satisfying the conditions $p > 0$, $0 \leqq N < 2^p - 1$. We have already established in (i) above that these latter intervals are the images of sets from $\overline{\mathcal{R}}'$ hence from

$BR(\bar{\mathcal{R}}')$. Since $\mathcal{J}$ is a Borel-ring it contains with these intervals the sum of any countable set of them; hence it contains the intersection of I with any open set. This establishes the second part of the theorem.

THEOREM 10.5.5. Any open, half-open, or closed interval of the form $a \underset{(=)}{<} x \underset{(=)}{<} b$ $(0 \leqq a \leqq b \leqq 1)$ in I has (is the image of a set having) the $\tilde{\bar{\mu}}'$-measure $b - a$.

Proof: The intervals in question belong to the domain of $\tilde{\bar{\mu}}'(M)$ by Theorem 10.5.4. If the theorem holds for all open intervals $a < x < b$ with $0 \leqq a < b \leqq 1$, then it carries over to all other cases, because from the proof of Theorem 10.5.3, 1) the measure of a set consisting of a single point is 0. Further, if the present theorem holds for all half-open intervals $\frac{N}{2^p} < x \leqq \frac{N+1}{2^p}$ (N and p as in Theorem 10.5.4), it will hold for open intervals because they can be represented as the sum of countably many disjunct such intervals. We thus need only to prove that a set M whose image in $\mathbf{I}$ is the interval

$$\frac{N}{2^p} = \sum_{n=1}^{p} \frac{x_n - 1}{2^n} < x \leqq \sum_{n=1}^{p} \frac{x_n - 1}{2^n} + \frac{1}{2^p} = \frac{N+1}{2^p}$$ has the $\tilde{\bar{\mu}}'$-measure $\frac{1}{2^p}$. As this M is equal, up to except possibly a subset of S°, to $\{x_1\} \times \ldots \times \{x_p\} \times \Sigma_2 \times \Sigma_2 \times \ldots$, this fact follows from the definition of $\tilde{\bar{\mu}}'(M)$.

THEOREM 10.5.6. Let $\bar{\mathcal{R}}_i$ be the set of all half-open intervals M: $a < x \leqq b$ $(0 \leqq a \leqq b \leqq 1)$ in I. Define for $M \in \bar{\mathcal{R}}_i$, $\bar{\mu}_i(M) = b-a$. Then

1) $\bar{\mathcal{R}}_i$ is a half-ring.

2) $\bar{\mu}_i(M)$ is a finite, non-negative, totally-additive measure function.

3) $\bar{\mu}_i(M)$ is equivalent to $\bar{\mu}'(M)$.

Proof: Conclusion 1) is evident. By Theorem 10.5.5, $\bar{\mu}'(M)$ is an extension of $\bar{\mu}_i(M)$ and therefore conclusion 2) holds. As for 3), we need to prove that $\overline{\mu}^*_{iI}(M) \equiv \overline{\mu}'^*_I(M)$. Since $\bar{\mu}_i(M)$ and $\bar{\mu}'(M)$ agree on $\bar{\mathcal{R}}_i$, it is sufficient by the Corollary to Theorem 10.3.1 to show that $\bar{\mathcal{R}}_i$ determines both outer measures. For $\overline{\mu}^*_{iI}(M)$ this is definitory; $\overline{\mu}'^*_I(M)$ is definitorily determined by $\bar{\mathcal{R}}'$ (more precisely in I by the image of $\bar{\mathcal{R}}'$). By Theorem 10.5.4 the image of $BR(\bar{\mathcal{R}}')$ is $BR(\bar{\mathcal{R}}_i)$. Thus 3) is established. Now simply define the Lebesgue measure $\bar{\mu}_\chi(M)$ as the maximal equivalent measure function to $\bar{\mu}_i(M)$ and $\bar{\mu}'(M)$

$$\mu_\chi(M) \equiv \tilde{\bar{\mu}}_i(M) \equiv \tilde{\bar{\mu}}'(M).$$

We can now discuss how the customary method of introducing Lebesgue measure --- the one used in Chapters 2-4 compares with the present method. These remarks will also apply to the situation to be encountered in Example 3. The customary method consists in forming $\mu_i(M)$, or its many dimensional analogue, proving its additivity and total additivity --- the latter by a topological process, cf. Example 4 Theorem 10.5.18 as well as Theorems 1.3 and 2.3 and Theorem 10.1.20 of the present chapter --- and then passing over from the outer measure $\overline{\mu}^*_{iI}(M)$ to $\tilde{\mu}_i(M)$. This is now all automatically taken care of by the general theorems of sections 1-4 of this chapter.

It may be remarked that we could have used the $\sum_m$'s of Example 1a with any $m > 1$. The results would have been similar to those obtained above for $m = 2$.

<u>Second Definition</u>. Form with the sets of Example 1c the space $\bar{S}'' = \sum_\omega \times \sum_\omega \times \ldots$ to a countably infinite number of factors. For each $\sum_\omega$ use the half-ring $\mathcal{P}(\sum_\omega)$ and the measure function $\mu_{\Delta_\omega}(M)$. As $\sum_\omega \in \mathcal{P}(\sum_\omega)$ and $\mu_{\Delta_\omega}(\sum_\omega) = 1$, Definition 10.4.3 applies with this as an admissible system and we can form in the product space $\bar{S}''$ the half-ring $\bar{\mathcal{R}}''$ and the measure func-

tion $\bar{\mu}''(M)$ (corresponding respectively to $\mathcal{S}$, $\mathcal{R}$, $\mu(\mathcal{M})$ in Definition 10.4.3). The general point ξ of $\bar{S}''$ is a sequence $\xi = [y_1, y_2, \ldots]$ where the y_n are arbitrary positive integers. Define

$$y(\xi) = \sum_{n=1}^{\infty} \frac{1}{2^{y_1+\ldots+y_n}},$$

then we have

<u>THEOREM 10.5.7.</u> <u>$\bar{S}''$ is mapped by $x = y(\xi)$ in a one-to-one way on the interval</u> I: $0 < x \leqq 1$.

Proof: $x = y(\xi) = \sum_{n=1}^{\infty} \frac{1}{2^{y_1+\ldots+y_n}}$ is clearly a possible way to describe the (unique) dyadic expansion of x in which infinitely many terms $\neq 0$ occur.

As was the case with Theorem 10.5.3, this result makes it possible to replace subsets of $\bar{S}''$ by subsets of I. Here, however, it is even simpler as there is nothing to take the place of the exceptional set $S°$. Hereafter we shall speak of the sets of $\bar{S}''$ in terms of their images in I.

<u>THEOREM 10.5.8.</u> <u>$BR(\bar{\mathcal{R}}'')$ in I is the set of all Borel sets.</u>

Proof: (i) Every $M \in BR(\bar{\mathcal{R}}'')$ is (has as its image in I) a Borel set. As in the proof of Theorem 10.5.4, we need only to consider the M of the form $M = M_1 \times \ldots \times M_p \times \sum_{\omega} \times \sum_{\omega} \times \ldots$ where M_n consists of a single element y_n. Such an M is simply the half-open interval

$$\sum_{n=1}^{p} \frac{1}{2^{y_1+\ldots+y_n}} < x \leqq \sum_{n=1}^{p} \frac{1}{2^{y_1+\ldots y_n}} + \frac{1}{2^{y_1+\ldots+y_p}},$$

and this is a Borel set.

(ii) Every Borel set belongs to (is the image of a set from) $BR(\bar{\mathcal{R}}'')$. As in the proof of Theorem 10.5.4, we need only to consider the half-open intervals of the form, $\frac{N}{2^q} < x \leqq \frac{N+1}{2^q}$ where $q = 1, 2, \ldots$, $N = 0, 1, \ldots, 2^q - 1$.

We have just established it for those of the form $\frac{2N'+1}{2^{q'+1}} < x \leqq \frac{2N'+2}{2^{q'+1}}$ where $q' = 0, 1, 2, \ldots, N' = 0, 1, \ldots, 2^{q'} - 1$. The former intervals are the sums of infinitely many intervals of the latter type: $\frac{N}{2^q} < x \leqq \frac{N+1}{2^q}$ is obtained by assigning to N', q' in turn values N, q; $2N, q+1$; $4N, q+2$; ... and adding. This completes the proof.

THEOREM 10.5.9. Any open, half-open, or closed interval of the form $a \underset{(=)}{<} x \underset{(=)}{<} b$ in I has the $\tilde{\bar{\mu}}''$-measure $b - a$.

Proof: The intervals in question belong to the domain of $\mu''(M)$ because by the preceding theorem they belong to $BR(\bar{\mathcal{R}}'')$. The same arguments as used in the second half of the proof of Theorem 10.5.4, in the proof of Theorem 10.5.5, and in the second part of the preceding theorem, show that it suffices to consider the half-open interval of the type

$$\sum_{n=1}^{p} \frac{1}{2^{y_1+\ldots+y_n}} < x \leqq \sum_{n=1}^{p} \frac{1}{2^{y_1+\ldots+y_n}} + \frac{1}{2^{y_1+\ldots+y_p}}.$$

This interval must have the measure $\frac{1}{2^{y_1+\ldots+y_p}}$; but as the interval is the image of $\{y_1\} \times \{y_2\} \times \ldots \times \{y_p\} \times \sum_\omega \times \sum_\omega \times \ldots$, this fact follows from the definition of the measure in $\bar{S}''$.

THEOREM 10.5.10. Let $\bar{\mathcal{R}}_i$, $\bar{\mu}_i(M)$ be as in Theorem 10.5.6. Then $\bar{\mu}_i(M)$ is equivalent to $\bar{\mu}''(M)$.

Proof: This follows from Theorems 10.5.8 and 10.5.9 in the same way that conclusion 3) of Theorem 10.5.6 followed from Theorems 10.5.4 and 10.5.5.

Thus all we said at the end of the first definition about the relationship between Lebesgue measure and $\bar{\mu}'(M)$ applies equally with $\bar{\mu}''(M)$ instead. Thus we have

$$\mu_{\mathcal{L}}(M) \equiv \tilde{\bar{\mu}}_i(M) \equiv \tilde{\bar{\mu}}''(M).$$

Example 3. Lebesgue measure in the s-dimensional Euclidean space R_s (s=1,2,...)

Consider first the case s = 1. Denote the S, $\mu_\alpha(M)$ of Example 1b respectively by $\sum_\infty$, $\mu_{\Delta\infty}(M)$. Note that it is not necessary to pass this time to $\tilde{\mu}_\alpha(M)$. Form the space $\overline{S}''' = \sum_\infty \times I$. For $\sum_\infty$ use the measure function $\mu_{\Delta\infty}(M)$ and its half-ring $\mathcal{R}_\alpha$ consisting of all sets having one or no elements; for I we could, recalling Theorem 10.4.8, take any one of the four equivalent measure functions $\bar{\mu}_\chi(M)$, $\bar{\mu}_i(M)$, $\bar{\mu}'(M)$, $\bar{\mu}''(M)$; we choose to take $\bar{\mu}_i(M)$. As we have only two factors, Definition 10.4.3 applies, and we can form the product space $\overline{S}'''$, the half-ring $\overline{\mathcal{R}}'''$, and the measure function $\bar{\mu}'''(M)$, corresponding respectively to $\mathcal{S}$, $\mathcal{R}$, $\mu(\mathcal{M})$ of Definition 10.4.3. The general point η of $\overline{S}'''$ is a pair $\eta = [\nu, x]$ where $\nu = 0, \pm 1, \pm 2, \ldots$, $0 < x \leqq 1$. Define the function

$$z(\eta) = \nu + x,$$

then we have

THEOREM 10.5.11. $\overline{S}'''$ is mapped by $z = z(\eta)$ in a one-to-one way upon the set of all real numbers z, that is on the R_1.

Proof: Clear.

As in past examples, we replace every set by its image in R_1; thus we pass from subsets of $\overline{S}'''$ to subsets of R_1. Analogous theorems hold:

THEOREM 10.5.12. $BR(\overline{\mathcal{R}}''')$ in R_1 is the set of all Borel sets.

Proof: (i) Every $M \in BR(\overline{\mathcal{R}}''')$ is a Borel set. As before it is sufficient to prove it for $M \in \overline{\mathcal{R}}'''$. But $M = \{\nu\} \times S_x[a < x \leqq b] = S_z[\nu + a < z \leqq \nu + b]$ is a Borel set.

(ii) Every Borel set belongs to $BR(\overline{\mathcal{R}}''')$. We need only to prove this for half-open intervals $a < z \leqq b$. We have already seen that this is true for the intervals $\nu + a < z < \nu + b$ $(0 \leqq a \leqq b \leqq 1)$, that is for the intervals $c < z \leqq d$ where $\nu \leqq c \leqq d \leqq \nu + 1$. But every interval of the form

$a < z \leqq b$ is a finite sum of those of the latter type.

THEOREM 10.5.13. *Every open, half open, or closed interval of the form* $a \underset{(=)}{<} z \underset{(=)}{<} b$ *in* $\mathcal{R}'''$ *has the* $\tilde{\bar{\mu}}'''$ *-measure* $b - a$.

Proof: The argument follows closely the lines the proofs of the two analogous Theorems 10.5.5 and 10.5.9. We observe that one point sets $M = \{z\} = \{[\nu, x]\} = \{\nu\} \times \{x\}$ have the measure zero and hence that it is sufficient to consider the half-open intervals $a < x \leqq b$. For these, as we saw in the last half of the proof of the preceding theorem, it is sufficient to consider $\nu + c < x \leqq \nu + d$, $0 \leqq c \leqq d \leqq 1$. But such an $M = \{\nu\} \times S_x[c < x \leqq d]$ and our original statement here follows from the definition of $\bar{\mu}'''(M)$.

THEOREM 10.5.14. *Let* $\bar{\mathcal{R}}_{\mathcal{J}}$ *be the set of all half-open intervals* M: $a < x \leqq b$ $(a \leqq b)$ *in* R_1. *Define for* $M \in \bar{\mathcal{R}}_{\mathcal{J}}$, $\bar{\mu}_{\mathcal{J}}(M) = b - a$. *Then*

1) $\bar{\mathcal{R}}_{\mathcal{J}}$ *is a half-ring.*

2) $\bar{\mu}_{\mathcal{J}}(M)$ *is a finite, non-negative, and totally additive measure function.*

3) $\bar{\mu}_{\mathcal{J}}(M)$ *is equivalent to* $\tilde{\bar{\mu}}'''(M)$.

Proof: Conclusion 1) is evident. Conclusion 2) follows from the fact that by Theorem 10.5.13 $\tilde{\bar{\mu}}'''(M)$ is an extension of $\bar{\mu}_{\mathcal{J}}(M)$. Conclusion 3) follows from Theorems 10.5.12 and 10.5.13 in the same way in which Theorem 10.5.6, conclusion 3) followed from Theorems 10.5.4 and 10.5.5.

Thus we have essentially the same relationship between $\bar{\mu}'''(M)$ and the Lebesgue measure in R_1 as we had between $\bar{\mu}'(M)$ or $\bar{\mu}''(M)$ and the Lebesgue - measure in I.

We can thus define the Lebesgue measure in R_1 as the maximal equivalent measure function to $\bar{\mu}_{\mathcal{J}}(M)$ and $\bar{\mu}'''(M)$:

$$\bar{\mu}_{\mathcal{L}}(M) \equiv \tilde{\bar{\mu}}_{\mathcal{J}}(M) \equiv \tilde{\bar{\mu}}'''(M).$$

Consider now an arbitrary $s = 1, 2, \ldots$. Form the space $R_s = R_1 \times \ldots \times R_1$ (with s factors). For each R_1 we could use --- recalling Theorem 10.4.8 --- any one of the three equivalent measure functions $\bar{\mu}_{\mathcal{L}}(M)$, $\bar{\mu}_{\mathcal{J}}(M)$ and $\bar{\mu}'''(M)$. We choose to take $\bar{\mu}_{\mathcal{J}}(M)$. As R_s is a direct product of a finite number of factors, the process of Definition 10.4.3 is applicable. We can thus define the space R_s, the half-ring $\bar{\mathcal{R}}_{\mathcal{J},s}$ and the measure function $\bar{\mu}_{\mathcal{J},s}(M)$ corresponding respectively to the $\mathcal{S}$, $\mathcal{R}$, $\mu(\mathcal{M})$ of Definition 10.4.3.

R_s is the s-dimensional Euclidean space with the general point $[z_1, \ldots, z_s]$ where $z_1, \ldots, z_s$ are arbitrary real numbers. $M \in \bar{\mathcal{R}}_{\mathcal{J},s}$ is any s-dimensional half-open interval: $a_\sigma < z_\sigma \leqq b_\sigma$, $a_\sigma \leqq b_\sigma$, $\sigma = 1, 2, \ldots, s$, and $\bar{\mu}_{\mathcal{J},s}(M) = \prod_{\sigma=1}^{s} (b_\sigma - a_\sigma)$. All this follows by direct application of the definitions.

We see that $\bar{\mu}_{\mathcal{J},s}(M)$ corresponds in the same way to Lebesgue measure in R_s as $\bar{\mu}_{\mathcal{J}}(M)$ and $\bar{\mu}_i(M)$ did in R_1 and I respectively. Thus we define --- avoiding the detailed discussions of Chapters 1-5 which are obviously contained in the general theorems of sections 1-3 of this chapter --- the Lebesgue measure as the maximal equivalent measure function to $\bar{\mu}_{\mathcal{J},s}(M)$:

$$\bar{\mu}_{\mathcal{L},s}(M) \equiv \tilde{\bar{\mu}}_{\mathcal{J},s}(M).$$

To sum up what has been done so far in this section: We have, by direct product processes, reached I, R_1, R_s

$I = \sum_2 \times \sum_2 \times \ldots$ or $\sum_\omega \times \sum_\omega \times \ldots$ (with a countably infinite number of factors)

$R_1 = \sum_\infty \times I$

$R_s = R_1 \times \ldots \times R_1$ (s factors)

Thus by rearranging the factors as a simple sequence we obtain

$$R_s = \sum_\infty \times \ldots \times \sum_\infty \times \sum_2 \times \sum_2 \times \ldots \text{ or } \sum_\infty \times \ldots \times \sum_\infty \times \sum_\omega \times \sum_\omega \times \ldots$$

(with s factors $\sum_{\infty}$ and a countably infinite number of factors $\sum_2$ or $\sum_{\omega}$).

By this process, which is the natural one to generate R_s from discrete spaces, we have obtained the Lebesgue measure from the discrete measures of Example 1.

Example 4. Lebesgue-Radon-Stieltjes measure in the s-dimensional Euclidean space R_s.

In Example 3 it was shown that the set $\bar{\mathcal{R}}_{\mathcal{Y},s}$ of all s-dimensional half-open intervals M: $a_\sigma < z_\sigma \leqq b_\sigma$, $\sigma = 1, 2, \ldots, s$, $a_\sigma \leqq b_\sigma$ is a half-ring and that it fulfills condition (b) of Definition 10.1.8, that is a countable set from $\bar{\mathcal{R}}_{\mathcal{Y},s}$ covers R_s. Lebesgue measure was seen to be a finite, non-negative and totally additive measure function which was the maximal equivalent one to a similar measure function with the domain $\bar{\mathcal{R}}_{\mathcal{Y},s}$. It is natural and significant to consider the problem of determining all such measure functions with $\bar{\mathcal{R}}_{\mathcal{Y},s}$ as their domain. Their maximal equivalents are the Stieltjes measures or more precisely the Lebesgue-Radon-Stieltjes measures. We proceed to carry out this determination.

Denote the interval $a_\sigma < z_\sigma \leqq b_\sigma$, $\sigma = 1, \ldots, s$, $a_\sigma \leqq b_\sigma$ by $\mathcal{Y}\begin{bmatrix} a_1 \ldots a_s \\ b_1 \ldots b_s \end{bmatrix}$. Theorems 10.4.1 and 10.4.2 show that either $\mathcal{Y}\begin{bmatrix} a_1 \ldots a_s \\ b_1 \ldots b_s \end{bmatrix}$ is empty, in which case we must have $a_\sigma = b_\sigma$ for some $\sigma = \sigma_1$ or else it is not empty and all the a_σ and b_σ satisfy $a_\sigma < b_\sigma$ and are uniquely determined by it. Thus any set function which vanishes for the empty set --- and every measure function is like that --- can, when considered on the half-open intervals, be thought of as a function $\mathcal{F}\begin{bmatrix} a_1 \ldots a_s \\ b_1 \ldots b_s \end{bmatrix}$ of the 2s real variables $a_1, \ldots, a_s, b_1, \ldots, b_s$ which is defined for $a_\sigma \leqq b_\sigma$ and which vanishes whenever for some σ $a_\sigma = b_\sigma$. We shall characterize completely in the next theorem those functions

$\mathcal{F}\begin{bmatrix} a_1 \dots a_s \\ b_1 \dots b_s \end{bmatrix}$ of this type which arise from finite measure functions.

THEOREM 10.5.15: *A function $\mathcal{F}\begin{bmatrix} a_1 \dots a_s \\ b_1 \dots b_s \end{bmatrix}$ belonging to a finite measure function $\mu(M)$ in $\overline{\mathcal{R}}_{\mathcal{Y},s}$ in the sense given above, i.e., with*

$$(1) \qquad \mu\left[\mathcal{Y}\begin{bmatrix} a_1 \dots a_s \\ b_1 \dots b_s \end{bmatrix}\right] = \mathcal{F}\begin{bmatrix} a_1 \dots a_s \\ b_1 \dots b_s \end{bmatrix} \qquad (a_\sigma \leqq b_\sigma, \quad \sigma = 1, \dots, s)$$

is characterized by the property that it is finite and satisfies

$$(2) \quad \mathcal{F}\begin{bmatrix} a_1 \dots a_\tau \dots a_s \\ b_1 \dots b_\tau \dots b_s \end{bmatrix} = \mathcal{F}\begin{bmatrix} a_1 \dots a_\tau \dots a_s \\ b_1 \dots e_\tau \dots b_s \end{bmatrix} + \mathcal{F}\begin{bmatrix} a_1 \dots e_\tau \dots a_s \\ b_1 \dots b_\tau \dots b_s \end{bmatrix}$$

where $a_\sigma \leqq b_\sigma$, $\sigma = 1, \dots, s$, and $a_\tau \leqq e_\tau \leqq b_\tau$. Such an $\mathcal{F}\begin{bmatrix} a_1 \dots a_s \\ b_1 \dots b_s \end{bmatrix}$ can be extended in a unique manner to all values of $a_1, \dots, a_s, b_1, \dots, b_s$, so that (2) holds without exception.

Proof: If $\mathcal{F}\begin{bmatrix} a_1 \dots a_s \\ b_1 \dots b_s \end{bmatrix}$ belongs to a measure function, then it obviously satisfies condition (2). On the other hand, if $\mathcal{F}\begin{bmatrix} a_1 \dots a_s \\ b_1 \dots b_s \end{bmatrix}$ is finite and satisfies (2), we can show that (1) may be used to define a finite measure function on $\overline{\mathcal{R}}_{\mathcal{Y},s}$. We have merely to show that if M, N, $P \in \overline{\mathcal{R}}_{\mathcal{Y},s}$, $N \cdot P = 0$ and $M = N + P$, then $\mu(M)$, $\mu(N)$, $\mu(P)$ defined by (1) satisfy $\mu(M) = \mu(N) + \mu(P)$. By Theorem 10.4.3 these last assumptions guarantee in the case $N \neq 0$, $P \neq 0$, that M, N, P have the respective forms

$$M = \mathcal{Y}\begin{bmatrix} a_1 \dots a_\tau \dots a_s \\ b_1 \dots b_\tau \dots b_s \end{bmatrix} \quad N = \mathcal{Y}\begin{bmatrix} a_1 \dots a_\tau \dots a_s \\ b_1 \dots e_\tau \dots b_s \end{bmatrix} \quad P = \mathcal{Y}\begin{bmatrix} a_1 \dots e_\tau \dots a_s \\ b_1 \dots b_\tau \dots b_s \end{bmatrix}$$

with $a_\sigma < b_\sigma$, $\sigma = 1, \dots, s$, $a_\tau \leqq e_\tau \leqq b_\tau$, or else these forms with the roles of N and P reversed. The condition that $\mu(M) = \mu(N) + \mu(P)$ then becomes simply the equation (2). To show the additivity when N or P is the empty set, it is clearly sufficient to show that this set has the measure zero. This may be shown by writing $a_\tau = e_\tau = b_\tau$ in equation (2).

As to the existence of an extension of $\mathcal{F}\begin{bmatrix} a_1 \dots a_s \\ b_1 \dots b_s \end{bmatrix}$ let $a_1, \dots, a_s$ $b_1, \dots, b_s$ be arbitrary real numbers. Denote by $\bar{a}_\sigma$, $\bar{b}_\sigma$ the pair a_σ, b_σ in case $a_\sigma \leqq b_\sigma$ and the pair b_σ, a_σ in case $a_\sigma > b_\sigma$, and let $\bar{e}$ be the number of indices σ for which $a_\sigma > b_\sigma$. Define

$$\mathcal{F}^0\begin{bmatrix} a_1 \dots a_s \\ b_1 \dots b_s \end{bmatrix} = (-1)^{\bar{e}}\, \mathcal{F}\begin{bmatrix} \bar{a}_1 \dots \bar{a}_s \\ \bar{b}_1 \dots \bar{b}_s \end{bmatrix}$$

We show that $\mathcal{F}^0\begin{bmatrix} a_1 \dots a_s \\ b_1 \dots b_s \end{bmatrix}$ is the unique desired extension. If for every $\sigma = 1, \dots, s$, $a_\sigma \leqq b_\sigma$, then $e = 0$, $a_\sigma = \bar{a}_\sigma$, $b_\sigma = \bar{b}_0$, and $\mathcal{F}^0\begin{bmatrix} a_1 \dots a_s \\ b_1 \dots b_s \end{bmatrix}$ agrees with $\mathcal{F}\begin{bmatrix} a_1 \dots a_s \\ b_1 \dots b_s \end{bmatrix}$ in the latter's domain of definition. To show that $\mathcal{F}^0\begin{bmatrix} a_1 \dots a_s \\ b_1 \dots b_s \end{bmatrix}$ satisfies equation (2) consider the expression

$$G \equiv \mathcal{F}^0\begin{bmatrix} a_1 \dots a_\tau \dots a_s \\ b_1 \dots e_\tau \dots b_s \end{bmatrix} + \mathcal{F}^0\begin{bmatrix} a_1 \dots e_\tau \dots a_s \\ b_1 \dots b_\tau \dots b_s \end{bmatrix} + \mathcal{F}^0\begin{bmatrix} a_1 \dots b_\tau \dots a_s \\ b_1 \dots a_\tau \dots b_s \end{bmatrix}.$$

G remains invariant under any cyclic permutation of a_τ, e_τ, b_τ, and changes its sign if we replace a_τ, e_τ, b_τ, by b_τ, e_τ, a_τ. Thus under any permutation of a_τ, e_τ, b_τ, G at most changes its sign. Assume therefore that $a_\tau \leqq e_\tau \leqq b_\tau$. Using the definition of $\mathcal{F}^0\begin{bmatrix} a_1 \dots a_s \\ b_1 \dots b_s \end{bmatrix}$ and equation (2) and denoting by $\bar{e}'$ the number of $\sigma \neq \tau$ for which $a_\sigma > b_\sigma$, we get

$$G = (-1)^{\bar{e}'}\mathcal{F}\begin{bmatrix} \bar{a}_1 \dots \bar{a}_\tau \dots \bar{a}_s \\ \bar{b}_1 \dots e_\tau \dots \bar{b}_s \end{bmatrix} + (-1)^{\bar{e}'}\mathcal{F}\begin{bmatrix} \bar{a}_1 \dots e_\tau \dots \bar{a}_s \\ \bar{b}_1 \dots b_\tau \dots \bar{b}_s \end{bmatrix} +$$

$$+ (-1)^{\bar{e}'+1}\, \mathcal{F}\begin{bmatrix} \bar{a}_1 \dots a_\tau \dots \bar{a}_s \\ \bar{b}_1 \dots b_\tau \dots \bar{b}_s \end{bmatrix} = 0$$

and this is equation (2) for the extended function. Finally, the extension is unique because if $\mathcal{F}'\begin{bmatrix} a_1 \dots a_s \\ b_1 \dots b_s \end{bmatrix}$ is any other extension, equation (2) for it implies $\mathcal{F}'\begin{bmatrix} a_1 \dots b_\tau \dots a_s \\ b_1 \dots a_\tau \dots b_s \end{bmatrix} = -\mathcal{F}'\begin{bmatrix} a_1 \dots a_\tau \dots a_s \\ b_1 \dots b_\tau \dots b_s \end{bmatrix}$ which clearly implies

$$\mathcal{F}'\begin{bmatrix} a_1 \cdots a_s \\ b_1 \cdots b_s \end{bmatrix} = (-1)^{\bar{e}}\, \mathcal{F}\begin{bmatrix} \bar{a}_1 \cdots \bar{a}_s \\ \bar{b}_1 \cdots \bar{b}_s \end{bmatrix} = \mathcal{F}^0\begin{bmatrix} a_1 \cdots a_s \\ b_1 \cdots b_s \end{bmatrix} .$$

THEOREM 10.5.16. *The extended functions* $\mathcal{F}\begin{bmatrix} a_1 \cdots a_s \\ b_1 \cdots b_s \end{bmatrix}$ *described in Theorem 10.5.15 coincide with the functions which can be written in the form*

$$\mathcal{F}\begin{bmatrix} a_1 \cdots a_s \\ b_1 \cdots b_s \end{bmatrix} = \sum_{r_1,\dots,r_s=1,2} (-1)^{r_1+\dots+r_s}\, \phi(c_1^{(r_1)} \cdots c_s^{(r_s)}) \tag{3}$$

where $c_\sigma^{(1)} = a_\sigma$ *and* $c_\sigma^{(2)} = b_\sigma$ *for* $\sigma = 1,\dots,s$, *and where* $\phi(c_1 \cdots c_s)$ *is an arbitrary finite (complex) function.*

Proof: Let $\tau = 1,\dots,s$. Equation (2) above gives

$$\begin{aligned} &\mathcal{F}\begin{bmatrix} 0 & \cdots & 0 & a_\tau \cdots a_s \\ c_1^{(r_1)} & \cdots & c_{\tau-1}^{(r_{\tau-1})} & b_\tau \cdots b_s \end{bmatrix} = \mathcal{F}\begin{bmatrix} 0 & \cdots & 0 & 0\ a_{\tau+1} \cdots a_s \\ c_1^{(r_1)} & \cdots & c_{\tau-1}^{(r_{\tau-1})} & b_\tau\ b_{\tau+1} \cdots b_s \end{bmatrix} - \\ &- \mathcal{F}\begin{bmatrix} 0 & \cdots & 0 & 0\ a_{\tau+1} \cdots a_s \\ c_1^{(r_1)} & \cdots & c_{\tau-1}^{(r_{\tau-1})} & a_\tau\ b_{\tau+1} \cdots b_s \end{bmatrix} = \sum_{r_\tau=1,2} (-1)^{r_\tau} \mathcal{F}\begin{bmatrix} 0 & \cdots & 0 & a_{\tau+1} \cdots a_s \\ c_1^{(r_1)} & \cdots & c_\tau^{(r_\tau)} & b_{\tau+1} \cdots b_s \end{bmatrix} . \end{aligned} \tag{4}$$

Multiplying (4) through by $(-1)^{r_1+\dots+r_{\tau-1}}$ and adding over $r_1,\dots,r_{\tau-1} = 1,2$ we get

$$\begin{aligned} &\sum_{r_1,\dots,r_{\tau-1}=1,2} (-1)^{r_1+\dots+r_{\tau-1}} \mathcal{F}\begin{bmatrix} 0 & \cdots & 0 & a_\tau \cdots a_s \\ c_1^{(r_1)} & \cdots & c_{\tau-1}^{(r_{\tau-1})} & b_\tau \cdots b_s \end{bmatrix} = \\ &= \sum_{r_1,\dots,r_\tau=1,2} (-1)^{r_1+\dots+r_\tau} \mathcal{F}\begin{bmatrix} 0 & \cdots & 0 & a_{\tau+1} \cdots a_s \\ c_1^{(r_1)} & \cdots & c_\tau^{(r_\tau)} & b_{\tau+1} \cdots b_s \end{bmatrix} \end{aligned} \tag{5}$$

Adding equations (5) for $\tau = 1,\dots,s$, we get (3) with

$$\phi(c_1 \cdots c_s) \equiv \mathcal{F}\begin{bmatrix} 0 \cdots 0 \\ c_1 \cdots c_s \end{bmatrix}$$

Conversely, every function of the form (3) satisfies equation (2). For simplicity we take $\tau = 1$. Then

$$\mathcal{F}\begin{bmatrix} a_1 \dots a_s \\ b_1 \dots b_s \end{bmatrix} = \sum_{r_1, \dots, r_s = 1,2} (-1)^{r_1 + \dots + r_s} \phi(c_1^{(r_1)} \dots c_s^{(r_s)}) =$$

$$= \sum_{r_2, \dots, r_s = 1,2} (-1)^{r_2 + \dots + r_s} \left(\phi(b_1 c_2^{(r_2)} \dots c_s^{(r_s)}) - \phi(e_1 c_2^{(r_2)} \dots c_s^{r_s}) + \right.$$

$$\left. + \phi(e_1 c_2^{(r_2)} \dots c_s^{(r_s)}) - \phi(a_1 c_2^{r_2} \dots c_s^{r_s})\right) = \mathcal{F}\begin{bmatrix} a_1 \dots a_s \\ e_1 \dots b_s \end{bmatrix} + \mathcal{F}\begin{bmatrix} e_1 \dots a_s \\ b_1 \dots b_s \end{bmatrix} .$$

<u>THEOREM 10.5.17.</u> $\mathcal{F}\begin{bmatrix} a_1 \dots a_s \\ b_1 \dots b_s \end{bmatrix}$ <u>does not determine</u> $\phi(c_1 \dots c_s)$ <u>uniquely, but</u>

1) <u>if one solution</u> $\phi(c_1 \dots c_s)$ <u>is known, all others are given by</u>

(6) $$\phi'(c_1 \dots c_s) = \phi(c_1 \dots c_s) + \sum_{\tau=1}^{s} \psi_\tau(c_1 \dots c_{\tau-1} c_{\tau+1} \dots c_s)$$

<u>where</u> $\psi_1, \dots, \psi_s$ <u>are arbitrary</u> s-1 <u>variable functions,</u>

2) <u>there is a unique solution</u> $\phi(c_1 \dots c_s)$ <u>with the property that</u> $\phi(c_1 \dots c_s) = 0$ <u>if any</u> $c_\sigma = 0$; <u>it is</u>

$$\phi(c_1 \dots c_s) = \phi^\circ(c_1 \dots c_s) \equiv \mathcal{F}\begin{bmatrix} 0 \dots 0 \\ c_1 \dots c_s \end{bmatrix} .$$

Proof: 1) We first observe that the equation (3) may be regarded as defining $\mathcal{F}\begin{bmatrix} a_1 \dots a_s \\ b_1 \dots b_s \end{bmatrix}$ by means of a homogeneous linear operation upon $\phi(c_1 \dots c_s)$. This is clear from the form of (3). If therefore two functions $\phi(c_1 \dots c_s)$ and $\phi'(c_1 \dots c_s)$ generate the same interval function $\mathcal{F}\begin{bmatrix} a_1 \dots a_s \\ b_1 \dots b_s \end{bmatrix}$, then their difference will generate an interval function identically zero, and conversely. We have thus reduced the question to one of determining all functions which generate zero interval functions. Let $\chi(c_1 \dots c_s)$ be such a function. We wish to prove that $\chi(c_1 \dots c_s) = \sum_{\tau=1}^{s} \psi_\tau(c_1 \dots c_{\tau-1} c_{\tau+1} \dots c_s)$. We prove it by induction on s. The result is clear for s = 1 for then χ must be a constant. Assume it for s - 1. Then we have

$$\sum_{r_1,\dots,r_s=1,2} (-1)^{r_1+\dots+r_s}\chi(c_1^{(r_1)}\dots c_s^{(r_s)}) =$$

$$= \sum_{r_1,\dots,r_{s-1}=1,2} (-1)^{r_1+\dots+r_{s-1}} \{\chi(c_1^{(r_1)}\dots c_{s-1}^{(r_{s-1})}\, b_s) - \chi(c_1^{(r_1)}\dots c_{s-1}^{(r_s)}\, a_s)\} = 0$$

We apply the induction hypothesis to the expression in brackets in this last equation (regarding a_s and b_s as parameters) and get

$$\chi(c_1 \dots c_{s-1}b_s) - \chi(c_1 \dots c_{s-1}a_s) = \tag{7}$$

$$= \sum_{\tau=1}^{s-1} \overline{\Psi}_\tau(c_1 \dots c_{\tau-1}c_{\tau+1}\dots c_{s-1};\, a_s\, b_s)$$

Now giving to a_s some fixed value and writing $b_s = c_s$, we obtain the completed induction by writing in (7)

$$\Psi_\tau(c_1 \dots c_{\tau-1}, c_{\tau+1}\dots c_s) = \overline{\Psi}_\tau(c_1 \dots c_{\tau-1}c_{\tau+1} \dots c_{s-1};\, a_s\, c_s)$$

$$(\tau=1,\dots,s-1)$$

$$\Psi_s(c_1 \dots c_{s-1}) = \chi(c_1 \dots c_{s-1}\, a_s)$$

Thus every function which generates a zero function of intervals is of the form $\sum_{\tau=1}^{s} \Psi_\tau$. The converse is clear because when the difference is formed with respect to the variable c_τ the function Ψ_τ is annihilated.

2) That $\phi^\circ(c_1 \dots c_s) \equiv \mathcal{F}\begin{bmatrix} 0 & \dots & 0 \\ c_1 & \dots & c_s \end{bmatrix}$ is a solution of equation (3) was shown in the proof of the preceding theorem. Also $\varphi(c_1 \dots 0 \dots c_s) = \mathcal{F}\begin{bmatrix} 0 & \dots & 0 & \dots & 0 \\ c_1 & \dots & 0 & \dots & c_s \end{bmatrix} = 0$. Conversely if $\phi(c_1 \dots c_s)$ is a solution having the given property, then upon using (3) with $a_1 = \dots a_s = 0$, $b_\sigma = c_\sigma$, it is readily verified that all terms in the sum of the right-hand side of (3) drop out

except that for which $r_1 = \ldots = r_s = 2$; and we get $\phi(c_1 \ldots c_s) = \phi^\circ(c_1 \ldots c_s)$. Hence $\phi^\circ(c_1 \ldots c_s)$ is unique.

The foregoing theorems show that any finite measure function $\mu(M)$ in $\bar{\mathcal{R}}_{\mathcal{Y},s}$ is characterized through equation (1) by an interval function $\mathcal{F}\begin{bmatrix} a_1 \ldots a_s \\ b_1 \ldots b_s \end{bmatrix}$ satisfying equation (2), or, through equation (3), by a point function $\phi(c_1 \ldots c_s)$. While $\mu(M)$ corresponds to exactly one $\mathcal{F}\begin{bmatrix} a_1 \ldots a_s \\ b_1 \ldots b_s \end{bmatrix}$ it corresponds to many $\phi(c_1 \ldots c_s)$. The family of all such $\phi(c_1 \ldots c_s)$ is characterized in Theorem 10.5.17,1), and a normalization which makes $\phi(c_1 \ldots c_s)$ unique is given in Theorem 10.5.17,2).

It is now significant to characterize those $\phi(c_1 \ldots c_s)$ which lead to non-negative and totally additive measure functions. The condition for non-negativity is clearly (cf.Theorem 10.5.16)

$$(8) \qquad \sum_{r_1,\ldots,r_s=1,2} (-1)^{r_1+\ldots+r_s} \phi(c_1^{(r_1)} \ldots c_s^{(r_s)}) \text{ is real and } \geqq 0$$

for all choices $a_1, \ldots, a_s, b_1, \ldots, b_s$, with $a_\sigma \leqq b_\sigma$, $(\sigma = 1,2,\ldots,s)$. If $s = 1$, this means $\phi(b_1) - \phi(a_1) \geqq 0$ for $a_1 \leqq b_1$; that is, $\phi(c)$ is monotonous. The monotonous functions were discussed in detail in Chapter IX. If $s > 1$, there is no such simple statement to characterize (8).

<u>Definition 10.5.3.</u> <u>A function</u> $\phi(c_1 \ldots c_s)$ <u>satisfying condition</u> (8) <u>for all choices of</u> $a_1,\ldots,a_s$, $b_1,\ldots,b_s$ <u>with</u> $a_\sigma \leqq b_\sigma$ $(\sigma = 1,2,\ldots,s)$ <u>is called quasi-monotonous.</u>

<u>Definition 10.5.4.</u> <u>A function</u> $\phi(c_1, \ldots, c_s)$ <u>is called right-continuous when for every system of sequence</u> $\{c_\sigma^{(\nu)}\}$ $(\sigma = 1, 2,\ldots,s; \nu = 1, 2, \ldots)$ <u>for which</u> $c_\sigma^{(\nu)} \geqq c_\sigma$ <u>and</u> $\lim_{\nu\to\infty} c_\sigma^{\nu} = c_\sigma$ <u>we have</u> $\lim_{\nu\to\infty} \phi(c_1^{(\nu)} \ldots c_s^{(\nu)}) = \phi(c_1,\ldots,c_s)$.

We are now in a position to prove an exhaustive theorem concerning non-negativity and total additivity.

THEOREM 10.5.18. *A finite measure function $\mu(M)$ in $\bar{\mathcal{R}}_{\mathcal{J},s}$ is non-negative and totally additive if and only if at least one of its $\phi(c_1,\dots,c_s)$'s is quasi-monotonous and right-continuous.*

Remark: We remark first that if one $\phi(c_1,\dots,c_s)$ of $\mu(M)$ is quasi-monotonous, all are; for quasi-monotony is by (8) really a property of $\mathcal{F}\begin{bmatrix} a_1 \dots a_s \\ b_1 \dots b_s \end{bmatrix} = \mu\left[\mathcal{J}\begin{bmatrix} a_1 \dots a_s \\ b_1 \dots b_s \end{bmatrix}\right]$. On the other hand, if one $\phi(c_1,\dots,c_s)$ of $\mu(M)$ is right-continuous, others will not be as the Ψ_τ of Theorem 10.7.17,1), are perfectly arbitrary. Nevertheless, if one $\phi(c_1,\dots,c_s)$ is right-continuous, then $\phi^\circ(c_1,\dots,c_s)$ will also be as the subsequent proof will show.

Proof: We have to prove that in the case of quasi-monotonous functions the right-continuity of one $\phi(c_1,\dots,c_s)$ is sufficient for the total additivity of $\mu(M)$ and that the right-continuity of $\phi^\circ(c_1,\dots,c_s)$ is necessary.

To prove the sufficiency we employ the sufficient topological criterium developed in Theorem 10.1.20. Assume that $\phi'(c_1,\dots,c_s)$ is a right-continuous point function associated with $\mu(M)$. Let $M = \mathcal{J}\begin{bmatrix} a_1 \dots a_s \\ b_1 \dots b_s \end{bmatrix}$ be any fixed interval in $\bar{\mathcal{R}}_{\mathcal{J},s}$; let ε be an arbitrary positive number. We show that there exist sets P, Q from $\bar{\mathcal{R}}_{\mathcal{J},s}$, a compact set C and an open set $\bar{\Phi}$ such that $P \subset C \subset M \subset \bar{\Phi} \subset Q$ and $\mu(P) > \mu(M) - \varepsilon$, $\mu(Q) < \mu(M) + \varepsilon$. Assume first that $M \neq 0$ so that $a_\sigma < b_\sigma$. Since $\mathcal{F}\begin{bmatrix} a_1 \dots a_s \\ b_1 \dots b_s \end{bmatrix}$ is a finite linear combinàtion of the $\phi'(c_1^{(r_1)} \dots c_s^{(r_s)})$, it is right-continuous in its 2s arguments. Now choose an integer ν so large that $a_\sigma + \frac{1}{\nu} < b_\sigma$ and that $\mathcal{F}\begin{bmatrix} a_1 + \frac{1}{\nu} \dots a_s + \frac{1}{\nu} \\ b_1 \dots b_s \end{bmatrix}$ and

$\mathcal{F}\begin{bmatrix} a_1 \dots a_s \\ b_1 \dots b_s \end{bmatrix}$ by an amount in absolute value less than ε. Let

$P \equiv \mathcal{Y}\begin{bmatrix} a_1+\frac{1}{\nu} \dots a_s+\frac{1}{\nu} \\ b_1 \quad \dots b_s \end{bmatrix}$ $Q \equiv \mathcal{Y}\begin{bmatrix} a_1 \quad \dots a_s \\ b_1+\frac{1}{\nu} \dots b_s+\frac{1}{\nu} \end{bmatrix}$. Let C be the closed interval $a_\sigma + \frac{1}{2\nu} \leqq z_\sigma \leqq b_\sigma$ and let Φ be the open interval $a_\sigma < z_\sigma < b_\sigma + \frac{1}{\nu}$ $(\sigma = 1,\dots,s)$. In the space R_s, C is compact and Φ is open. Furthermore we have $P \subset C \subset M \subset \Phi \subset Q$ where P, $Q \in \widetilde{\mathcal{R}}_{\mathcal{Y},s}$ and $\mu(P) = \begin{bmatrix} a_1+\frac{1}{\nu} \dots a_s+\frac{1}{\nu} \\ b_1 \quad \dots b_s \end{bmatrix} >$ $> \mathcal{F}\begin{bmatrix} a_1 \dots a_s \\ b_1 \dots b_s \end{bmatrix} - \varepsilon = \mu(M) - \varepsilon$. Similarly $\mu(Q) < \mu(M) + \varepsilon$. If $M = 0$ take $a_1 = \dots = a_s = b_1 = \dots = b_s = 0$, $P = C = 0$, Φ and Q as before, then the same inequalities hold. Theorem 10.1.20 therefore applies to prove that $\mu(M)$ is totally additive.

We now proceed to prove that if $\mu(M)$ is (non-negative and) totally additive then $\Phi^\circ(c_1,\dots,c_s)$ is right-continuous. For this it is obviously more then sufficient to prove that the extended $\mathcal{F}\begin{bmatrix} a_1 \dots a_s \\ b_1 \dots b_s \end{bmatrix}$ of $\mu(M)$ is right-continuous in its 2s arguments. Let $a_1, \dots, a_s, b_1, \dots, b_s$ be any fixed set of arguments, $\{a_1^{(\nu)}\}, \dots \{a_s^{(\nu)}\}, \{b_1^{(\nu)}\}, \dots, \{b_s^{(\nu)}\}$ sequences such that for $\sigma = 1,\dots,s$, $\nu = 1, 2, \dots$ we have $a_\sigma \leqq a_\sigma^{(\nu)}$, $b_\sigma \leqq b_\sigma^{(\nu)}$, $\lim_{\nu\to\infty} a_\sigma^{(\nu)} = a_\sigma$, $\lim_{\nu\to\infty} b_\sigma^{(\nu)} = b_\sigma$. We have to prove that

$$\lim_{\nu\to\infty} \mathcal{F}\begin{bmatrix} a_1^{(\nu)} \dots a_s^{(\nu)} \\ b_1^{(\nu)} \dots b_s^{(\nu)} \end{bmatrix} = \mathcal{F}\begin{bmatrix} a_1 \dots a_s \\ b_1 \dots b_s \end{bmatrix}.$$

Define $\bar{a}_\sigma^{(\nu)}, \bar{b}_\sigma^{(\nu)}, \bar{e}_\nu$ in terms of $a_\sigma^{(\nu)}, b_\sigma^{(\nu)}$, and $\bar{a}_\sigma, \bar{b}_\sigma$, e, in terms of a_σ, b_σ, exactly as was done in the proof of Theorem 10.5.15. Now $\bar{a}_\sigma = \min(a_\sigma, b_\sigma) \leqq \min(a_\sigma^{(\nu)}, b_\sigma^{(\nu)}) = \bar{a}_\sigma^{(\nu)}$ and $\bar{b}_\sigma = \max(a_\sigma, b_\sigma) \leqq$ $\leqq \max(a_\sigma^{(\nu)}, b_\sigma^{(\nu)}) = \bar{b}_\sigma$; also $\lim_{\nu\to\infty} \bar{a}_\sigma^{(\nu)} = \bar{a}_\sigma$, $\lim_{\nu\to\infty} \bar{b}_\sigma^{(\nu)} = \bar{b}_\sigma$. We now define for $\nu = 1,2,\ldots$, the intervals $M \equiv \mathcal{Y}\begin{bmatrix}\bar{a}_1 \ldots \bar{a}_s \\ \bar{b}_1 \ldots \bar{b}_s\end{bmatrix}$ $M_\nu \equiv \mathcal{Y}\begin{bmatrix}\bar{a}_1^{(\nu)} \ldots \bar{a}_s^{(\nu)} \\ \bar{b}_1^{(\nu)} \ldots \bar{b}_s^{(\nu)}\end{bmatrix}$.

Consider in R_s the sets $\bar{M}_\nu = \sum_{n=0}^{\infty} M_{\nu+n}$, $\bar{\bar{M}}_\nu = \prod_{n=0}^{\infty} M_{\nu+n}$ ($\nu = 1, 2, \ldots$). These belong certainly to $BR(\bar{\mathcal{R}}_{\mathcal{Y},s})$ and we have the relations $\bar{M}_1 \supset \bar{M}_2 \supset \ldots \supset M = \lim_{\nu\to\infty} \bar{M}_\nu$ and $\bar{\bar{M}}_1 \subset \bar{\bar{M}}_2 \subset \ldots \subset M = \lim_{\nu\to\infty} \bar{\bar{M}}_\nu$ as well as $\bar{M}_\nu \supset M_\nu$, $\bar{\bar{M}}_\nu \subset M_\nu$. Now pass to the maximal extension $\tilde{\mu}(M)$ of $\mu(M)$. We have, on applying Theorems 10.2.8, 10.2.9,

$$\overline{\lim_{\nu\to\infty}}\, \mu(M_\nu) = \overline{\lim_{\nu\to\infty}}\, \tilde{\mu}(M_\nu) \leqq \lim_{\nu\to\infty} \tilde{\mu}(\bar{M}\ ) = \tilde{\mu}(M) = \mu(M)$$

$$\underline{\lim_{\nu\to\infty}}\, \mu(M\ ) = \underline{\lim_{\nu\to\infty}}\, \tilde{\mu}(M_\nu) \geqq \lim \tilde{\mu}(\bar{\bar{M}}_\nu) = \tilde{\mu}(M) = \mu(M)$$

from which

$$\lim_{\nu\to\infty} \mu(M_\nu) = \mu(M) .$$

Thus we have

$$(9) \qquad \lim_{\nu\to\infty} \mathcal{F}\begin{bmatrix}\bar{a}_1^{(\nu)} \ldots \bar{a}_s^{(\nu)} \\ \bar{b}_1^{(\nu)} \ldots \bar{b}_s^{(\nu)}\end{bmatrix} = \mathcal{F}\begin{bmatrix}\bar{a}_1 \ldots \bar{a}_s \\ \bar{b}_1 \ldots \bar{b}_s\end{bmatrix}$$

Now if for all $\sigma = 1,\ldots,s$ we have $a_\sigma \neq b_\sigma$, that is $\bar{a}_\sigma < \bar{b}_\sigma$, then it is clear that for sufficiently large ν the value of $\bar{e}_\nu$ remains fixed and equal to $\bar{e}$. Hence in this case

$$(10) \qquad \begin{aligned} \lim_{\nu\to\infty} \mathcal{F}\begin{bmatrix}a_1^{(\nu)} \ldots a_s^{(\nu)} \\ b_1^{(\nu)} \ldots b_s^{(\nu)}\end{bmatrix} &= \lim_{\nu\to\infty} (-1)^{\bar{e}_\nu} \mathcal{F}\begin{bmatrix}\bar{a}_1^{(\nu)} \ldots \bar{a}_s^{(\nu)} \\ \bar{b}_1^{(\nu)} \ldots \bar{b}_s^{(\nu)}\end{bmatrix} = \\ &= (-1)^{\bar{e}} \mathcal{F}\begin{bmatrix}\bar{a}_1 \ldots \bar{a}_s \\ \bar{b}_1 \ldots \bar{b}_s\end{bmatrix} = \mathcal{F}\begin{bmatrix}a_1 \ldots a_s \\ b_1 \ldots b_s\end{bmatrix} . \end{aligned}$$

If for some $\sigma = 1,\ldots,s$, $a_\sigma = b_\sigma$, which means $\bar{a}_\sigma = \bar{b}_\sigma$, then the value of $\bar{e}_\nu$ does not matter, for we then have zero on the right-hand side of (9); hence the limit in the second member of equation (10) exists and is equal to zero independently of the $\bar{e}_\nu$. Hence (10) holds in all cases. This completes the proof.

Lebesgue measure is by definition a special case of these Lebesgue-Radon-Stieltjes measures. We have in this instance

$\mathcal{F}\begin{bmatrix} a_1 \dots a_s \\ b_1 \dots b_s \end{bmatrix} = \mu\left[\mathcal{Y}\begin{bmatrix} a_1 \dots a_s \\ b_1, \dots b_s \end{bmatrix}\right] = \prod_{\sigma=1}^{s} (b_\sigma - a_\sigma)$ if $a_\sigma \leqq b_\sigma$, and the extension Theorem 10.5.15 gives $\mathcal{F}\begin{bmatrix} a_1 \dots a_s \\ b_1 \dots b_s \end{bmatrix} = \prod_{\sigma=1}^{s} (b_\sigma - a_\sigma)$ in general. Therefore

$$\phi^\circ(c_1 \dots c_s) = \mathcal{F}\begin{bmatrix} 0 \dots 0 \\ c_1 \dots c_s \end{bmatrix} = c_1 \dots c_s.$$

The $\phi^\circ(c)$'s of the general one-dimensional case are simply the monotonous, right-continuous functions which vanish at 0 --- a class which has been discussed in Chapter IX.

That the general Stieltjes measure $\mu(M)$ may differ widely in behavior from the Lebesgue measure is best seen by considering the fact that one-point sets may have positive measure in the general case. For instance in one dimension let M_n be the half-open interval $c - \frac{1}{n} < x \leqq c$; then

$M_1 \supset M_2 \supset \dots \supset \{c\} = \lim_{n\to\infty} M_n$. Now $\mu(M_n) = \phi(c) - \phi(c - \frac{1}{n})$. Therefore

$\mu(\{c\}) = \lim_{n\to\infty} \mu(M_n) = \phi(c) - \lim_{n\to\infty} \phi(c - \frac{1}{n}) = \phi(c) - \phi_-(c)$. Owing to the right continuity of $\phi(c)$, $\phi(c) = \phi_+(c)$; thus $\mu(\{c\}) = \phi_+(c) - \phi_-(c) =$

$= \text{osc}\ \phi(c)$. Hence $\mu(\{c\}) > 0$ if and only if $\phi(c)$ is (left) discontinuous at c.

A common feature of all Stieltjes measures is this:

THEOREM 10.5.19. *If a measure function $\mu(M)$ defined on $\overline{\mathcal{R}}_{\mathcal{Y},s}$ is restricted to the set $\overline{\mathcal{R}}'_{\mathcal{Y},s}$ of all intervals $\mathcal{Y}\begin{bmatrix} a_1 \dots a_s \\ b_1 \dots b_s \end{bmatrix}$ with rational*

$a_1, \ldots, a_s, b_1, \ldots, b_s$, <u>it remains equivalent to itself</u>.

Proof: We need only show $BR(\overline{\mathcal{R}}_{\mathcal{Y},s}) = BR(\overline{\mathcal{R}}'_{\mathcal{Y},s})$ and as $\overline{\mathcal{R}}_{\mathcal{Y},s} \supset \overline{\mathcal{R}}'_{\mathcal{Y},s}$, this follows if $\overline{\mathcal{R}}_{\mathcal{Y},s} \subset BR(\overline{\mathcal{R}}'_{\mathcal{Y},s})$ is established.

Let $\mathcal{Y}\begin{bmatrix} a_1 \ldots a_s \\ b_1 \ldots b_s \end{bmatrix}$ be an arbitrary element of $\overline{\mathcal{R}}_{\mathcal{Y},s}$; as the case is clear for 0, we may assume it to be $\neq 0$, that is $a_\sigma < b_\sigma$. Choose 2s sequences of rational numbers $a_\sigma^{(\nu)}$, $b_\sigma^{(\nu)}$, $\sigma = 1, \ldots, s$, $\nu = 1, 2, \ldots$, with $a_\sigma \leqq a_\sigma^{(\nu)} \leqq b_\sigma \leqq b_0^{(\nu)}$, and $\lim_{\nu\to\infty} a_\sigma^{(\nu)} = a_\sigma$, $\lim_{\nu\to\infty} b_\sigma^{(\nu)} = b_\sigma$. Then

$$\mathcal{Y}\begin{bmatrix} a_1 \ldots a_s \\ b_1 \ldots b_s \end{bmatrix} = \prod_{\nu=1,2,\ldots} \sum_{\mu=1,2,\ldots} \mathcal{Y}\begin{bmatrix} a_1^{(\mu)} \ldots a_s^{(\mu)} \\ b_1^{(\nu)} \ldots b_s^{(\nu)} \end{bmatrix}$$

which proves $\mathcal{Y}\begin{bmatrix} a_1 \ldots a_s \\ b_1 \ldots b_s \end{bmatrix} \in BR(\overline{\mathcal{R}}'_{\mathcal{Y},s})$.

We finally prove a theorem on the behavior of $\Phi(c_1, \ldots, c_s)$ under direct multiplication.

Put $s = s' + s''$ so that $R_s = R_{s'} \times R_{s''}$. Consider two finite, non-negative and totally additive measure functions $\mu'(M)$ and $\mu''(M)$ defined on $\overline{\mathcal{R}}_{\mathcal{Y},s'}$ and $\overline{\mathcal{R}}_{\mathcal{Y},s''}$ respectively. As we have only two factors, Definition 10.4.3 applies and we can form the product space R_s, the half-ring $\overline{\mathcal{R}}_{\mathcal{Y},s}$ and the measure function $\mu(M)$ corresponding respectively to $\mathcal{S}$, $\mathcal{R}$ and $\mu(\mathcal{M})$ in Definition 10.4.3.

<u>THEOREM 10.5.20. Let the</u> $\Phi^\circ(c_1, \ldots, c_s)$<u>'s of the</u> $\mu(M)$, $\mu'(M)$, $\mu''(M)$ <u>above be respectively</u> $\Phi^\circ(c_1, \ldots, c_s)$, $\Phi'^\circ(c_1, \ldots, c_{s'})$, $\Phi''^\circ(c_1, \ldots, c_{s''})$. <u>Then</u>

$$\Phi^\circ(c_1, \ldots, c_s) = \Phi'(c_1, \ldots, c_{s'})\, \Phi''^\circ(c_{s'+1}, \ldots, c_{s'+s''})$$

Proof: We have by definition

$$\mu\left[\mathcal{Y}\begin{bmatrix} a_1 \dots a_s \\ b_1 \dots b_s \end{bmatrix}\right] = \mu'\left[\mathcal{Y}\begin{bmatrix} a_1 \dots a_{s'} \\ b_1 \dots b_{s'} \end{bmatrix}\right]\mu''\left[\mathcal{Y}\begin{bmatrix} a_{s'+1} \dots a_{s'+s''} \\ b_{s'+1} \dots b_{s'+s''} \end{bmatrix}\right]$$

and thus

$$\mathcal{F}\begin{bmatrix} a_1 \dots a_s \\ b_1 \dots b_s \end{bmatrix} = \mathcal{F}'\begin{bmatrix} a_1 \dots a_{s'} \\ b_1 \dots b_{s'} \end{bmatrix} \cdot \mathcal{F}''\begin{bmatrix} a_{s'+1} \dots a_{s'+s''} \\ b_{s'+1} \dots b_{s'+s''} \end{bmatrix}$$

if $a_\sigma \leqq b_\sigma$ for $\sigma = 1, \dots, s$. The construction in Theorem 10.5.15 extends this to all $a_1, \dots, a_s, b_1, \dots, b_s$; putting $a_1 = \dots = a_s = 0$ and $b_1 = c_1, \dots, b_s = c_s$ gives the desired result.

Example 5. Infinite-dimensional spaces.

Let us return to the Lebsgue measure $\bar{\mu}_\mathcal{L}(M)$ in I: $0 < x \leqq 1$, or better still to the equivalent $\bar{\mu}_i(M)$ (cf. Example 2). Since $I \in \bar{\mathcal{R}}_i$, $\bar{\mu}_i(I) = 1$, Definition 10.4.3 applies to the product $I \times I \times \dots$ (with a countably infinite number of factors) using in each I the half-ring $\bar{\mathcal{R}}_i$ and the measure function $\bar{\mu}_i(M)$. Thus we may take the product space $I_\infty = I \times I \times \dots$, the half-ring $\bar{\mathcal{R}}_{i,\infty}$ and the measure function $\bar{\mu}_{i,\infty}(M)$ corresponding respectively to $\mathcal{S}$, $\mathcal{R}$, $\mu(\mathcal{M})$ of Definition 10.4.3.

Thus $[z_1, z_2, \dots]$ with $0 < z_\sigma \leqq 1$ for $\sigma = 1, 2, \dots$ is the general point of I_∞ and $\bar{\mathcal{R}}_{i,\infty}$ consists of all sets $\mathcal{Y}\begin{bmatrix} a_1 \dots a_t \\ b_1 \dots b_t \end{bmatrix}$: $0 \leqq a_\sigma < z_\sigma \leqq b_\sigma \leqq 1$ for $\sigma = 1, \dots, t$, and $0 < z_\sigma \leqq 1$ for $\sigma = t+1, t+2, \dots$, where t is an arbitrary positive integer. For $M = \mathcal{Y}\begin{bmatrix} a_1 \dots a_t \\ b_1 \dots b_t \end{bmatrix}$ $\bar{\mu}_{i,\infty}(M) = \prod_{\sigma=1}^{t} (b_\sigma - a_\sigma)$ by definition.

Lebesgue measure in this case is defined to be the maximal equivalent extension of $\bar{\mu}_{i,\infty}(M)$:

$$\bar{\mu}_{\mathcal{L},\infty}(M) = \tilde{\bar{\mu}}_{i,\infty\, I}(M).$$

As $I = \sum_2 \times \sum_2 \times \ldots$ or $\sum_\omega \times \sum_\omega \times \ldots$, and as doubly infinite sequence can be written as a simple sequence $I_\infty = I \times I \times \ldots = \sum_2 \times \sum_2 \times \ldots$ or $\sum_\omega \times \sum_\omega \times \ldots$ it follows that I_∞ can be mapped in a measure-conserving way upon I. (Steinhaus.)

$R_1 \times R_1 \times \ldots$ to a countably infinite number of factors cannot be formed directly using $\bar{\mu}_{\mathcal{L}}(M)$ or $\bar{\mu}_{\mathcal{J}}(M)$ because $\bar{\mu}_{\mathcal{L}}(R_1) = +\infty$ and $\bar{\mu}_{\mathcal{J}}(R_1)$ is undefined. There are, however, various indirect ways to reach such spaces.

For instance, if $z = \varphi(x)$ is a one-to-one mapping of $0 < x < 1$ on $-\infty < z < +\infty$ --- that is, of I (without the one-point set $\{1\}$ which does not matter as its measure is zero) on R_1 --- then we can map $I \times I \times \ldots$ on $R_1 \times R_1 \times \ldots$ by letting correspond to an $[x_1, x_2, \ldots] \in I \times I \times \ldots$ the $[z_1, z_2, \ldots] \in R_1 \times R_1 \times \ldots$ with $z_\sigma = \varphi(x_\sigma)$, $\sigma = 1, 2, \ldots$. This carries the measure $\bar{\mu}_{\mathcal{L},\infty}(M)$ from $I \times I \times \ldots$ over to $R_1 \times R_1 \times \ldots$.

Very interesting applications have been made of the measure which results from the $\varphi(x)$ inverse to

$$\psi(z) = \frac{\int_{-\infty}^{z} e^{-u^2}\,du}{\int_{-\infty}^{+\infty} e^{-u^2}\,du} = \frac{1}{2} + \frac{1}{\sqrt{\pi}} \int_0^z e^{-u^2}\,du$$

(N. Wiener)

CHAPTER XI

PROPERTIES OF THE GENERAL INTEGRAL

§1. Bounded variation and the extension of finite measure functions

If we examine the extension theorems in section 3 of the last chapter, there is suggested at once a class of measure functions which are not necessarily non-negative but to which it seems likely similar extension theorems ought to apply. We refer to the class of measure functions which are finite linear combinations of the non-negative totally-additive ones. It will turn out in the course of the argument that similar extension theorems do hold for these functions, and that these are the only functions for which there exist totally additive extensions to a restricted Borel ring.

We now proceed to introduce the notion of bounded variation. A later theorem (Theorem 11.1.14) will make it clear that the concept is by no means extraneous and is in fact connected in an essential manner with the possibility of making totally additive extensions of not necessarily non-negative measure functions.

The nature of these preliminary remarks makes it clear that we shall be concerned with measure functions expressible as finite linear combinations of others. We make the following observation: If $\mu(M)$ and $\nu(M)$ are finite measure functions on a half-ring $\mathcal{R}$ then so are $\mu(M) + \nu(M)$ and c $\mu(M)$ where c is any real or complex constant. This is an immediate consequence of the definition of measure function. The statement may be put in other words: "The finite measure functions on $\mathcal{R}$ form a linear set." In particular the finite real measure functions form a linear set with respect to the real numbers. The finite and totally additive measure functions also form a linear set.

This is merely a consequence of the fact that the term-by-term sum of two absolutely convergent series is absolutely convergent and that a series remains absolutely convergent if its terms are all multiplied by a fixed number.

Definition 11.1.1. A real or complex valued measure function $\mu(M)$ _defined over a ring_ $\mathcal{R}$ _is said to be of bounded variation over a set_ $M \in \mathcal{R}$ _if the set of values_ $\mu(N)$ _where_ $N \in \mathcal{R}$ _and_ $N \subset M$ _is bounded; it is said to be of bounded variation over_ $\mathcal{R}$ _if it is of bounded variation over every_ $M \in \mathcal{R}$.

A real or complex valued measure function $\mu(M)$ _defined on a half-ring_ $\mathcal{R}$ _is said to be of bounded variation over a set_ $M \in \mathcal{R}$ _if its extension_ $\mu_I(M)$ _to_ $R(\mathcal{R})$ _is of bounded variation over_ M; _it is said to be of bounded variation over_ $\mathcal{R}$ _if it is of bounded variation over every_ $M \in \mathcal{R}$.

The form of Definition 11.1.1 makes it clear that the concept of bounded variation is simpler when the domain of the measure function is a ring rather than merely a half-ring. It will therefore be more convenient to make most manipulations by first extending to rings all the measure functions that occur. Nevertheless, the definition when $\mathcal{R}$ is a half-ring can be put in a form which is very reminiscent of the usual definition of bounded variation. We do this in Theorem 11.1.1.

THEOREM 11.1.1. _A measure function_ $\mu(M)$ _defined over a half-ring_ $\mathcal{R}$ _is of bounded variation over a set_ $M \in \mathcal{R}$ _if and only if the sum_ $\sum_i |\mu(M_i)|$ _is bounded over all finite systems of disjunct sets_ M_i _where_ $M_i \in \mathcal{R}$ _and_ $\sum_i M_i \subset M$.

Proof: Let $\mu_I(M)$ be the extension of $\mu(M)$ from $\mathcal{R}$ to $R(\mathcal{R})$. Let $M \in \mathcal{R}$ and $M_1, \ldots, M_n$ be disjunct sets from $\mathcal{R}$ each contained in M. Then $\sum_i M_i \in R(\mathcal{R})$ and $\mu_I(\sum_i M_i) = \sum_i \mu(M_i)$. On the assumption that $\mu(M)$ is of bounded variation over M we have the existence of a $C(M) > 0$ such that for

$N \in R(\mathcal{R})$, $N \subset M$ it is true that $|\mu(N)| < C(M)$. Thus we have $|\sum_i \mu(M_i)| < C(M)$. Since this inequality persists when $M_1, \ldots, M_n$ are replaced by any subset of them, we have

$$\sum_i |\mu(M_i)| \leqq \sum_i |\mathcal{R}\mu(M_i)| + \sum_i |\mathcal{J}\mu(M_i)| = \sum_{\mathcal{R}\mu(M_i) \geqq 0} \mathcal{R}\mu(M_i) +$$

$$+ \Big|\sum_{\mathcal{R}\mu(M_i)<0} \mathcal{R}\mu(M_i)\Big| + \sum_{\mathcal{J}\mu(M_i) \geqq 0} \mathcal{J}\mu(M_i) + \Big|\sum_{\mathcal{J}\mu(M_i)<0} \mathcal{J}\mu(M_i)\Big| \leqq$$

$$\leqq \Big|\sum_{\mathcal{R}\mu(M_i) \geqq 0} \mu(M_i)\Big| + \Big|\sum_{\mathcal{R}\mu(M_i)<0} \mu(M_i)\Big| + \Big|\sum_{\mathcal{J}\mu(M_i) \geqq 0} \mu(M_i)\Big| +$$

$$+ \Big|\sum_{\mathcal{J}\mu(M_i)<0} \mu(M_i)\Big| < 4\,C(M).$$

The converse is trivial on account of the inequality $|\sum_i \mu(M_i)| \leqq \sum_i |\mu(M_i)|$.

We remark, in disposal of the connection between bounded variation on a half-ring and on a ring, that it is obvious that if a measure function is of bounded variation over two sets of a ring it is of bounded variation over their sum. Hence it follows at once that a measure function defined on a half-ring $\mathcal{R}$ is of bounded variation over $\mathcal{R}$ if and only if its extension $\mu_I(M)$ is of bounded variation over $R(\mathcal{R})$.

Any finite non-negative measure function $\mu(M)$ on a half-ring $\mathcal{R}$ is clearly of bounded variation over $\mathcal{R}$, for its extension $\mu_1(M)$ on $R(\mathcal{R})$ is also finite and non-negative and for $M, N \in R(\mathcal{R})$, $N \subset M$ we have $\mu_I(N) \leqq \mu_I(M)$.

The measure functions of bounded variation on a half-ring $\mathcal{R}$ form a linear set. For if $\mu(M)$ and $\nu(M)$ are of bounded variation over $\mathcal{R}$, then by the definition for a given $M \in \mathcal{R}$ the sets of values $\mu_I(N)$ and $\nu_I(N)$ where $N \in R(\mathcal{R})$ and $N \subset M$ are bounded. Hence so also are the sets of values

$\mu_I(N) + \nu_I(N)$ and $c\,\mu_I(N)$.

Definition 11.1.2. *Let $\mu(M)$ be a real measure function on a half-ring $\mathcal{R}$. Let $\mu_I(M)$ be the extension of $\mu(M)$ to $R(\mathcal{R})$. The positive variation function $\mu_+(M)$ of $\mu(M)$ is defined for $M \in \mathcal{R}$ by means of*

$$(1) \qquad \mu_+(M) \equiv \underset{\substack{N \in R(\mathcal{R}) \\ N \subset M}}{\text{l.u.b.}} \mu_I(N) = \underset{P \in R(\mathcal{R})}{\text{l.u.b.}} \mu_I(MP).$$

The negative variation function $\mu_-(M)$ of $\mu(M)$ is defined as the positive variation function of $-\mu(M)$; i.e.,

$$(1') \qquad \mu_-(M) \equiv \underset{\substack{N \in R(\mathcal{R}) \\ N \subset M}}{\text{l.u.b.}} (-\mu_I(N)) = -\underset{\substack{N \in R(\mathcal{R}) \\ N \subset M}}{\text{g.l.b.}} \mu_I(N) = -\underset{P \in R(\mathcal{R})}{\text{g.l.b.}} \mu_I(MP).$$

Evidently in order that a real measure function $\mu(M)$ be of bounded variation over $\mathcal{R}$ it is necessary and sufficient that $\mu_+(M)$ and $\mu_-(M)$ be finite functions over $\mathcal{R}$.

THEOREM 11.1.2. *If $\mu(M)$ is a real measure function of bounded variation over a half-ring $\mathcal{R}$ then $\mu_+(M)$ and $\mu_-(M)$ are finite non-negative measure functions over $\mathcal{R}$, and we have for any $M \in \mathcal{R}$,*

$$(2) \qquad \mu(M) = \mu_+(M) - \mu_-(M).$$

If furthermore $\mu(M)$ is totally additive, then so are $\mu_+(M)$ and $\mu_-(M)$.

Proof: It is sufficient to make the proof under the assumption that $\mathcal{R}$ is a ring. For if $\mathcal{R}$ is only a half-ring we can form the extension $\mu_I(M)$ of $\mu(M)$ from $\mathcal{R}$ to $R(\mathcal{R})$. Then we can form the positive and negative variation functions $\mu_{I+}(M)$ and $\mu_{I-}(M)$ of $\mu_I(M)$ over $R(\mathcal{R})$. Equations (1) and (1') show that $\mu_{I+}(M)$ and $\mu_{I-}(M)$ agree respectively with $\mu_+(M)$ and $\mu_-(M)$ for $M \in \mathcal{R}$. Hence if the theorem is established for rings it will follow that $\mu_{I+}(M)$ and

$\mu_{I-}(M)$ are finite non-negative measure functions on $R(\mathcal{R})$ and satisfy the equation $\mu_I(M) = \mu_{I+}(M) - \mu_{I-}(M)$. If further $\mu(M)$ is totally additive on $\mathcal{R}$ then $\mu_I(M)$ and hence $\mu_{I+}(M)$ and $\mu_{I-}(M)$ will be totally additive. These results remain true if we now restrict ourselves to $\mathcal{R}$, and on $\mathcal{R}$ $\mu_I(M)$, $\mu_{I+}(M)$, and $\mu_{I-}(M)$ are respectively equal to $\mu(M)$, $\mu_+(M)$ and $\mu_-(M)$.

Assume therefore that $\mathcal{R}$ is a ring.

That $\mu_+(M)$ is finite and non-negative is a consequence of its definition, and the fact that the empty set has $\mu(M)$ measure zero and is contained in every $M \in \mathcal{R}$.

To prove that $\mu_+(M)$ is a measure function, let M_1 and M_2 be disjunct sets from $\mathcal{R}$ and let ε be a positive number. Let N_1, N_2 be sets from $\mathcal{R}$ such that $N_1 \subset M_1$, $N_2 \subset M_2$ and $\mu_+(M_1) < \mu(N_1) + \frac{\varepsilon}{2}$, $\mu_+(M_2) < \mu(N_2) + \frac{\varepsilon}{2}$. Now

$$\mu_+(M_1 + M_2) = \underset{P \in \mathcal{R}}{\text{l.u.b.}}\ \mu((M_1 + M_2)P) = \underset{P \in \mathcal{R}}{\text{l.u.b.}}\ (\mu(M_1P) + \mu(M_2P)) \leqq$$

$$\leqq \underset{P \in \mathcal{R}}{\text{l.u.b.}}\ \mu(M_1P) + \underset{P \in \mathcal{R}}{\text{l.u.b.}}\ \mu(M_2P) = \mu_+(M_1) + \mu_+(M_2) <$$

$$< \mu(N_1) + \mu(N_2) + \varepsilon = \mu(N_1 + N_2) + \varepsilon \leqq \mu_+(M_1 + M_2) + \varepsilon .$$

If now ε approaches zero we get

$$\mu_+(M_1 + M_2) = \mu_+(M_1) + \mu_+(M_2),$$

so that $\mu_+(M)$ is a measure function.

To prove the total additivity of $\mu_+(M)$ on the assumption of that of $\mu(M)$, proceed as follows: Let $M \in \mathcal{R}$ be the sum of an infinite sequence of disjunct sets from $\mathcal{R}$, say $M = \sum_i M_i$. Let ε be a positive number and select N from $\mathcal{R}$ such that $N \subset M$ and $\mu_+(M) - \varepsilon < \mu(N)$. Then we have

$$\mu_+(M) - \varepsilon < \mu(N) = \mu(\sum_i M_iN) = \sum_i \mu(M_iN) \leqq \sum_i \mu_+(M_i) \leqq \mu_+(M),$$

so that $\mu_+(M)$ is totally additive on the assumption that $\mu(M)$ is.

To establish equation (2) we simply observe that

$$\mu_+(M) - \mu(M) = \underset{P \in \mathcal{R}}{\text{l.u.b.}}\ \mu(MP) - \mu(M) = \underset{P \in \mathcal{R}}{\text{l.u.b.}}(\mu(MP) - \mu(M)) =$$

$$= \underset{P \in \mathcal{R}}{\text{l.u.b.}} - \mu(M - MP) = \underset{Q \in \mathcal{R}}{\text{l.u.b.}} - \mu(MQ) = \mu_-(M).$$

As for $\mu_-(M)$ its definition shows that it is finite and non-negative; the equation (2) shows everything else that we need. The proof is therefore complete.

Corollary: Even if $\mu(M)$ is not of bounded variation but is simply finite, the functions $\mu_+(M)$ and $\mu_-(M)$ are non-negative measure functions on $\mathcal{R}$. The equation (2) holds for every set M for which $\mu_+(M)$ or $\mu_-(M)$ is finite and these sets M are those over which $\mu(M)$ is of bounded variation. If further $\mu(M)$ is totally additive, then so are $\mu_+(M)$ and $\mu_-(M)$.

Proof: The proof of the additivity and total additivity of $\mu_+(M)$ and $\mu_-(M)$ involved in no essential fashion the finiteness of $\mu_+(M)$ and $\mu_-(M)$. The argument used to establish (2) is valid up to and including the statement that $\mu_+(M) - \mu(M) = \mu_-(M)$. This statement on the basis of the finiteness of $\mu(M)$ shows that $\mu_+(M)$ and $\mu_-(M)$ are either both finite or both infinite. If for a set M both are finite, equation holds and $\mu(M)$ is of bounded variation over M.

This theorem shows that every real measure function of bounded variation over a half ring $\mathcal{R}$ is expressible as the difference of two finite non-negative measure functions. Of course it is obvious that any function so expressible is of bounded variation. The decomposition is by no means unique. For example, if $\mu(M) = \mu_1(M) - \mu_2(M)$ is a decomposition and $\rho(M)$ is any finite non-negative measure function over $\mathcal{R}$, then $\mu(M) = (\mu_1(M)+\rho(M))-(\mu_2(M)+\rho(M))$ is another decomposition. It is therefore of interest to characterize all decompositions. With this in mind we introduce:

Definition 11.1.3. Let $\mu_1(M)$ and $\mu_2(M)$ be real finite measure functions on a half-ring $\mathcal{R}$. Let $\mu_{1I}(M)$ and $\mu_{2I}(M)$ be their respective extensions from $\mathcal{R}$ to $R(\mathcal{R})$. For $M \in \mathcal{R}$ define

$$(\text{Max } \mu_1, \mu_2)(M) \equiv \underset{P \in R(\mathcal{R})}{\text{l.u.b.}} (\mu_{1I}(MP) + \mu_{2I}(M-MP))$$

$$(\text{Min } \mu_1, \mu_2)(M) = \underset{P \in R(\mathcal{R})}{\text{g.l.b.}} (\mu_{1I}(MP) + \mu_{2I}(M-MP))$$

THEOREM 11.1.3. Let $\mu_1(M)$, $\mu_2(M)$ be as in Definition 11.1.3. Denote by $\mu(M)$ the difference $\mu_1(M) - \mu_2(M)$. The functions $(\text{Max } \mu_1, \mu_2)(M)$ and $(\text{Min } \mu_1, \mu_2)(M)$ satisfy the relations

$$(3) \qquad (\text{Max } \mu_1, \mu_2)(M) = \mu_1(M) + \mu_-(M) = \mu_2(M) + \mu_+(M)$$

$$(3') \qquad (\text{Min } \mu_1, \mu_2)(M) = \mu_1(M) - \mu_+(M) = \mu_2(M) - \mu_-(M) .$$

Proof: By an argument similar to that employed in the first part of the proof of Theorem 11.1.2 we can show that it is permissible to assume that $\mathcal{R}$ is a ring. We therefore take $\mathcal{R}$ as a ring. Now we have

$$\begin{aligned}(\text{Max } \mu_1, \mu_2)(M) &= \underset{P \in \mathcal{R}}{\text{l.u.b.}} (\mu_1(MP) + \mu_2(M-MP)) = \\ &= \underset{P \in \mathcal{R}}{\text{l.u.b.}} (\mu_1(MP) - \mu_2(MP) + \mu_2(M)) = \\ &= \underset{P \in \mathcal{R}}{\text{l.u.b.}} (\mu(MP) + \mu_2(M)) = \underset{P \in \mathcal{R}}{\text{l.u.b.}} \mu(MP) + \mu_2(M) = \\ &= \mu_+(M) + \mu_2(M).\end{aligned}$$

The function $(\text{Max } \mu_1, \mu_2)(M)$ is symmetric in μ_1 and μ_2 as is clear from its definition. If therefore we interchange $\mu_1(M)$ and $\mu_2(M)$, $(\text{Max } \mu_1, \mu_2)(M)$ goes over into itself, $\mu(M)$ goes over into $-\mu(M)$, and hence $\mu_+(M)$ becomes $\mu_-(M)$. The above equation therefore becomes

$$(\text{Max } \mu_1, \mu_2)(M) = \mu_-(M) + \mu_1(M).$$

The equation (3) is thus established. Equation (3') follows from (3) and the identity

$$(\text{Min } \mu_1, \mu_2)(M) = -(\text{Max} - \mu_1, -\mu_2)(M)$$

which is obtained directly from the definitions.

Corollary 1. *If $\mu_1(M)$ and $\mu_2(M)$ are real measure functions of bounded variation over a half-ring $\mathcal{R}$, then so are $(\text{Max } \mu_1, \mu_2)(M)$ and $(\text{Min } \mu_1, \mu_2)(M)$. If further they are totally additive, then so are $(\text{Max } \mu_1, \mu_2)(M)$ and $(\text{Min } \mu_1, \mu_2)(M)$.*

Proof: Equations (3), (3') and Theorem 11.1.2.

Corollary 2. *In order that two finite real measure functions $\mu_1(M)$ and $\mu_2(M)$ on $\mathcal{R}$ be respectively the positive and negative variation functions of their difference $\mu(M) = \mu_1(M) - \mu_2(M)$ it is necessary and sufficient that $(\text{Min } \mu_1, \mu_2)(M) = 0$.*

Proof: Equation (3').

THEOREM 11.1.4. *A necessary and sufficient condition that $\mu(M) = \mu_1(M) - \mu_2(M)$ be a decomposition of a function $\mu(M)$ of bounded variation on a half-ring $\mathcal{R}$ into the difference of two finite and non-negative measure functions $\mu_1(M)$, $\mu_2(M)$ is that*

$$\mu_1(M) = \mu_+(M) + \rho(M)$$
$$\mu_2(M) = \mu_-(M) + \rho(M)$$

where $\rho(M)$ is a finite non-negative measure function on $\mathcal{R}$.

Proof: Since $\mu_1(M)$ and $\mu_2(M)$ are non-negative, it follows from the definition of $(\text{Min } \mu_1, \mu_2)(M)$ that it cannot be negative. We now simply use equation (3') of Theorem 11.1.3, take $\rho(M) = (\text{Min } \mu_1, \mu_2)(M)$ and apply Corollary 1 above. This proves the necessity. The sufficiency is obvious.

This last theorem gives us a complete characterization of the decompositions of real measure functions of bounded variation into the difference of two finite non-negative measure functions. It is to be observed that if $\mu(M)$ is totally additive $\mu_1(M)$ and $\mu_2(M)$ need not be since $\rho(M)$ is arbitrary. If, however, $\mu_1(M)$ and $\mu_2(M)$ are totally additive $\mu(M)$ and hence $\mu_+(M)$ and $\mu_-(M)$ will also be.

Definition 11.1.4. If $\mu(M)$ is a finite real measure function over a half-ring $\mathcal{R}$, we shall call the function $\mu_+(M) + \mu_-(M)$ the total variation function of $\mu(M)$, and denote it by $\overline{\mu}(M)$.

Clearly by Theorem 11.1.2 and its corollary, $\overline{\mu}(M)$ is a non-negative measure function and is totally additive if $\mu(M)$ is. $\mu(M)$ is of bounded variation if and only if $\overline{\mu}(M)$ is finite.

We can now easily obtain analogous results on the decomposition of complex-valued measure functions of bounded variation. This is essentially done by splitting the function into its real and imaginary parts and treating them separately.

THEOREM 11.1.5. Let $\mathcal{R}$ be a half-ring and $\mu(M)$ a (finite) complex-valued measure function on $\mathcal{R}$. The functions $\lambda(M) \equiv \mathcal{R}\mu(M)$ and $\nu(M) \equiv \mathcal{I}\mu(M)$ are (finite) real measure functions on $\mathcal{R}$. $\mu(M)$ is of bounded variation (respectively totally additive) if and only if $\lambda(M)$ and $\nu(M)$ are.

Proof: That $\lambda(M)$ and $\nu(M)$ are measure functions (i.e. are additive) is obvious. To show that $\mu(M)$ is of bounded variation when and only when $\lambda(M)$ and $\nu(M)$ are, we observe that for any $M, M_1, \ldots, M_n \in \mathcal{R}$ with the M_i disjunct and contained in M, we have

$$\sum_i |\lambda(M_i)| + \sum_i |\nu(M_i)| \geqq \sum_i |\mu(M_i)| \geqq \text{Max}\ (\sum_i |\lambda(M_i)|, \sum_i |\nu(M_i)|).$$

Thus $\sum_i |\mu(M_i)|$ is bounded for all such systems $M_1, \ldots, M_n$ if and only if $\sum_i |\lambda(M_i)|$ and $\sum_i |\nu(M_i)|$ are. Theorem 11.1.1 then gives what we want. The equivalence for the total-additivity is a consequence of the fact that a series of complex terms converges absolutely if and only if the corresponding series for its real and imaginary parts do.

THEOREM 11.1.6. *Let $\mu(M)$ be a (finite) complex-valued measure function on a half-ring $\mathcal{R}$. Then*

1) *In order that $\mu(M)$ be of bounded variation over $\mathcal{R}$, it is necessary and sufficient that it be expressible in the form $\mu(M) = \sum_{\rho=0}^{3} i^\rho \mu_\rho(M)$ where $\mu_0(M), \ldots, \mu_3(M)$ are finite, non-negative measure functions on $\mathcal{R}$. We shall refer to such an expression as a decomposition of $\mu(M)$.*

2) *If $\mu(M)$ is of bounded variation on $\mathcal{R}$ there is a (unique) special decomposition $\mu(M) = \sum_{\rho=0}^{3} i^\rho \mu_\rho^{(o)}(M)$ having the property that in order that $\mu(M) = \sum_{\rho=0}^{3} i^\rho \mu_\rho(M)$ be another decomposition of $\mu(M)$, it is necessary and sufficient that*

$$\mu_\rho(M) = \begin{cases} \mu_\rho^{(o)}(M) + \alpha(M) & (\rho \text{ even}) \\ \mu_\rho^{(o)}(M) + \beta(M) & (\rho \text{ odd}) \end{cases} \tag{4}$$

where $\alpha(M)$ and $\beta(M)$ are finite non-negative measure functions on $\mathcal{R}$.

3) *If the $\mu_\rho(M)$ for any decomposition of $\mu(M)$ are totally additive, then so is $\mu(M)$; if $\mu(M)$ is totally additive, so are the $\mu_\rho^{(o)}(M)$.*

Proof: The sufficiency in statement 1) is trivial. As for the necessity, let $\lambda(M)$ and $\nu(M)$ be as in Theorem 11.1.5, and define $\mu_o^{(o)}(M) = \lambda_+(M)$,

$\mu_1^{(o)}(M) = \nu_+(M)$, $\mu_2^{(o)}(M) = \lambda_-(M)$, $\mu_3^{(o)}(M) = \nu_-(M)$. On the assumption that $\mu(M)$, and hence $\nu(M)$ and $\lambda(M)$, is of bounded variation, the $\mu_\rho^{(o)}(M)$ are finite non-negative measure functions. In view of Theorem 11.1.2 we have $\mu(M) = \lambda(M) + i\,\nu(M) = \sum_{\rho=o}^{3} i^\rho\,\mu_\rho^{(o)}(M)$. Hence a decomposition exists.

We now show that the $\mu_\rho^{(o)}(M)$ just defined satisfy the conditions of statement 2). In the first place it is obvious that the equation (4) for any finite non-negative $\alpha(M)$ and $\beta(M)$ defines a decomposition of $\mu(M)$. Conversely if $\mu(M) = \sum_{\rho=o}^{3} i^\rho\,\mu_\rho(M)$ is a decomposition of $\mu(M)$, we have $\lambda(M) = \mu_o(M) - \mu_2(M)$ and $\nu(M) = \mu_1(M) - \mu_3(M)$. Theorem 11.1.4 then gives us equation (4) with $\alpha(M) = (\text{Min}\ \mu_o, \mu_2)(M)$, $\beta(M) = (\text{Min}\ \mu_1, \mu_3)(M)$. The uniqueness of the special decomposition is clear from the equations (4) because if $\mu_\rho'^{(o)}(M)$ were another, (4) would give $\mu_\rho^{(o)}(M) \geqq \mu_\rho'^{(o)}(M)$ and $\mu_\rho'^{(o)}(M) \geqq \mu_\rho^{(o)}(M)$.

The first part of statement 3) is trivial. The second part is a consequence of Theorems 11.1.5 and 11.1.2. The proof is complete.

It is to be noted that the expressions by which the $\mu_\rho^{(o)}(M)$ were defined have a sense even if $\mu(M)$ is not of bounded variation. They will not of course in this case be finite. Nevertheless for any (finite) complex-valued measure function $\mu(M)$ we use the notation $\mu_\rho^{(o)}(M)$ to denote the four functions defined by the equations of the first paragraph of the proof of Theorem 11.1.6.

Definition 11.1.5. Let $\mu(M)$ be a complex-valued measure function on a half-ring $\mathcal{R}$. Define the functions $\mu_\rho^{(o)}(M)$ as above. The function $\bar{\mu}(M) \equiv \sum_{\rho=o}^{3} \mu_\rho^{(o)}(M)$ is called the total variation function of $\mu(M)$.

It is evident that in case $\mu(M)$ happens to be a real measure function, the $\bar{\mu}(M)$ of Definition 11.1.5 coincides with the $\bar{\mu}(M)$ of Definition 11.1.4, for in this case we have $\mu_1^{(o)}(M) = \mu_3^{(o)}(M) = 0$, $\mu_o^{(o)}(M) = \mu_+(M)$, $\mu_2^{(o)}(M) = \mu_-(M)$ As in the remarks following Definition 11.1.4, it is clear that $\bar{\mu}(M)$ is a non-negative measure function, that $\mu(M)$ is of bounded variation if and only if $\bar{\mu}(M)$ is finite, and that $\bar{\mu}(M)$ is totally additive if $\mu(M)$ is. As a matter of fact it can readily be shown that if $\bar{\mu}(M)$ is finite and totally additive, $\mu(M)$ is totally additive, but we shall not need this result.

Note. There is another and perhaps more customary definition of the total variation function of a complex measure function. This total variation function which we may denote by $\bar{\bar{\mu}}(M)$ is defined as l.u.b. $\sum_i |\mu(M_i)|$ over all finite systems of disjunct sets $M_i \in \mathcal{R}$, $M_i \subset M$. The two functions $\bar{\mu}(M)$ and $\bar{\bar{\mu}}(M)$ coincide when $\mu(M)$ is real, but are not in general identical; they do, however, enjoy very similar properties with respect to the results we expect to obtain. Each of the definitions has its advantages and the reason we choose to stick to Definition 11.1.5 is that it lends itself more readily to the program of treating complex measure functions by splitting them into their real and imaginary parts.

We now turn our attention to the problem of extending the domain of definition of the measure functions of bounded variation.

It is convenient first to develop a lemma concerning the measurability of a set simultaneously with respect to a number of outer measures. For use later we state the lemma in a somewhat more general form than is immediately necessary.

Lemma 1. *Let S be a space and $\{\mu_n^*(M)\}$ a (finite or infinite) sequence of outer measures each determined by a fixed half-ring $\mathcal{R}$. Let $\{a_n\}$ be a* corresponding sequence of positive numbers such that for every $M \in \mathcal{R}$, $\sum_n a_n \mu_n^*(M)$ is convergent and finite. Consider the function

$$\mu^*(M) = \sum_n a_n \mu_n^*(M).$$

(This is defined for every $M \in \mathcal{P}(S)$, the sum being either finite or $+\infty$.) Then

1) The function $\mu^*(M)$ is a regular outer measure function determined by $\mathcal{R}$.
2) A set M is measurable with respect to $\mu^*(M)$ if and only if it is measurable with respect to every $\mu_n^*(M)$. In other words, $MS_{\mu^*} = \prod_n MS_{\mu_n^*}$.

Proof: First, $\mu^*(M)$ is an outer measure. It obviously satisfies postulates I and II for outer measures. As for III, let $M = \sum_i M_i$, then

$$\mu^*(M) = \sum_n a_n \mu_n^*(M) \leq \sum_n a_n \sum_i \mu_n^*(M_i) = \sum_i \sum_n a_n \mu_n^*(M_i) = \sum_i \mu^*(M_i).$$

The interchange in order of summation is permissible because all the terms are non-negative.

Next, every set measurable with respect to every $\mu_n^*(M)$ is also measurable with respect to $\mu^*(M)$. Let M be such a set and P be any set. Then $\mu^*(P) = \sum_n a_n \mu_n^*(P) = \sum_n a_n(\mu_n^*(MP) + \mu_n^*(P - MP)) = \mu^*(MP) + \mu^*(P - MP)$. In particular it follows that the sets of $BR(\mathcal{R})$ are measurable with respect to $\mu^*(M)$.

Now let M be any set. Since $\mu_n^*(M)$ is determined by $\mathcal{R}$, there is for each n a $N_n \in BR(\mathcal{R})$ such that $N_n \supset M$ and $\mu_n^*(N_n) = \mu_n^*(M)$. If now $N = \prod_n N_n$, then $N \in BR(\mathcal{R})$, $N \supset M$, and $\mu_n^*(N) = \mu_n^*(M)$. Hence $\mu^*(N) = \mu^*(M)$. From this result and that of the preceding paragraph together with the finiteness of

$\mu^*(M)$ for $M \in \mathcal{R}$, it follows that $\mathcal{R}$ determines $\mu^*(M)$. We have thus proved statement 1) and half of 2).

To complete the proof we need only show that a set M measurable with respect to $\mu^*(M)$ is measurable with respect to every $\mu_n^*(M)$. If M is measurable, the we use the Corollary to Theorem 10.2.12, write $M = N + Z$ where $N \in BR(\mathcal{R})$ and $\mu^*(Z) = 0$. Since $\mu^*(Z) = 0$ it follows that for every n, $\mu_n^*(Z) = 0$. Hence M is measurable with respect to $\mu_n^*(M)$.

Corollary 1. If the functions $\mu_n^*(M)$ are finite in number, then $MS'_{\mu^*} = \prod_n MS'_{\mu_n^*}$.

Proof: The proof is clear.

Corollary 2. If $\mu^*(M)$ and $\nu^*(M)$ are outer measures determined by a half-ring $\mathcal{R}$ and if on $\mathcal{R}$, $\mu(M) \geqq \nu(M)$, then for all M, $\mu^*(M) \geqq \nu^*(M)$ and every set measurable with respect to $\mu^*(M)$ is also measurable with respect to $\nu^*(M)$.

Proof: That $\mu^*(M) \geqq \nu^*(M)$ is evident for example from the explicit definition, given in Theorem 10.3.2, of the outer measures determined by $\mu(N)$ and $\nu(N)$ on $\mathcal{R}$. The measurability property is a consequence of the lemma by writing $\mu(M) = \nu(M) + (\mu(M) - \nu(M))$.

THEOREM 11.1.7. Let $\mathcal{R}$ be a half-ring and on it let $\mu_1(M)$, $\mu_2(M)$, ..., $\mu_m(M)$ be finite non-negative and totally additive measure functions; let $a_1, \ldots, a_m$ be non-negative numbers and form the function $\mu(M) = \sum_1^m a_n \mu_n(M)$. Form the corresponding outer measures $\mu_I^*(M)$, $\mu_{1I}^*(M)$, $\mu_{2I}^*(M)$, ..., $\mu_{mI}^*(M)$. Then

1) $\mu_I^*(M) = \sum_1^m a_n \mu_{nI}^*(M)$.

2) For every half-ring $\mathcal{T}$ such that $\mathcal{R} \subset \mathcal{T} \subset MS'_{\mu_I^*}$ there exist uniquely $\tilde{\mu}(M)$, $\tilde{\mu}_1(M)$, ..., $\tilde{\mu}_n(M)$ equivalent exten-

sions of the corresponding functions from $\mathcal{R}$ to $\mathcal{T}$. These satisfy the identity $\bar{\mu}(M) = \sum_1^m a_n \bar{\mu}_n(M)$.

3) If for a half-ring $\mathcal{R}'$ the functions $\mu_1'(M), \mu_2'(M), \ldots, \mu_m'(M)$ are respectively equivalent to $\mu_1(M), \ldots, \mu_m(M)$, then $\mu'(M) = \sum_1^m a_n \mu_n'(M)$ is equivalent to $\mu(M)$.

Proof: We may assume that no a_n is zero otherwise the corresponding terms simply do not occur. It is clear from the explicit definition (cf. Theorem 10.3.2) of the outer measure $\nu_I^*(M)$ determined by a totally additive non-negative $\nu(M)$ on $\mathcal{R}$ that for a positive a the function $a\,\nu(M)$ determines the (unique) outer measure $a\,\nu_I^*(M)$. Thus two functions $\nu(M)$ on $\mathcal{R}$ and $\nu'(M)$ on $\mathcal{R}'$ determine the same outer measure if and only if $a\,\nu(M)$ and $a\,\nu'(M)$ do; i.e., $\nu(M)$ is equivalent to $\nu'(M)$ if and only if $a\,\nu(M)$ is equivalent to $a\,\nu'(M)$. We may as well therefore assume in the proof of the theorem that $a_1 = \ldots = a_m = 1$. Now by Lemma 1, the function $\sum_1^m \mu_{nI}^*(M)$ is an outer measure determined by $\mathcal{R}$. It obviously agrees with $\mu_I^*(M)$ for $M \in \mathcal{R}$ and therefore by Theorem 10.3.2 is equal to it. Statement 1) therefore holds. If $\mu_1'(M), \ldots, \mu_m'(M)$ are respectively equivalent to $\mu_1(M), \ldots, \mu_m(M)$, and $\mu'(M) = \sum_1^m \mu_n'(M)$, then the use of statement 1) gives $\mu_I'^*(M) = \mu_I^*(M)$ and therefore the equivalence of $\mu(M)$ and $\mu'(M)$. Thus 3) is established. By the use of Corollary 1 to Lemma 1 we see that it is permissible to make the extension of the $\mu_n(M)$ to $\mathcal{T}$ and hence 2) follows from 3).

THEOREM 11.1.8. Let $\mu_1(M), \mu_2(M)$, finite, non-negative and totally additive on a half-ring $\mathcal{R}$, be equivalent (cf. Definition 10.3.1) respectively to $\mu_1'(M), \mu_2'(M)$ on a half-ring $\mathcal{R}'$. Then if $(\text{Min}\ \mu_1, \mu_2) = 0$ for $M \in \mathcal{R}$ we have $(\text{Min}\ \mu_1', \mu_2')(M) = 0$ for $M \in \mathcal{R}$.

Proof: Let $\bar{\mu}(M) = \mu_1(M) + \mu_2(M)$; form the outer measure $\bar{\mu}_I^*(M)$ and the set $\mathcal{T} = MS'_{\bar{\mu}_1^*}$. Let $\tilde{\mu}_1(M)$, $\tilde{\mu}_2(M)$ be the equivalent extensions of $\mu_1(M)$, $\mu_2(M)$ from $\mathcal{R}$ to $\mathcal{T}$. These are by the equivalence also the equivalent extensions of $\mu_1'(M)$, $\mu_2'(M)$ from $\mathcal{R}'$ to $\mathcal{T}$. Form on $\mathcal{T}$ the function $(\text{Min}\ \tilde{\mu}_1, \tilde{\mu}_2)(M)$, Recalling Definition 11.1.3 we see that for $M \in \mathcal{R}$ this function is zero (since in defining it we take a g.l.b. over a wider class of numbers). Since every set from $\mathcal{T}$ is contained in the sum of a sequence of sets from $\mathcal{R}$, the non-negative function $(\text{Min}\ \tilde{\mu}_1, \tilde{\mu}_2)(M)$ is zero for every $M \in \mathcal{T}$. Now define on $\mathcal{R}'$ the function $\rho'(M) = (\text{Min}\ \mu_1', \mu_2')(M)$ and the functions $\nu_1'(M) = \mu_1'(M) - \rho'(M)$, $\nu_2'(M) = \mu_2'(M) - \rho'(M)$. The functions $\nu_1'(M)$, $\nu_2'(M)$ are finite non-negative and totally additive (for in fact they are respectively the positive and negative variation functions of $\mu'(M) = \mu_1'(M) - \mu_2'(M)$ as is seen from Theorem 11.1.4). We now form the extensions $\tilde{\mu}_1'(M)$, $\tilde{\mu}_2'(M)$, $\rho'(M)$, $\tilde{\nu}_1'(M)$, $\tilde{\nu}_2'(M)$. By Theorem 11.1.7 we have $\tilde{\mu}_1'(M) = \tilde{\nu}_1'(M) + \tilde{\rho}'(M)$ and $\tilde{\mu}_2'(M) = \tilde{\nu}_2'(M) + \tilde{\rho}'(M)$. Now for $M \in \mathcal{T}$

$$0 \leq \tilde{\rho}'(M) \leq (\text{Min}\ \tilde{\mu}_1', \tilde{\mu}_2')(M) = (\text{Min}\ \tilde{\mu}_1, \tilde{\mu}_2)(M) = 0 .$$

Hence for $M \in \mathcal{R}'$, $\rho'(M) = 0$.

Definition 11.1.6. *Let $\mu(M)$ be a real (finite) totally additive measure function of bounded variation over a half-ring $\mathcal{R}$, and $\mu'(M)$ a similar function over a half ring $\mathcal{R}'$. $\mu(M)$ and $\mu'(M)$ are called equivalent if $\mu_+(M)$ and $\mu_-(M)$ are respectively equivalent to $\mu_+'(M)$ and $\mu_-'(M)$.*

If $\mu(M)$, $\mu'(M)$ are complex they are called equivalent if the $\mu_\rho^{(o)}(M)$ of $\mu(M)$ are respectively equivalent to the $u_\rho'^{(o)}(M)$ of $\mu'(M)$.

Since the relation of equivalence for non-negative measure functions is a transitive one, it is clear that it will also be transitive in the present more general situation. It is also evident that two complex measure functions are equivalent if and only if their real and imaginary parts are.

THEOREM 11.1.9. *Let $\mu(M)$ be real, of bounded variation, and totally additive on a half-ring $\mathcal{R}$; $\mu'(M)$ the same on a half-ring $\mathcal{R}'$. If there exist* *a decomposition* $\mu(M) = \mu_1(M) - \mu_2(M)$ *into finite non-negative totally additive functions, and a similar decomposition* $\mu'(M) = \mu_1'(M) - \mu_2'(M)$ *such that* $\mu_1(M)$, $\mu_2(M)$ *are respectively equivalent to* $\mu_1'(M)$, $\mu_2'(M)$, *then* $\mu(M)$ *and* $\mu'(M)$ *are equivalent.*

A corresponding statement holds when $\mu(M)$ and $\mu'(M)$ are complex.

Proof: First assume that $\mathcal{R} \subset \mathcal{R}'$. Then on $\mathcal{R}$, $\mu(M)$ and $\mu'(M)$ agree. On $\mathcal{R}'$ form the functions $\mu_+'(M)$ and $\mu_-'(M)$; for $M \in \mathcal{R}$ define $\nu_1(M) = \mu_+'(M)$ and $\nu_2(M) = \mu_-'(M)$. If we knew now that $\mu_+'(M)$, $\mu_-'(M)$ and $\nu_1(M)$, $\nu_2(M)$ were respectively equivalent, the Theorem 11.1.8 would show that on $\mathcal{R}$ $(\text{Min } \nu_1, \nu_2)(M) = 0$, and hence that $\nu_1(M)$, $\nu_2(M)$ are respectively $\mu_+(M)$ and $\mu_-(M)$; therefore that $\mu(M)$ and $\mu'(M)$ are equivalent. To show that $\mu_+'(M)$ is equivalent to $\nu_1(M)$ it is sufficient, since the first is a totally additive extension of the second, to show that every set of $\mathcal{R}'$ is measurable with respect to the outer measure determined by $\nu_1(M)$ (by Theorem 10.3.3 and Corollary to Theorem 10.3.7).

To prove the last statement, form the common outer measure determined by $\mu_1(M)$ on $\mathcal{R}$ and $\mu_1'(M)$ on $\mathcal{R}'$, and call it $\mu_{1I}^*(M)$; form the outer measure $\nu_{1I}^*(M)$ determined by $\nu_1(M)$ on $\mathcal{R}$. Now on $\mathcal{R}'$, $\mu_1'(M) \geqq \mu_+'(M)$ by Theorem 11.1.4. Therefore on $\mathcal{R}$, $\mu_1(M) \geqq \nu_1(M)$. Hence, since $\mathcal{R}$ determines both outer measures, we have everywhere $\mu_{1I}^*(M) \geqq \nu_{1I}^*(M)$. Since $\mathcal{R}'$ is a determining set for $\mu_{1I}^*(M)$, every set from $\mathcal{R}'$ is measurable with respect to it, and by Corollary 2 to Lemma 1, is therefore measurable with respect to $\nu_{1I}^*(M)$. Thus $\mu_+'(M)$ and $\nu_1(M)$ are equivalent. Similarly for $\mu_-'(M)$ and $\nu_2(M)$.

We now dispense with the requirement that $\mathcal{R} \subset \mathcal{R}'$. Form the outer measures $\mu_{1I}^*(M)$, $\mu_{2I}^*(M)$ determined respectively by $\mu_1(M)$ and $\mu_2(M)$, and the system $\mathcal{T} = MS'_{\mu_{1I}^*} \cdot MS'_{\mu_{2I}^*}$. Let $\tilde{\mu}_1(M)$, $\tilde{\mu}_2(M)$ be respectively the equivalent extensions of $\mu_1(M)$, $\mu_2(M)$ from $\mathcal{R}$ to $\mathcal{T}$; they are also the equivalent extensions of $\mu_1'(M)$,

$\mu_2'(M)$ from $\mathcal{R}'$ to $\mathcal{T}$. Thus by what we have proved above, $\mu(M)$ and $\mu'(M)$ are both equivalent to $\tilde{\mu}(M) = \tilde{\mu}_1(M) - \tilde{\mu}_2(M)$ and therefore to each other.

For the complex case the same proof works if real and imaginary parts are considered separately.

THEOREM 11.1.10. Let $\mu(M)$ be a real or complex (finite) totally-additive measure function of bounded variation over a half-ring $\mathcal{R}$. Form the total variation function $\bar{\mu}(M)$ of $\mu(M)$, the outer measure $\bar{\mu}_I^*(M)$, and the system $\mathcal{T} = MS'_{\bar{\mu}_1}{}^{-*}$. There is a unique equivalent extension $\tilde{\mu}(M)$ of $\mu(M)$ from $\mathcal{R}$ to $\mathcal{T}$ which is maximal in the sense that if $\mu'(M)$ is totally additive and of bounded variation over a half-ring $\mathcal{R}'$ and is equivalent to $\mu(M)$, then $\mathcal{R}' \subset \mathcal{T}$.

Proof: We go through the proof for the real case. Form the functions $\mu_1(M) = \mu_+(M)$, $\mu_2(M) = \mu_-(M)$ on $\mathcal{R}$; form their respective equivalent extensions $\tilde{\mu}_1(M)$, $\tilde{\mu}_2(M)$ from $\mathcal{R}$ to $\mathcal{T}$. (They exist by Corollary 1 to Lemma 1 above, and Theorem 10.3.3.) Define $\tilde{\mu}(M) = \tilde{\mu}_1(M) - \tilde{\mu}_2(M)$. By Theorem 11.1.9, $\tilde{\mu}(M)$ is equivalent to $\mu(M)$.

Now let $\mu'(M)$ and $\mathcal{R}'$ be as in the statement of the theorem. By hypothesis the outer measures determined by $\mu_1'(M)$, $\mu_2'(M)$ are identical with those determined by $\mu_1(M)$ and $\mu_2(M)$. Hence every set of $\mathcal{R}'$ is measurable and of finite measure with respect to both of them. Thus $\mathcal{R}' \subset \mathcal{T}$.

Any other equivalent extension to $\mathcal{T}$ would be equivalent to $\tilde{\mu}(M)$ on $\mathcal{T}$ and thus identical with it. Therefore $\tilde{\mu}(M)$ is unique.

The argument for the complex case is essentially the same.

THEOREM 11.1.11. Let $\mu(M)$, $\tilde{\mu}(M)$, $\mathcal{R}$, $\mathcal{T}$ be as in the preceding theorem. For every half-ring $\mathcal{R}''$ satisfying $\mathcal{R} \subset \mathcal{R}'' \subset \mathcal{T}$ there is a unique equivalent extension $\mu''(M)$ of $\mu(M)$ from $\mathcal{R}$ to $\mathcal{R}''$.

Proof: If $\mu''(M)$ exists, it is unique for it is equivalent to $\mu(M)$ and hence to $\tilde{\mu}(M)$, and hence agrees with $\tilde{\mu}(M)$ on $\mathcal{R}''$.

Preserve the notations of the proof of the preceding theorem. On $\mathcal{R}''$ define $\mu_1''(M) = \tilde{\mu}_1(M)$, $\mu_2''(M) = \tilde{\mu}_2(M)$. By the Corollary to Theorem 10.3.7, $\mu_1''(M)$, $\mu_2''(M)$ are respectively equivalent extensions of $\mu_1(M)$, $\mu_2(M)$. By Theorem 11.1.9 the function $\mu''(M) = \mu_1''(M) - \mu_2''(M)$ has the desired properties.

We obtain the complex case from this in the usual fashion.

THEOREM 11.1.12. *Let $\mathcal{R}$ be a half-ring, $\mu(M)$ and $\nu(M)$ finite real or complex measure functions of bounded variation on $\mathcal{R}$; let $\mu'(M)$ and $\nu'(M)$ be similar functions on a half-ring $\mathcal{R}'$. Assume that $\mu'(M)$, $\nu'(M)$ are respectively equivalent to $\mu(M)$, $\nu(M)$. Then for any constant a, $\mu(M) + \nu(M)$ and $a \cdot \mu(M)$ are respectively equivalent to $\mu'(M) + \nu'(M)$ and $a \cdot \mu'(M)$.*

Proof: We prove the equivalence of $\mu(M) + \nu(M)$ and $\mu'(M) + \nu'(M)$ when $\mu(M)$ and $\nu(M)$ are real. We have

$$\mu(M) + \nu(M) = [\mu_+(M) + \nu_+(M)] - [\mu_-(M) + \nu_-(M)]$$

$$\mu'(M) + \nu'(M) = [\mu'_+(M) + \nu'_+(M)] - [\mu'_-(M) + \nu'_-(M)]$$

Now by Theorem 11.1.7, the first bracketed terms on the right side of these equations are equivalent, and similarly the second. Thus by Theorem 11.1.9 the left-hand sides are equivalent. If $\mu(M)$ and $\nu(M)$ are complex we apply the process just used to the real and imaginary parts. If $\mu(M)$, $\mu'(M)$ are real and a is non-negative, then $a\,\mu(M) = a\,\mu_+(M) - a\,\mu_-(M)$, and $a\,\mu'(M) = a\,\mu'_+(M) - a\,\mu'_-(M)$ are totally additive decompositions. By Theorem 11.1.7, $a\,\mu_+(M)$, $a\,\mu_-(M)$ are respectively equivalent to $a\,\mu'_+(M)$, $a\,\mu'_-(M)$. Theorem 11.1.9 then shows that $a\,\mu(M)$ is equivalent to $a\,\mu'(M)$. Finally, let $\mu(M)$ and $\mu'(M)$ be complex: $\mu(M) = \sum_{\rho=0}^{3} i^\rho \mu_\rho^{(o)}(M)$ and $\mu'(M) = \sum_{\rho=0}^{3} i^\rho \mu_\rho^{\prime(o)}(M)$, and let $a = i$. The multiplication by i effects a cyclic permutation of the $\mu_\rho^{(o)}(M)$ and the same cyclic permutation of the $\mu_\rho^{\prime(o)}(M)$. Hence $i\,\mu(M)$ is equivalent to $i\,\mu'(M)$. What we have just proved is sufficient to give the result for arbitrary complex a.

Corollary: If $\mu_1(M)$, ..., $\mu_m(M)$ are arbitrary real or complex totally-additive measure functions of bounded variation on $\mathcal{R}$, $\mu_1'(M)$, ..., $\mu_m'(M)$ corresponding equivalent functions on $\mathcal{R}'$, and a_1, ..., a_m complex numbers; then $\sum_1^m a_n\mu_n(M)$ and $\sum_1^m a_n\mu_n'(M)$ are equivalent.

Proof: Induction.

THEOREM 11.1.13. Let $\mu(M)$ be a real or complex valued measure function, totally additive and of bounded variation over a half-ring $\mathcal{R}$. Let $\mu'(M)$ be a totally additive extension of $\mu(M)$ from $\mathcal{R}$ to $\mathcal{R}'$ where $\mathcal{R} \subset \mathcal{R}' \subset BR'(\mathcal{R})$. Then if $\mu'(M)$ is of bounded variation over $\mathcal{R}'$ it is equivalent to $\mu(M)$.

Proof: As before we assume that $\mu(M)$ is real. On $\mathcal{R}$ define $\mu_1(M) = \mu_+'(M)$, $\mu_2(M) = \mu_-'(M)$. Form the equivalent extensions $\tilde{\mu}_1(M)$ and $\tilde{\mu}_2(M)$ of $\mu_1(M)$, $\mu_2(M)$ respectively from $\mathcal{R}$ to $BR'(\mathcal{R})$. As extensions of $\mu_+'(M)$, $\mu_-'(M)$ from $\mathcal{R}'$ to $BR'(\mathcal{R}') = BR'(\mathcal{R})$ the functions $\tilde{\mu}_1(M)$, $\tilde{\mu}_2(M)$ are respectively equivalent to $\mu_+'(M)$ and $\mu_-'(M)$. Hence $\mu_+'(M)$, $\mu_-'(M)$ are respectively equivalent to $\mu_1(M)$ and $\mu_2(M)$. Now on $\mathcal{R}'$, $(\mathrm{Min}\ \mu_+', \mu_-')(M) = 0$, and therefore by Theorem 11.1.8 we have $(\mathrm{Min}\ \mu_1, \mu_2)(M) = 0$ on $\mathcal{R}$; i.e. $\mu_1(M) = \mu_+(M)$ and $\mu_2(M) = \mu_-(M)$. Hence $\mu'(M)$ and $\mu(M)$ are equivalent.

Remark: We cannot in general replace $BR'(\mathcal{R})$ in the foregoing theorem by an arbitrary wider domain $\mathcal{T}$ contained in the domain of the maximal equivalent extension of $\mu(M)$. This is due to the fact that in order to carry through the proof we would need to know that the sets $\mathcal{T}$ were measurable and of finite measure with respect to the outer measures determined by $\mu_1(M)$ and $\mu_2(M)$ on $\mathcal{R}$. We would of course replace $BR'(\mathcal{R})$ by a somewhat wider restricted Borel-ring, for example the Borel-ring of "analytic sets with respect to $\mathcal{R}$" which are contained in some set of $R(\mathcal{R})$.

The following example serves to show that the difficulty just mentioned is not illusory. Let S consist of the two points a, b. Let $\mathcal{R}$ consist of S and the empty set 0. We clearly have $\mathcal{R} = R(\mathcal{R}) = BR'(\mathcal{R}) = BR(\mathcal{R})$. On $\mathcal{R}$ define $\mu(M) = 0$. Now on $\mathcal{P}(S)$ define $\mu'(M)$ as follows: $\mu'(0) = 0$; $\mu'(\{a\}) = 1$; $\mu'(\{b\}) = -1$; $\mu'(S) = 0$. $\mu'(M)$ is obviously of bounded variation and totally additive. Furthermore, it is an extension of $\mu(M)$. Any equivalent extension of $\mu(M)$ is identically zero; hence $\mu(M)$ and $\mu'(M)$ are inequivalent. We may observe that if we form the outer measure $\mu^*_{1I}(M)$, $\mu^*_{2I}(M)$ determined respectively by $\mu_1(M)$, $\mu_2(M)$ defined in the proof of Theorem 11.1.13, we have: $\mu^*_{1I}(0) = \mu^*_{2I}(0) = 0$; $\mu^*_{1I}(\{a\}) = \mu^*_{2I}(\{a\}) = 1$; $\mu^*_{1I}(\{b\}) = \mu^*_{2I}(\{b\}) = 1$; $\mu^*_{1I}(S) = \mu^*_{2I}(S) = 1$. From which it follows that neither $\{a\}$ nor $\{b\}$ is measurable with respect to either outer measure (cf. Corollary to Theorem 10.2.4).

In order to complete this section it is desirable to show that the concept of bounded variation is actually essential to the problem of extending the general finite totally-additive measure functions. What we show is that if we are to hope for totally additive extension of a finite measure function from a half-ring $\mathcal{R}$ to $BR'(\mathcal{R})$, then the function must be assumed of bounded variation (and totally additive) on $\mathcal{R}$. First we prove a lemma for which we shall have use later.

<u>Lemma 2. If a real (or complex) measure function</u> $\mu(M)$ <u>defined on a ring</u> $\mathcal{R}$ <u>is not of bounded variation over</u> $\mathcal{R}$, <u>then for every positive constant</u> c <u>there exists an infinite sequence of disjunct sets</u> $M_1, M_2, \ldots$ <u>from</u> $\mathcal{R}$, <u>all contained in a set</u> $M' \in \mathcal{R}$, <u>such that</u> $|\mu(M_i)| \geqq c$.

Proof: If $\mu(M)$ is of bounded variation over each of two sets from $\mathcal{R}$, then it is of bounded variation over their sum. If therefore $\mu(M)$ is not of bounded variation over $M = N + P$, it is either of unbounded variation over N or else over P. Let M' be a set from $\mathcal{R}$ over which $\mu(M)$ is not of bounded variation. M' contains a set $\bar{M}'$ such that $|\mu(\bar{M}')| \geqq |\mu(M')| + c$. Now $|\mu(M' - \bar{M}')| = |\mu(M') - \mu(\bar{M}')| \geqq |\mu(\bar{M}')| - |\mu(M')| \geqq c$, so that $\bar{M}'$ and $M' - \bar{M}'$

both have μ-measures in absolute value greater than c. Denote by M_1' the first of $\bar{M}'$ and $M' - \bar{M}'$ over which $\mu(M)$ is not of bounded variation (there is one such by the Remark at the outset of the proof), and by M_1 the remaining set. Thus we have $M' = M_1 + M_1'$, $|\mu(M_1)| \geqq c$, and $\mu(M)$ of unbounded variation over M_1'. Apply a similar argument to M_1' as applied to M', and obtain $M_1' = M_2 + M_2'$, $|\mu(M_2)| \geqq c$, $\mu(M)$ of unbounded variation over M_2'. Repeat the process indefinitely, defining M_{n+1}, M_{n+1}' in terms of M_n'. The sets M_i are disjunct, contained in M' and $|\mu(M_i)| \geqq c$.

THEOREM 11.1.14. *If $\mu(M)$ is a finite real (or complex) measure function over a restricted Borel-ring $\mathcal{R}$, and is totally additive over $\mathcal{R}$, then it is of bounded variation over $\mathcal{R}$.*

Proof: Assume the contrary. Let M' be a set over which $\mu(M)$ is of unbounded variation, and let $\{M_i\}$ be as defined in the above Lemma with $c = 1$. Then $\sum_i M_i \in \mathcal{R}$ and the series $\sum_i \mu(M_i)$ is divergent.

Remark: Theorem 11.1.14 does not remain true if the hypotheses are weakened by simply requiring that $\mathcal{R}$ be a ring. This can be shown by the following example. Let the space S be the real interval $0 \leqq x < 1$, and let $\mathcal{R}$ be the ring generated from the half-open intervals $\mathcal{J}\binom{a}{b}$: $0 \leqq a \leqq x < b \leqq 1$.

Let $\varphi(x) = \begin{cases} x \cos \frac{\pi}{x} & (0 < x < 1) \\ 0 & (x = 0) \end{cases}$ Let $\mu(\mathcal{J}\binom{a}{b}) = \varphi(b) - \varphi(a)$. Define

$$\bar{M}_n = \mathcal{J}\left(\begin{matrix}\frac{1}{2n+1}\\ \frac{1}{2n}\end{matrix}\right), \; n = 1, 2, \ldots . \quad \text{Thus } \mu(\bar{M}_n) = \frac{1}{2n} + \frac{1}{2n+1} \geqq \frac{1}{n+1} .$$

Now $\mu(M)$ is certainly not of bounded variation over $S = \mathcal{J}\binom{0}{1}$. For we have

$$\underset{\substack{N \in \mathcal{R} \\ N \subset S}}{\text{l.u.b.}} \mu(N) \geqq \underset{m}{\text{l.u.b.}}\, \mu\Big(\sum_1^m \bar{M}_n\Big) = \underset{m}{\text{l.u.b.}} \sum_1^m \mu(\bar{M}_n) \geqq \underset{m}{\text{l.u.b.}} \sum_1^m \frac{1}{n+1} = +\infty .$$

On the other hand $\mu(M)$ is totally additive over $\mathcal{R}$. Let $M \in \mathcal{R}$ and $M = \sum_i M_i$

where the M_i form a sequence of disjunct sets from $\mathcal{R}$. If the point 0 does not belong to M, then obviously M lies in a closed interval having a positive distance from the point 0. In this case, since $x \cos \frac{\pi}{x}$ is analytic in such an interval, it is the difference of two continuous non-decreasing functions. Therefore $\mu(M)$ is totally additive over M, the series $\sum_i \mu(M_i)$ is absolutely convergent, and its sum is $\mu(M)$. If M contains the point 0, then 0 is contained in exactly one of the intervals M_i, let us say M_1. By the preceding argument applied to $M - M_1$, the series $\sum_{i \neq 1} \mu(M_i)$ is absolutely convergent and its sum is $\mu(M - M_1)$. Hence $\sum_i \mu(M_i)$ is absolutely convergent and its sum is $\mu(M)$.

We finally establish a result whose applications will not occur until the next section, but whose subject matter seems properly to belong here.

THEOREM 11.1.15. *Let $\mu(M)$ be a real (finite) totally-additive measure function over a restricted Borel-ring $\mathcal{R}$. For every $\bar{M} \in \mathcal{R}$ there is an $\bar{N} \in \mathcal{R}$ contained in $\bar{M}$ such that $\mu_+(\bar{N}) = 0$, and $\mu_-(\bar{M} - \bar{N}) = 0$.*

Proof: By Corollary 2 to Theorem 11.1.3 and the definition of $(\text{Min}\ \mu_+, \mu_-)(M)$ we have

$$\underset{N \in \mathcal{R}}{\text{g.l.b.}}\ \mu_+(N) + \mu_-(\bar{M} - N)) = 0;$$

thus for each integer n there is an $N_n \in \mathcal{R}$ such that

$$\mu_+(N_n) + \mu_-(\bar{M} - N_n) \leqq \frac{1}{2^n}$$

Now define $\bar{N} = \overline{\lim_{n\to\infty}}\ N_n = \prod_{n=1}^{\infty} \sum_{m=n}^{\infty} N_m$. Thus $\bar{M} - \bar{N} = \underline{\lim}_{n\to\infty} (\bar{M} - N_n) = \sum_{n=1}^{\infty} \prod_{m=n}^{\infty} (\bar{M} - N_m)$. It is clear that $\bar{N} \in \mathcal{R}$. Now, since $\mu_+(N_n) \leqq \frac{1}{2^n}$, we have $\mu_+(\sum_{m=n}^{\infty} N_m) \leqq$ $\leqq \sum_{m=n}^{\infty} \frac{1}{2^m} = \frac{1}{2^{n-1}}$. But since for every n, $\bar{N} \subset \sum_{m=n}^{\infty} N_m$ we get $\mu_+(\bar{N}) = 0$. On the other hand, since $\prod_{m=n}^{\infty} (\bar{M} - N_m) \subset \bar{M} - N_k$ for every $k > n$, it follows that

$\mu_-(\prod_{m=n}^{\infty}(\bar{M} - N_m)) \leq \frac{1}{2^k}$ for every $k > n$, and is therefore zero. Hence

$$\mu_-(\bar{M} - \bar{N}) \leq \sum_{n=1}^{\infty} \mu_-(\prod_{m=n}^{\infty} (\bar{M} - N_m)) = 0.$$

Corollary: For all $P \in \mathcal{R}$, $P \subset \bar{N}$, $\mu(P) \leq 0$, and for all $Q \in \mathcal{R}$, $Q \subset \bar{M} - \bar{N}$, $\mu(Q) \geq 0$.

§2. Absolute continuity and the integral

It is a well known result of Lebesgue that the class of absolutely continuous functions in an interval $a \leq a \leq b$ is identical with the class of functions of the form $F(x) = \int_a^x f(x)dx + c$ where c is a constant and f(x) is summable over (a,b). If f(x) and g(x) generate by this process the same F(x), then f(x) = g(x) almost everywhere in (a, b). An analogous connection exists between the notions of absolute continuity and integral which we shall define in this section.

The integral we shall define contains as special cases the Lebesgue integral, the Lebesgue-Stieltjes integral, and the Radon-Stieltjes integral.

We begin with absolute continuity.

Unless mention is made to the contrary, we fix on the following assumptions:

$\mathcal{R}$ is a half-ring in $\mathcal{P}(S)$.

$\nu(M)$ is a finite non-negative and totally additive measure function on $\mathcal{R}$.

Definition 11.2.1. Let $\mu(M)$ be a finite (real or complex) measure function on $\mathcal{R}$. Form the extensions $\mu_1(M)$, $\nu_1(M)$ of $\mu(M)$, $\nu(M)$ from $\mathcal{R}$ to $R(\mathcal{R})$. $\mu(M)$ is called absolutely continuous with respect to $\nu(M)$ if for every $M \in \mathcal{R}$ and every $\varepsilon > 0$ there is a $\delta(\varepsilon, M) > 0$ such that if $N \in R(\mathcal{R})$, $N \subset M$, and $\nu_1(N) < \delta(\varepsilon, M)$ then $|\mu_1(M)| < \varepsilon$. If $\mu(M)$ has this property for some particular $M \in \mathcal{R}$ (but not necessarily for others) it is said to be absolutely con-

tinuous over M.

THEOREM 11.2.1. *Any finite linear combination of measure functions absolutely continuous with respect to* $\nu(M)$ *is also absolutely continuous with respect to* $\nu(M)$.

Proof: We omit the proof. It is analogous to the proof that a finite linear combination of continuous functions is continuous.

THEOREM 11.2.2. *If* $\mu(M)$ *is absolutely continuous with respect to* $\nu(M)$, *then it is totally additive and of bounded variation on* $\mathcal{R}$.

Proof: $\mathcal{R}$ can be assumed a ring, for the properties of total additivity and bounded variation are undisturbed in the extension from $\mathcal{R}$ to $R(\mathcal{R})$, and if they hold on $R(\mathcal{R})$ they hold on $\mathcal{R}$.

The total additivity of $\mu(M)$ is proved as follows: Let $\{M_i\}$ be a sequence of disjunct sets from $\mathcal{R}$ whose sum $M = \sum_i M_i \in \mathcal{R}$, and let ε be a positive number. Because of the total additivity of $\nu(M)$ we can choose n_0 so great that for $n > n_0$, $\nu(M) - \sum_1^n \nu(M_i) = \nu(M - \sum_1^n M_i) < \delta(\varepsilon, M)$. Therefore

$$|\mu(M) - \sum_1^n \mu(M_i)| = |\mu(M - \sum_1^n M_i)| < \varepsilon$$

and hence $\mu(M) = \sum_i \mu(M_i)$.

If $\mu(M)$ were not of bounded variation there would be by Lemma 2 of the last section a set $M \in \mathcal{R}$ and a sequence of disjunct sets $M_i \in \mathcal{R}$ contained in M and such that $|\mu(M_i)| > 1$. For each M_i we should then have $\nu(M_i) \geqq \delta(1,M)$. Thus $\nu(M) \geqq \sum_i \nu(M_i) = +\infty$. Since $\nu(M)$ is finite, this contradiction proves that $\mu(M)$ is of bounded variation.

THEOREM 11.2.3. *A real* $\mu(M)$ *is absolutely continuous with respect to* $\nu(M)$ *if and only if* $\mu_+(M)$ *and* $\mu_-(M)$ *are.*

Proof: As before, we may assume that $\mathcal{R}$ is a ring. If $M, N \in \mathcal{R}$, $N \subset M$ and $\nu(N) < \delta(\varepsilon, M)$ then for any $P \in \mathcal{R}$, $\nu(NP) < \delta(\varepsilon, M)$, and $|\mu(NP)| < \varepsilon$.

Thus $|\mu_+(N)| = \mu_+(N) = \underset{P \in \mathcal{R}}{\text{l.u.b.}}\ \mu(NP) \leq \varepsilon$. A similar argument shows (under the same assumptions) that $|\mu_-(N)| = \mu_-(N) \leq \varepsilon$.

The converse statement is a consequence of $\mu(M) = \mu_+(M) - \mu_-(M)$ and Theorem 11.2.1.

Corollary: *A complex* $\mu(M)$ *is absolutely continuous with respect to* $\nu(M)$ *if and only if* $\mathcal{R}\mu(M)$ *and* $\mathcal{J}\mu(M)$ *are. Thus if and only if the functions* $\mu_\rho^{(o)}(M)$ are.

Proof: The absolute continuity of $\mathcal{R}\mu(M)$ and $\mathcal{J}\mu(M)$ on the assumption of that of $\mu(M)$ follows from the inequalities $|\mathcal{R}\mu(M)| \leq |\mu(M)|$ and $|\mathcal{J}\mu(M)| \leq |\mu(M)|$. The converse is contained in Theorem 11.21.1.

The second statement is a consequence of the definition of $\mu_\rho^{(o)}(M)$ and the present theorem.

THEOREM 11.2.4. *If* $\mathcal{R}$ *is a restricted Borel ring and* $\mu(M)$ *is non-negative and totally additive, then a necessary and sufficient condition that* $\mu(M)$ *be absolutely continuous with respect to* $\nu(M)$ *is that for every* $M \in \mathcal{R}$ *for which* $\nu(M) = 0$, *we have* $\mu(M) = 0$.

Proof: The necessity of the condition is obvious, for if $\nu(M) = 0$ then $\nu(M) < \delta(\varepsilon, M)$ for all $\varepsilon > 0$, thus $|\mu(M)| < \varepsilon$, and therefore $\mu(M) = 0$.

To prove sufficiency assume the converse. Then there exists a set $M \in \mathcal{R}$ and a constant $c > 0$ such that to every $\delta > 0$ corresponds a set $N_\delta \in \mathcal{R}$, $N_\delta \subset M$ for which $\nu(N_\delta) < \delta$ and $\mu(N_\delta) \geq c$. Write $N_n = N_\delta$ for $\delta = \frac{1}{2^n}$ $(n=1,2,\ldots)$. Define $\bar{N}_m = \sum_1^m N_n$, $\bar{N} = \prod_m \bar{N}_m$. Now

$$\nu(\bar{N}_m) \leq \sum_{n=1}^{m} \nu(N_n) < \sum_{n=1}^{m} \frac{1}{2^n} = \frac{1}{2^{m-1}} .$$

Since for any m, $\bar{N} \subset \bar{N}_m$, it follows that $\nu(\bar{N}) = 0$, and thus we should have $\mu(\bar{N}) = 0$.

But since $\bar{N}_m \supset N_m$, we have $\mu(\bar{N}_m) \geqq c$. The total additivity of $\mu(M)$ now gives

$$\mu(\bar{N}) = \lim_{n\to\infty} \mu(\bar{N}_m) \geqq c > 0$$

and we have the desired contradiction.

THEOREM 11.21.5. *If $\mu(M)$ is absolutely continuous with respect to $\nu(M)$, and if $\mu'(M)$, $\nu'(M)$ defined on a half-ring $\mathcal{R}'$ are respectively equivalent to $\mu(M)$, $\nu(M)$, then $\mu'(M)$ is absolutely continuous with respect to $\nu'(M)$.*

Proof: In view of Theorem 11.2.3 and its Corollary, it is sufficient to prove the statement when $\mu(M)$ is non-negative. Also, since the equivalence of $\mu(M)$, $\nu(M)$ with $\mu'(M)$, $\nu'(M)$ persists if we replace $\mu(M)$, $\nu(M)$ by their extensions from $\mathcal{R}$ to $R(\mathcal{R})$, we may assume that $\mathcal{R}$ is a ring.

Form the outer measures $\mu_I^*(M)$ and $\nu_I^*(M)$ determined respectively by $\mu(M)$ and $\nu(M)$. Assume that $M \in \mathcal{R}$, $\bar{M} \subset M$, $\nu_I^*(\bar{M}) = 0$, and $\varepsilon > 0$. There exists (cf. Theorem 10.3.2) a sequence $\{M_i\}$ of sets from $\mathcal{R}$ such that $\bar{M} \subset \sum_i M_i$ and $\sum_i \nu(M_i) \leqq \nu_I^*(\bar{M}) + \delta(\varepsilon, M) = \delta(\varepsilon, M)$. Therefore for any n we have

$$\nu(M \sum_1^n M_i) \leqq \nu(\sum_1^n M_i) \leqq \sum_1^n \nu(M_i) \leqq \sum_i \nu(M_i) < \delta(\varepsilon, M)$$

and therefore by the absolute continuity in $\mathcal{R}$, $\mu(M \sum_1^n M_i) < \varepsilon$. Thus we get

$$\mu_I^*(\bar{M}) \leqq \mu_I^*(M \sum_1^\infty M_i) = \lim_{n\to\infty} \mu_I^*(M \sum_1^n M_i) = \lim_{n\to\infty} \mu(M \sum_1^n M_i) < \varepsilon ,$$

and since ε is arbitrary, $\mu_I^*(\bar{M}) = 0$. We have thus proved that for any set $\bar{M} \subset M \in \mathcal{R}$ for which $\nu_I^*(\bar{M}) = 0$ we have $\mu_I^*(M) = 0$. The restriction that $\bar{M} \subset M$ can be removed by covering the set $\bar{M}$ with a sequence of disjunct sets from $\mathcal{R}$ and applying what we have just proved to the part of $\bar{M}$ lying in each of them.

Now form the systems $MS'_{\nu^*_I}$, $MS'_{\mu^*_I}$, write $\mathcal{T} = MS'_{\nu^*_I} \cdot MS'_{\mu^*_I}$, and let $\tilde{\mu}(M)$ and $\tilde{\nu}(M)$ be the equivalent extensions of $\mu(M)$, $\nu(M)$ from $\mathcal{R}$ to $\mathcal{T}$. $\mathcal{T}$ is a restricted Borel-ring and for every $M \in \mathcal{T}$ for which $\tilde{\nu}(M) = 0$ we have $\mu(M) = 0$. By Theorem 11.2.4, $\tilde{\mu}(M)$ is absolutely continuous with respect to $\tilde{\nu}(M)$.

Since $\mu'(M)$, $\nu'(M)$ are respectively equivalent to $\mu(M)$, $\nu(M)$, we have $\mathcal{R}' \subset MS_{\mu^*_I}$ and $\mathcal{R}' \subset MS_{\nu^*_I}$, hence $\mathcal{R}' \subset \mathcal{T}$, and on $\mathcal{R}'$, $\mu'(M)$, $\nu'(M)$ agree respectively with $\tilde{\mu}(M)$, $\tilde{\nu}(M)$. Therefore $\mu'(M)$ is absolutely continuous with respect to $\nu'(M)$.

Corollary 1: Among all half-rings satisfying the conditions imposed upon $\mathcal{R}'$ in the statement of the theorem there is a maximal half-ring $\mathcal{T}$. $\mathcal{T}$ is a restricted Borel-ring and in particular it contains every zero set with respect to $\nu^*_I(M)$.

Proof: It is obvious that the intersection of $MS'_{\nu^*_I}$ with the domain of the maximal equivalent extension of $\mu(M)$ is such an $\mathcal{T}$, for it contains every $\mathcal{R}'$ satisfying the conditions of the theorem. As the intersection of two restricted Borel-rings it is one.

If we replace $\mu(M)$ in the argument successively by $\mu_+(M)$ and $\mu_-(M)$ (respectively $\mu^{(o)}_\rho(M)$), and form the outer measures $\mu^*_{+I}(M)$ and $\mu^*_{-I}(M)$, then the same reasoning that showed $\mu^*_I(M) = 0$ for every zero set with respect to $\nu^*_I(M)$ shows it for $\mu^*_{-I}(M)$ and $\mu^*_{+I}(M)$. Thus if $\nu^*_I(M) = 0$, then $M \in MS'_{\nu^*_I} \cdot MS'_{\mu^*_{+I}} \cdot MS'_{\mu^*_{-I}}$.

Corollary 2: If $\mu^*_I(M)$ and $\nu^*_I(M)$ are as in the proof of the theorem, then $MS_{\nu^*_I} \subset MS_{\mu^*_I}$.

Proof: Apply the Corollary to Theorem 10.2.12.

Throughout the next portion of the discussion $\nu^*(M)$ will denote a fixed regular outer measure function. By the term "measurable" we shall always mean measurable with respect to $\nu^*(M)$; by "zero set" we shall mean zero set with respect to $\nu^*(M)$.

Definition 11.2.2. *A real function $f(x)$ defined for $x \in S$ is called measurable if for every real α the set $S_x\{f(x) < \alpha\}$ is measurable. A complex valued function is called measurable if its real and imaginary parts are measurable. A function $f(x)$ defined over a measurable subset M of S is called measurable if the function* $f_1(x) = \begin{cases} f(x) & (x \in M) \\ 0 & (x \in S-M) \end{cases}$ *is measurable.*

Definition 11.2.3. *Two functions $f(x)$ and $g(x)$ are called equivalent with respect to $\nu^*(M)$, or simply equivalent, if the set of points where they differ is a zero set. If $f(x)$ and $g(x)$ are equivalent we write $f(x) \sim g(x)$.*

The equivalence relation here defined is clearly an equivalence relation in the ordinary sense, i.e. reflexive, symmetric, and transitive. It is also clear that if $f(x)$ is measurable and $f(x) \sim g(x)$ then $g(x)$ is measurable, because the sets $S_x\{f(x) < \alpha\}$ and $S_x\{g(x) < \alpha\}$ will differ at most on a zero set and thus be measurable or non-measurable together.

THEOREM 11.2.6. *If $f(x)$ is measurable, then for every α each of the four sets $S_x\{f(x) < \alpha\}$, $S_x\{f(x) \leqq \alpha\}$, $S_x\{f(x) \geqq \alpha\}$, $S_x\{f(x) > \alpha\}$ is measurable. If any one of the four is measurable for an everywhere dense set of α's, then $f(x)$ is measurable.*

Proof: As for the first statement, the third and fourth sets are respectively the complements of the first and second. Thus it is sufficient to prove the measurability of the first two. For the first the property is definitory; this leaves $S_x\{f(x) \leqq \alpha\}$. Now $S_x\{f(x) \leqq \alpha\} = \prod_n S_x\{f(x) < \alpha + \frac{1}{n}\}$ and is therefore measurable.

For the second statement, let D be an everywhere dense set of α's, let

$\bar{\alpha}$ be any number, and select a sequence $\{\alpha_n\} \subset D$ such that $\alpha_1 < \alpha_2 < \dots < \bar{\alpha} = \lim_{n\to\infty} \alpha_n$. Then $S_x\{f(x) < \bar{\alpha}\} = \sum_1^\infty S_x\{f(x) < \alpha_n\} = \sum_1^\infty S_x\{f(x) \leqq \alpha_n\}$. This shows that $S_x\{f(x) < \bar{\alpha}\}$ is measurable if for all $\alpha \in D$, $S_x\{f(x) < \alpha\}$ is measurable, or if for all $\alpha \in D$, $S_x\{f(x) \leqq \alpha\}$ is measurable. As complements of the first two sets, the third and fourth are already taken care of.

THEOREM 11.2.7. *If $f(x)$ and $g(x)$ are real and measurable, and if c is any real number, then $cf(x)$, $|f(x)|^c$, $f(x) + g(x)$, $f(x) \circ g(x)$, Max $(f(x), g(x))$, and Min $(f(x), g(x))$ are measurable. If $f_1(x), f_2(x), \dots$ is a sequence of measurable functions, then $\overline{\lim}_{n\to\infty} f_n(x)$ and $\underline{\lim}_{n\to\infty} f_n(x)$ are measurable.*

Proof: Assume the notation: $F_\alpha = S_x\{f(x) < \alpha\}$, $G_\alpha = S_x\{g(x) < \alpha\}$, $F_\alpha^n = S_x\{f_n(x) < \alpha\}$. Now

(i) $cf(x)$ is measurable; for, if $c > 0$, then $S_x\{cf(x) < \alpha\} = S_x\{f(x) < \frac{\alpha}{c}\}$ which is measurable; if $c = 0$, then $S_x\{cf(x) < \alpha\} = \begin{cases} S(\alpha > 0) \\ 0(\alpha \leqq 0)\end{cases}$ which is measurable, and if $c < 0$, then $S_x\{cf(x) < \alpha\} = S_x\{f(x) > \frac{\alpha}{c}\}$ which by Theorem 11.2.6 is measurable.

(ii) To prove the measurability of $|f(x)|^c$ we observe that if $c > 0$ then $S_x\{|f(x)|^c < \alpha\} = S_x\{-\alpha^{\frac{1}{c}} < f(x) < \alpha^{\frac{1}{c}}\} = S_x\{f(x) > -\alpha^{\frac{1}{c}}\} \circ S_x\{f(x) < \alpha^{\frac{1}{c}}\}$ for $\alpha > 0$, and $S_x\{|f(x)|^c < \alpha\} = 0$ for $\alpha \leqq 0$. If $c < 0$, then $S_x\{|f(x)|^c < \alpha\} = S_x\{f(x) > \alpha^{\frac{1}{c}}\} + S_x\{f(x) < -\alpha^{\frac{1}{c}}\}$ for $\alpha > 0$, and is empty for $\alpha \leqq 0$. The statement is trivial for $c = 0$.

(iii) To prove $f(x) + g(x)$ is measurable, let D be a countable everywhere dense set of α's. Form for a fixed α the set E of all pairs (β, γ) where $\beta, \gamma \in D$ and $\beta + \gamma < \alpha$. E is countable; let $\{(\beta_1, \gamma_1), (\beta_2, \gamma_2), \dots\}$ be an enumeration of it. If now $x_0 \in S_x\{f(x) + g(x) < \alpha\}$ then there is an n_0 such that

$f(x_0) < \beta_{n_0}$ and $g(x_0) < \gamma_{n_0}$; hence $x_0 \in F_{\beta_{n_0}} G_{\gamma_{n_0}}$. If $x_1 \in F_{\beta_n} G_{\gamma_n}$ then $f(x_1) + g(x_1) < \beta_n + \gamma_n < \alpha$, and hence $x_1 \in S_x\{ f(x) + g(x) < \alpha \}$. This proves that

$$S_x\{ f(x) + g(x) < \alpha \} = \sum_n F_{\beta_n} G_{\gamma_n}$$

As the sum of a sequence of measurable sets it is measurable.

(iv) The measurability of $f(x) \circ g(x)$ is a consequence of (i), (ii), (iii), and the identity $f(x)\, g(x) = \frac{1}{4} \circ (|f(x) + g(x)|^2 - |f(x) - g(x)|^2)$.

(v) $\mathrm{Max}(f(x), g(x))$ and $\mathrm{Min}\,(f(x), g(x))$ are measurable because $S_x\{ \mathrm{Max}\,(f(x), g(x)) < \alpha \} = F_\alpha G_\alpha$ and $S_x\{ \mathrm{Min}\,(f(x), g(x)) < \alpha \} = F_\alpha + G_\alpha$.

(vi) We prove that $\overline{\lim}_{n\to\infty} f_n(x)$ and $\underline{\lim}_{n\to\infty} f_n(x)$ are measurable by simply observing that $S_x\{ \overline{\lim}_{n\to\infty} f_n(x) < \alpha \} = \overline{\lim}_{n\to\infty} F^n_\alpha = \prod_{n=1}^{\infty} \sum_{m=n}^{\infty} F^m_\alpha$ and that $\underline{\lim}_{n\to\infty} f_n(x) = - \overline{\lim}_{n\to\infty} - f_n(x)$.

Corollary 1: If except on a zero set $\lim_{n\to\infty} f_n(x) = f_0(x)$, then $f_0(x)$ is measurable.

Proof: Because then $f_0(x) \sim \overline{\lim}_{n\to\infty} f_n(x)$.

Corollary 2: If $f(x), f_n(x), g(x)$ are complex measurable functions, and c is a complex number, then $cf(x)$ and $f(x) + g(x)$ are measurable, and if except on a zero set $\lim_{n\to\infty} f_n(x) = f_0(x)$, then $f_0(x)$ is measurable.

Proof: By applying the theorem separately to the real and imaginary parts.

Corollary 3: If for every α , $S_x\{f(x) < \alpha\}$, $S_x\{g(x) < \alpha\}$, belong to some fixed half-ring $\mathfrak{T}$, then the sets $S_x\{ cf(x) < \alpha \}$, $S_x\{f(x) + g(x) < \alpha\}$,... etc., belong to $BR(\mathfrak{T})$.

Proof: The processes used in the proof of the theorem were processes under which Borel rings are closed.

THEOREM 11.2.18. Assume that $\mathcal{R}$ is a restricted Borel-ring, that $\nu^*(M)$ is the outer measure determined by $\nu(M)$ on $\mathcal{R}$, and that $\mu(M)$ is real and absolutely continuous with respect to $\nu(M)$ on $\mathcal{R}$. Form the intersection of MS'_{ν^*} with the domain of the maximal equivalent extension of $\mu(M)$; call it $\mathcal{T}$. Let $\tilde{\mu}(M)$, $\tilde{\nu}(M)$ be the equivalent extensions of $\mu(M)$, $\nu(M)$ from $\mathcal{R}$ to $\mathcal{T}$. Then for any fixed $\bar{M} \in \mathcal{R}$ and any real number α there is a set $M_\alpha \in \mathcal{R}$ contained in $\bar{M}$ having the property that

$$(1) \qquad \begin{cases} P \in \mathcal{T} \text{ and } P \subset M \text{ imply } \tilde{\mu}(P) \leqq \alpha\, \tilde{\nu}(P) \\ Q \in \mathcal{T} \text{ and } Q \subset \bar{M} - M \text{ imply } \tilde{\mu}(Q) \geqq \alpha\, \tilde{\nu}(P) \end{cases}$$

The system of sets M_α may be so chosen that

(2) If $\alpha < \beta$, then $M_\alpha \subset M_\beta$

(3) If $\alpha_1 < \alpha_2 < \dots < \alpha = \lim_{n\to\infty} \alpha_n$, then $M_\alpha = \lim_{n\to\infty} M_{\alpha_n}$

(4) If $\lim \alpha_n = -\infty$, then $\prod_n M_{\alpha_n} = 0$; if $\lim_{n\to\infty} \alpha_n = +\infty$, then $\sum_n M_{\alpha_n} = \bar{M}$.

Proof: Define on $\mathcal{R}$ the function $\lambda^\alpha(M) = \mu(M) - \alpha\,\nu(M)$. This is finite and absolutely continuous with respect to $\nu(M)$. Apply Theorem 11.1.14 to $\lambda^\alpha(M)$ and get an $\bar{M}_\alpha \in \mathcal{R}$ (the $\bar{N}$ of Theorem 11.1.14) contained in $\bar{M}$ such that $\lambda^\alpha_+(\bar{M}_\alpha) = 0$ and $\lambda^\alpha_-(\bar{M} - \bar{M}_\alpha) = 0$. Now form an equivalent extension $\tilde{\lambda}^\alpha(M)$ of $\lambda^\alpha(M)$ from $\mathcal{R}$ to $\mathcal{T}$. $\tilde{\lambda}^\alpha(M) = \tilde{\mu}(M) - \alpha\,\tilde{\nu}(M)$ and the positive and negative variation functions of $\tilde{\lambda}^\alpha(M)$ are precisely the respective extensions of $\lambda^\alpha_+(M)$ and $\lambda^\alpha_-(M)$ (Theorems 11.1.9, 11.1.10) from $\mathcal{R}$ to $\mathcal{T}$. Thus the properties $\tilde{\lambda}^\alpha_+(\bar{M}_\alpha) = 0$, $\lambda^\alpha_-(\bar{M} - \bar{M}_\alpha) = 0$ are preserved in the extension. Hence for $P \subset \bar{M}_\alpha$, $\tilde{\mu}(P) - \alpha\,\tilde{\nu}(P) \leqq 0$ and for $Q \subset \bar{M} - \bar{M}$, $\tilde{\mu}(Q) - \alpha\,\tilde{\nu}(P) = $ $= \tilde{\lambda}^\alpha(Q) \leqq 0$; that is, the $\bar{M}_\alpha$ have the property (1);

We now modify the $\bar{M}_\alpha$ in such a way that (1) is not disturbed and (2), (3) and (4) are obtained.

Select an everyhwere dense countable set D of α's. For any α define $\bar{\bar{M}}_\alpha \equiv \sum_{\bar{\alpha} \in D} \bar{M}_{\bar{\alpha}}$. Now fix on a particular α and let $\{\alpha_1, \alpha_2, \ldots\}$ be an enumeration of the elements of D which are less than α. For simplicity write $\bar{M}_i = \bar{M}_{\alpha_i}$ so that $\bar{\bar{M}}_\alpha = \sum_i \bar{M}_i$. Define $N_i = \bar{M}_i - \bar{M}_i(\sum_{j=1}^{i-1} \bar{M}_j)$. The N_i are disjunct, their sum is $\bar{\bar{M}}_\alpha$ and for each i, $N_i \subset \bar{M}_i$. Now if $P \subset \bar{\bar{M}}_\alpha$, then observing that $PN_i \subset \bar{M}_i$,

$$\tilde{\mu}(P) = \tilde{\mu}(P\bar{\bar{M}}_\alpha) = \sum_i \tilde{\mu}(PN_i) \leqq \sum_i \alpha_i \tilde{\nu}(PN_i) \leqq \alpha \sum_i \nu(PN_i) = \alpha \tilde{\nu}(P)$$

On the other hand, if $Q \subset \bar{M} - \bar{\bar{M}}_\alpha = \prod (\bar{M} - \bar{M}_i)$ we have for every i, $Q \subset \bar{M} - \bar{M}_i$, and thus $\tilde{\mu}(Q) \geqq \alpha_i \ \tilde{\nu}(Q)$. As α = l.u.b. α_i, this gives $\tilde{\mu}(Q) \geqq \alpha \tilde{\nu}(Q)$. Therefore the $\bar{\bar{M}}_\alpha$ have the property (1). Property (2) is obvious for the $\bar{\bar{M}}_\alpha$ from their definition. As for property (3), let $\{\alpha_n\}$ be a sequence such that $\alpha_1 < \alpha_2 < \ldots < \alpha = \lim_{n\to\infty} \alpha_n$; then $\lim_{n\to\infty} \bar{\bar{M}}_{\alpha_n} = \sum_n \bar{\bar{M}}_{\alpha_n} = \sum_n \sum_{\substack{\bar{\alpha}\in D \\ \bar{\alpha}<\alpha_n}} \bar{M}_{\bar{\alpha}} = \sum_{\substack{\bar{\alpha}\in D \\ \bar{\alpha}<\alpha}} \bar{M}_{\bar{\alpha}} = \bar{\bar{M}}_\alpha$.

We now have a system $\bar{\bar{M}}_\alpha$ with properties (1), (2), (3). We proceed with one final modification. Retain the dense set D from the last paragraph. Form the sets $F = \prod_{\bar{\alpha}\in D} \bar{\bar{M}}_{\bar{\alpha}}$; $G \equiv \bar{M} - \sum_{\bar{\alpha}\in D} \bar{\bar{M}}_{\bar{\alpha}} = \prod_{\bar{\alpha}\in D} (\bar{M} - \bar{\bar{M}}_{\bar{\alpha}})$. Since $F \subset \bar{\bar{M}}_{\bar{\alpha}}$, for every $\bar{\alpha} \in D$ we have $\tilde{\mu}(F) \leqq \bar{\alpha} \tilde{\nu}(F)$, and this is impossible unless $\tilde{\nu}(F) = 0$. On the other hand $G \subset \bar{M} - \bar{\bar{M}}_{\bar{\alpha}}$ for all $\bar{\alpha} \in D$, and thus $\tilde{\mu}(G) \geqq \bar{\alpha} \tilde{\nu}(G)$, $(\bar{\alpha} \in D)$ which implies $\tilde{\nu}(G) = 0$. Now define $M_\alpha = \bar{\bar{M}}_\alpha + F$ for $\alpha > 0$, $M_\alpha = \bar{\bar{M}}_\alpha - G$ for $\alpha \leqq 0$. The sets M_α have all four properties: (1) because they differ from $\bar{\bar{M}}_\alpha$ at most on zero sets (remember that the absolute continuity implies that $\mu(M)$ vanishes on these sets also), and (2), (3) and (4) because of the particular way we chose to add F and subtract G.

Since the original $\bar{M}_\alpha$ were in $\mathcal{R}$ and since all sets considered were contained in $\bar{M}$, the M_α are in $\mathcal{R}$.

The property (1) is of sufficient importance in the sequel to formulate it alone:

Definition 11.2.4. Preserving the denotations of Theorem 11.2.8 we say that with respect to the functions $\tilde{\mu}(M)$, $\tilde{\nu}(M)$ and a set $\bar{M} \in \mathcal{T}$, a family of sets $M_\alpha \in \mathcal{T}$, $M_\alpha \subset \bar{M}$ has the property (1) if

$$(1) \qquad \begin{cases} P \in \mathcal{T} \text{ and } P \subset M_\alpha \text{ imply } \tilde{\mu}(P) \leqq \alpha \tilde{\nu}(P) \\ Q \in \mathcal{T} \text{ and } Q \subset \bar{M} - M_\alpha \text{ imply } \tilde{\mu}(Q) \geqq \alpha \tilde{\nu}(Q). \end{cases}$$

THEOREM 11.2.9. Assume that $\nu^*(M)$ is determined by $\nu(M)$ on $\mathcal{R}$ ($\mathcal{R}$ is a half-ring), and that $\mu(M)$ is absolutely continuous with respect to $\nu(M)$ on $\mathcal{R}$. Let $\mathcal{T}$ be the common part of the domains of the maximal equivalent extensions of $\mu(M)$ and $\nu(M)$ respectively, and let $\tilde{\mu}(M)$, $\tilde{\nu}(M)$ be equivalent extensions of $\mu(M)$, $\nu(M)$ from $\mathcal{R}$ to $\mathcal{T}$. Then there exists a function $f(x)$ such that:

(a) $f(x)$ is everywhere finite.

(b) The sets $S_x\{f(x) < \alpha\}$ belong to $BR(\mathcal{R})$. ($f(x)$ is thus a fortiori measurable).

(c) For any $\bar{M} \in \mathcal{T}$ the system of sets $M_\alpha = \bar{M} \cdot S_x\{f(x) < \alpha\}$ has the property (1).

Proof: Cover the space with a sequence $\{\bar{M}_i\}$ of disjunct sets from $\mathcal{R}$. To each $\bar{M}_i$ apply Theorem 11.2.8 with the $\mathcal{R}$, $\bar{M}$ of that theorem taken as $BR'(\mathcal{R})$, $\bar{M}_i$ here. We get a family of sets $M^i_\alpha \subset \bar{M}_i$ having properties (1), (2), (3) and (4) of Theorem 11.21.8. All the sets $M^i_\alpha \in BR'(\mathcal{R})$. Now define

$$f(x) \equiv \underset{x \in M^i_\alpha}{\text{g.l.b.}}\ \alpha \qquad (x \in \bar{M}_i)$$

If for a particular $\alpha = \alpha_0$, $x_0 \in M^i_{\alpha_0}$, then property (3) shows that there must be an $\alpha_1 < \alpha_0$ such that $x_0 \in M^i_{\alpha_1}$. Therefore $f(x_0) \leqq \alpha_1 < \alpha_0$. This shows that $M^i_\alpha \subset \bar{M}_i \cdot S_x\{f(x) < \alpha\}$. Conversely, if x_0 is a point for which $f(x_0) < \alpha_0$, there must be an $\alpha_1 < \alpha_0$ such that $x_0 \in M^i_{\alpha_1}$; property (2) then gives $x_0 \in M^i_{\alpha_0}$. Therefore $M^i_\alpha = \bar{M}_i \cdot S_x\{f(x) < \alpha\}$. From this we obtain $S_x\{f(x) < \alpha\} = \sum_i M^i_\alpha$; so that $S_x\{f(x) < \alpha\} \in BR(\mathcal{R})$.

Now for any $\bar{M} \in \mathcal{T}$ form the sets $M_\alpha = \bar{M} \cdot S_x\{f(x) < \alpha\}$. We have $M_\alpha = \sum_i \bar{M}M^i_\alpha$ and, since the M^i are disjunct, $\bar{M} - M_\alpha = \sum_i \bar{M}(\bar{M}_i - M^i_\alpha)$. From this it follows that the system of sets M_α has the property (1), for if $P \in \mathcal{T}$ and $P \subset M_\alpha$

$$\tilde{\mu}(P) = \tilde{\mu}(M_\alpha P) = \tilde{\mu}(\sum_i PM^i_\alpha) = \sum_i \tilde{\mu}(PM^i_\alpha) \leqq \alpha \sum_i \tilde{\nu}(PM^i_\alpha) = \alpha\, \tilde{\nu}(P),$$

and if $Q \in \mathcal{T}$ and $Q \subset \bar{M} - M_\alpha$, we get similarly

$$\tilde{\mu}(Q) \geqq \alpha\, \tilde{\nu}(Q).$$

Finally, the function $f(x)$ is finite. Suppose $f(x_0) = -\infty$. This x_0 belongs to some $\bar{M}_{i_0}$ and the definition of $f(x)$ in $\bar{M}_{i_0}$ shows that x_0 belongs to every $M^{i_0}_\alpha$, which contradicts property (4). Similarly the assumption that $f(x_0) = +\infty$ gives an x_0 belonging to no M^i_α, which fact again contradicts property (4).

<u>Definition 11.2.5. Let</u> $\mu(M)$, $\nu(M)$, $\tilde{\mu}(M)$, $\tilde{\nu}(M)$, $\mathcal{R}$, $\mathcal{T}$, <u>be as in the statement of Theorem</u> 11.2.9. <u>Any function</u> $f(x)$ <u>which is measurable and satisfies condition</u> (c) <u>of the statement of that theorem is called a derivative of</u> $\mu(M)$ <u>with respect to</u> $\nu(M)$ <u>and we write</u> $f(x) \cong \frac{d\mu(M_x)}{d\nu(M_x)}$.

The content of Theorem 11.2.9 is that absolute continuity implies the existence of a derivative function; the theorem, however, states more: it says that a derivative exists which in addition satisfies conditions (a) and (b).

<u>THEOREM 11.2.10: If</u> $\mu'(M)$ <u>and</u> $\nu'(M)$ <u>are respectively equivalent to</u> $\mu(M)$, $\nu(M)$ <u>and if</u> $f(x) \cong \frac{d\mu(M_x)}{d\nu(M_x)}$, <u>then</u> $f(x) \cong \frac{d\mu'(M_x)}{d\nu'(M_x)}$.

Proof: A derivative was defined as any function satisfying the condition (c) in Theorem 11.2.9. This condition does not involve $\mu(M)$, $\nu(M)$ at all except through their maximal simultaneous equivalent extensions. These are the same for $\mu'(M)$ and $\nu'(M)$.

THEOREM 11.2.11. *If $\mu(M)$ is absolutely continuous with respect to $\nu(M)$ on $\mathcal{R}$, and if $f(x)$ is a measurable function such that for every $\bar{M} \in \mathcal{R}$ the family of sets $M_\alpha = \bar{M} \cdot S_x\{f(x) < \alpha\}$ has the property (1')*

$$(1') \qquad \begin{cases} P \in BR'(\mathcal{R}) \text{ and } P \subset M_\alpha \text{ imply } \tilde{\mu}(P) \leqq \alpha\tilde{\nu}(P) \\ Q \in BR'(\mathcal{R}) \text{ and } Q \subset \bar{M} - M_\alpha \text{ imply } \tilde{\mu}(Q) \geqq \alpha\tilde{\nu}(Q) \end{cases}$$

where $\tilde{\mu}(M)$, $\tilde{\nu}(M)$ are any equivalent extensions of $\mu(M)$, $\nu(M)$ to a half-ring containing $BR'(\mathcal{R})$, then $f(x) = \frac{d\mu(M_x)}{d\nu(M_x)}$.

Proof: The domain of $\tilde{\mu}(M)$, $\tilde{\nu}(M)$ contains $BR'(\mathcal{R})$ and is contained in $\mathcal{V}$ of Theorem 11.2.9. The functions $\tilde{\mu}(M)$, $\tilde{\nu}(M)$ of Theorem 11.2.9 are equivalent extensions of the $\tilde{\mu}(M)$, $\tilde{\nu}(M)$ here. Since the property (1') makes assertions only about P, Q $\in BR'(\mathcal{R})$ it is a matter of indifference what domain our present $\tilde{\mu}(M)$, $\tilde{\nu}(M)$ have. We may therefore assume it to be the $\mathcal{V}$ of Theorem 11.2.9.

Now we show that the system of sets M_α above also has the property (1). If $P \in \mathcal{V}$ then as a measurable set with respect to $\nu^*(M)$ it is a set P' from $BR'(\mathcal{R})$ plus a zero set N. By the Corollary to Theorem 11.2.5, the zero set N is contained in $\mathcal{V}$, and by Theorem 11.2.4, $\tilde{\mu}(N) = 0$. Therefore $\tilde{\mu}(P) = \tilde{\mu}(P')$ and $\tilde{\nu}(P) = \tilde{\nu}(P')$. Property (1') above, then gives $\tilde{\mu}(P) = \tilde{\mu}(P') \leqq \alpha\,\tilde{\nu}(P') = \alpha\,\tilde{\nu}(P)$. A similar argument goes for Q.

This shows that property (1) holds for families M_α associated with an $\bar{M} \in \mathcal{R}$; this can be extended to any $\bar{M} \in \mathcal{V}$ exactly as was done in the proof of Theorem 11.2.9.

THEOREM 11.2.12. Let $f(x) \cong \frac{d\mu(M_x)}{d\nu(M_x)}$. Then $g(x) \cong \frac{d\mu(M_x)}{d\nu(M_x)}$ if and only if $f(x) \sim g(x)$.

Proof: Preserve the notations of Theorem 11.2.10 where possible. Assume first that $g(x) \cong \frac{d\mu(M_x)}{d\nu(M_x)}$. Let $F \equiv S_x\{f(x) \neq g(x)\}$. It is obviously sufficient to prove that the portion of F in any set of $\mathcal{R}$ is a zero set. Let $\bar{M} \in \mathcal{R}$ and define $M_\alpha = \bar{M} \cdot S_x\{f(x) < \alpha\}$, $N_\alpha = \bar{M} \cdot S_x\{g(x) < \alpha\}$; define $J_{\alpha,\beta} = M_\alpha - M_\alpha N_\beta$, $K_{\alpha,\beta} = N_\alpha - N_\alpha M_\beta$. If $\alpha < \beta$ then since $J_{\alpha,\beta} \subset M_\alpha$ and $J_{\alpha,\beta} \subset \bar{M} - N_\beta$ we have $\tilde{\mu}(J_{\alpha,\beta}) \leqq \alpha\, \tilde{\nu}(J_{\alpha,\beta})$ and $\tilde{\mu}(J_{\alpha,\beta}) \geqq \beta\, \tilde{\nu}(J_{\alpha,\beta})$ which are incompatible unless $\tilde{\nu}(J_{\alpha,\beta}) = 0$. Similarly when $\alpha < \beta$, $\tilde{\nu}(K_{\alpha,\beta}) = 0$.

Let $F_1 = \bar{M} \cdot S_x\{f(x) > g(x)\}$, $F_2 = \bar{M} \cdot S_x\{f(x) < g(x)\}$; so that $F\bar{M} = F_1 + F_2$.

As before, let D be a countable everywhere dense set of α's. Form the set E of all pairs (α,β) where $\alpha, \beta \in D$ and $\alpha < \beta$. E is countable and can be written as a sequence $\{(\alpha_1, \beta_1), (\alpha_2, \beta_2), \ldots\}$.

Now if $x \in F_1$, then $f(x) > g(x)$, and there is a pair $(a_k, b_k) \in E$ such that $g(x) < \alpha_k < \beta_k < f(x)$. Hence this $x \in N_{\alpha_k} - N_{\alpha_k} M_{\beta_k} = K_{\alpha_k,\beta_k}$. This proves that $F_1 \subset \sum_k K_{\alpha_k,\beta_k}$, and is therefore a zero set. By a similar argument we get $F_2 \subset \sum_k J_{\alpha_k,\beta_k}$ and so F_2 also is a zero set. Hence so is $F\bar{M} = F_1 + F_2$.

The statement that $f(x) \sim g(x)$ implies $g(x) \cong \frac{d\mu(M_x)}{d\nu(M_x)}$ is easy; for the sets M_α , N_α above can differ only on zero sets; so that if one system has the property (1), so does the other.

Corollary: If $\mu(M)$ is non-negative, then $f(x) \geqq 0$ except possibly on a zero set.

Proof: If $\alpha < 0$, then $\tilde{\mu}(M_\alpha) > \alpha\, \tilde{\nu}(M_\alpha)$ unless M_α is a zero set. Since $M_0 = \sum_n M_{-\frac{1}{n}}$, M_0 is a zero set.

THEOREM 11.2.13. Let $f(x) \cong \frac{d\lambda(M_x)}{d\nu(M_x)}$, $g(x) \cong \frac{d\mu(M_x)}{d\nu(M_x)}$, and c be a real number. Then:

1) $cf(x) \cong \frac{d(c\lambda(M_x))}{d\,\nu(M_x)}$.

2) $f(x) + g(x) \cong \frac{d(\lambda(M_x) + \mu(M_x))}{d\,\nu(M_x)}$.

3) $\text{Min}\,(f(x),\ g(x)) \cong \frac{d((\text{Min}\,\lambda,\mu)(M_x))}{d\,\nu(M_x)}$.

4) $\text{Max}\,(f(x),\ g(x)) \cong \frac{d((\text{Max}\,\lambda,\mu)(M_x))}{d\,\nu(M_x)}$.

5) $1 \cong \frac{d\nu(M_x)}{d\nu(M_x)}$.

Proof: We may assume that $f(x)$ and $g(x)$ satisfy conditions (a) and (b) of Theorem 11.2.9. If not, we could by Theorems 11.2.9, 11.2.12 find a $f_1(x) \sim f(x)$ and a $g_1(x) \sim g(x)$ which did. If the theorem held for $f_1(x)$, $g_1(x)$, then since $cf_1(x) \sim cf(x)$, $f_1(x) + g_1(x) \sim f(x) + g(x)$, ..., etc., Theorem 11.2.12 would show it held for $f(x)$ and $g(x)$.

Let $\mathcal{R}$ be the half-ring on which $\nu(M)$ and hence $\lambda(M)$, $\mu(M)$ are defined. Let $\tilde{\nu}(M)$, $\tilde{\lambda}(M)$, $\tilde{\mu}(M)$ be the equivalent extensions from $\mathcal{R}$ to $BR'(\mathcal{R})$.

Let $\bar{M}$ be a set from $\mathcal{R}$, and define $L_\alpha \equiv \bar{M} \cdot S_x\{f(x) < \alpha\}$, $M_\alpha \equiv \bar{M} \cdot S_x\{g(x) < \alpha\}$. Since $f(x)$ and $g(x)$ are assumed to satisfy condition (b) of Theorem 11.2.9, we have L_α, $M_\alpha \in BR'(\mathcal{R})$. In what follows, P and Q will always be sets from $BR'(\mathcal{R})$.

For 1) it is obviously sufficient to consider three cases $c > 0$, $c = 0$, and $c = -1$. Form the sets $K_\alpha \equiv \bar{M} \cdot S_x\{cf(x) < \alpha\}$. If $c > 0$, then $K_\alpha = L_{\frac{\alpha}{c}}$. Hence for $P \subset K_\alpha$, $Q \subset \bar{M} - K_\alpha$ we have $\tilde{\lambda}(P) \leqq \frac{\alpha}{c}\,\tilde{\nu}(P)$, $\tilde{\lambda}(Q) \geqq \frac{\alpha}{c}\,\tilde{\nu}(Q)$, and thus $c\,\tilde{\lambda}(P) \leqq \alpha\,\tilde{\nu}(P)$, $c\,\tilde{\lambda}(Q) \geqq \alpha\,\tilde{\nu}(Q)$. Thus the sets L_α have the property (1') of Theorem 11.2.11, and we have established 1) for $c > 0$.

If $c = 0$, then $K_\alpha = \begin{cases} \bar{M} & (\alpha \geq 0) \\ 0 & (\alpha \leqq 0) \end{cases}$. Hence for $P \subset K_\alpha$, $Q \subset \bar{M} - K_\alpha$, $0 \leqq \alpha \tilde{\nu}(P)$, $0 \geqq \alpha \tilde{\nu}(Q)$, and thus 1) holds for $c = 0$.

If $c = -1$, then $K_\alpha = \bar{M}S_x\{f(x) > -\alpha\} \subset \bar{M}S_x\{f(x) \geqq -\alpha\} = \bar{M} - L_{-\alpha}$. Therefore for $P \subset K_\alpha$, $\tilde{\lambda}(P) \geqq (-\alpha)\tilde{\nu}(P)$, or $-\lambda(P) \leqq \alpha\tilde{\nu}(P)$. Also, since $\bar{M} - K_\alpha = \bar{M}S_x\{f(x) \leqq -\alpha\} = \prod_{n=1}^{\infty} L_{-\alpha+\frac{1}{n}}$, it follows that if $Q \subset \bar{M} - K_\alpha$, then for any n, $\tilde{\lambda}(Q) \leqq (-\alpha + \frac{1}{n})\tilde{\nu}(Q)$. Allowing n to become infinite, we get $-\tilde{\lambda}(Q) \geqq \alpha\tilde{\nu}(Q)$. Statement 1) is now completely established.

For 2) form the system of sets $K_\alpha = \bar{M} \circ S_x\{f(x) + g(x) < \alpha\}$. Let D be a countable everywhere dense set of α's. For a fixed α form the set E of all pairs (β, γ) where $\beta, \gamma \in D$ and $\beta + \gamma < \alpha$. E is countable; let $\{(\beta_1, \gamma_1), (\beta_2, \gamma_2), \ldots$ be an enumeration. If now for any n, $x \in L_{\beta_n} M_{\gamma_n}$, then $f(x) + g(x) < \beta_n + \gamma_n < \alpha$, and thus $x \in K_\alpha$; hence $K_\alpha \supset \sum_n L_{\beta_n} M_{\gamma_n}$. If conversely $x \in K_\alpha$ then there exists n such that $f(x) < \beta_n$, $g(x) < \gamma_n$, and thus $x \in L_{\beta_n} M_{\gamma_n}$; hence $K_\alpha = \sum_n L_{\beta_n} M_{\gamma_n}$. Write $J_n = L_{\beta_n} M_{\gamma_n} - L_{\beta_n} M_{\gamma_n} \cdot (\sum_{m=1}^{n-1} L_{\beta_m} M_{\gamma_m})$. The J_n are disjunct, $J_n \subset L_{\beta_n} M_{\gamma_n}$ and $K_\alpha = \sum_n J_n$. Now if $P \subset K$

$$\tilde{\lambda}(P) + \tilde{\mu}(P) = \tilde{\lambda}(PK_\alpha) + \tilde{\mu}(PK_\alpha) = \sum_n (\tilde{\lambda}(PJ_n) + \tilde{\mu}(PJ_n)) \leqq$$

$$\leqq \sum_n (\beta_n \tilde{\nu}(PJ_n) + \gamma_n \tilde{\nu}(PJ_n)) \leqq \alpha \sum_n \tilde{\nu}(PJ_n) = \alpha\tilde{\nu}(P).$$

The first part of property (1') of Theorem 11.2.10 therefore holds.

To show the second, write $K'_\alpha = \bar{M} \cdot S_x\{f(x) + g(x) > \alpha\}$. For a fixed α form the set E' of all pairs (β', γ') where $\beta', \gamma' \in D$ and $\beta' + \gamma' > \alpha$; enumerate them: $\{(\beta'_1, \gamma'_1), (\beta'_2, \gamma'_2), \ldots\}$. By an analogous argument to that used above, we get $K'_\alpha = \sum_n (\bar{M} - L_{\beta'_n}) \cdot (\bar{M} - M_{\gamma'_n})$. We can now define J'_n analogously to the J_n above in such a way that they are disjunct, $J'_n \subset (\bar{M} - L_{\beta'_n}) \cdot (\bar{M} - M_{\gamma'_n})$ and $K'_\alpha = \sum_n J'_n$. Now if $Q' \in BR'(\mathcal{R})$ and $Q' \subset K'_\alpha$ we have

$$\tilde{\lambda}(Q') + \tilde{\mu}(Q') = \sum_n (\tilde{\lambda}(Q'J'_n) + \tilde{\mu}(Q'J'_n) \geqq \sum_n (\beta'_n + \gamma'_n)\tilde{\nu}(Q'J'_n) \geqq \alpha\,\tilde{\nu}(Q').$$

Now clearly $\bar{M} - K_\alpha = \prod_n K'_{\alpha - \frac{1}{n}}$. Hence if $Q \subset \bar{M} - K_\alpha$ we have for all n, $\tilde{\lambda}(Q) + \tilde{\mu}(Q) \geqq (\alpha - \frac{1}{n})\tilde{\nu}(Q)$. As n makes the transition to infinity, this last inequality completes the proof of 2).

For 3) assumes first that (Min λ, μ) (M) = 0. Thus we have $\lambda(M) \geqq 0$, $\mu(M) \geqq 0$, and by Theorem 11.1.14 the existence of an $\bar{N} \in BR'(\mathcal{R})$, $\bar{N} \subset \bar{M}$ such that for $P \subset \bar{N}$, $\tilde{\lambda}(P) = 0$, and for $Q \subset \bar{M} - \bar{N}$, $\tilde{\mu}(Q) = 0$. It follows from an argument exactly like that applied to the case c = 0 in 1) above, that in $\bar{N}$, $f(x) \sim 0$ and in $\bar{M} - \bar{N}$, $g(x) \sim 0$. By the Corollary to Theorem 11.21.12, f(x), and g(x) are non-negative except on a zero set. Therefore Min (f(x), g(x)) = 0. This proves 3) under the assumption (Min λ, μ)(M) = 0.

In the general case let $\lambda_1(M) = \lambda(M) - (\text{Min } \lambda, \mu)(M)$, $\mu_1(M) = \mu(M) - (\text{Min } \lambda, \mu)(M)$; let $h(x) \cong \frac{d((\text{Min } \lambda, \mu)(M_x))}{d\nu(M_x)}$, $f_1(x) \cong \frac{d\lambda_1(M_x)}{d\nu(M_x)}$, $g_1(x) \cong \frac{d\mu_1(M_x)}{d\nu(M_x)}$. Then from 1) and 2) above, $f_1(x) \sim f(x) - h(x)$, $g_1(x) \sim g(x) - h(x)$. Since (Min λ_1, μ_1)(M) = 0, we have Min (f(x), g(x)) - - h(x) $\sim$ Min ($f_1(x)$, $g_1(x)$) $\sim$ 0. Statement 3) therefore holds.

For 4) we simply use 1), 3) and the fact that

$$(\text{Max } \lambda, \mu)(M) = -(\text{Min } -\lambda, -\mu)(M)$$

$$\text{Max } (f(x), g(x)) = -\text{Min}(-f(x), -g(x)).$$

For 5) we form $K_\alpha = \bar{M} \cdot S_x\{1 < \alpha\}$. Thus $K_\alpha = \begin{cases} \bar{M} & (\alpha \geqq 1) \\ 0 & (\alpha \leqq 1) \end{cases}$ and the conditions of property (1') are trivially satisfied.

<u>THEOREM 11.2.14</u>. <u>If</u> $f(x) \cong \frac{d\lambda(M_x)}{d\nu(M_x)}$, $g(x) \cong \frac{d\mu(M_x)}{d\nu(M_x)}$ <u>and</u> $f(x) \sim g(x)$, <u>then</u> $\lambda(M) = \mu(M)$.

Proof: Let $\rho(M) = \lambda(M) - \mu(M)$. From 1) and 2) of the preceding theorem $\frac{d\rho(M_x)}{d\nu(M_x)} \cong h(x) \sim f(x) - g(x) \sim 0$. We must prove that $\rho(M) = 0$. Let $\bar{M} \in \mathcal{R}$. Then relative to $\rho(M)$, $\nu(M)$, $\bar{M}$ the system of sets $K_\alpha = \bar{M} \cdot S_x\{0 < \alpha\}$ must have the property (1) (or (1')). But $K_\alpha = \bar{M}$ if $0 < \alpha$ and $K_\alpha = 0$ if $0 \geqq \alpha$. In particular if $P \subset \bar{M}$, $P \subset K_{\frac{1}{n}}$ and therefore $\tilde{\rho}(P) \leqq \frac{1}{n} \tilde{\nu}(P)$; but $P \subset \bar{M} - K_{-\frac{1}{n}}$, and therefore $\tilde{\rho}(P) \geqq -\frac{1}{n}\tilde{\nu}(P)$. If n tends to $+\infty$ we get $\tilde{\rho}(P) = 0$. Hence $\rho(M) = 0$.

If in the results preceding we regard $\nu(M)$ as a fixed non-negative and totally additive function on $\mathcal{R}$, we have associated with every real $\mu(M)$ which is absolutely continuous with respect to $\nu(M)$, a measurable function $f(x) \cong \frac{d\mu(M_x)}{d\nu(M_x)}$ -- in fact, the whole class of functions $g(x) \sim f(x)$. The operation $\frac{d\mu(M_x)}{d\nu(M_x)}$ on $\mu(M)$ is linear and homogeneous if we regard equivalent functions $f(x)$ as identical. Also we have shown that if $f(x) \cong \frac{d\mu(M_x)}{d\nu(M_x)}$ then $\mu(M)$ is uniquely determined. We now ask what $f(x)$'s are derivatives of some real $\mu(M)$ absolutely continuous with respect to $\nu(M)$. Obviously not all $f(x)$'s, for the $f(x)$ has at least to be measurable with respect to $\nu^*(M)$. Not even all measurable $f(x)$'s have the property, however. We shall call those which do have it "summable".

Definition 11.2.6. A real function $f(x)$ defined everywhere in S is called summable (with respect to $\nu(M)$) if there is a real $\mu(M)$ absolutely continuous with respect to $\nu(M)$ such that $f(x) \cong \frac{d\mu(M_x)}{d\nu(M_x)}$.

Definition 11.2.7. Let $f(x)$ be summable with respect to $\nu(M)$. The $\mu(M)$ of Definition 11.2.6 we shall call the primitive $f(x)$ and denote it by $\mu(M) = \int_M f(x)d\nu(M_x)$. For a particular $\bar{M} \in \mathcal{R}$ we call the expression $\int_{\bar{M}} f(x)d\nu(M_x)$ the integral of $f(x)$ (with respect to $\nu(M)$) over $\bar{M}$.

This method of defining the integral is due in essence to H. Hahn. Perhaps the most striking advantage of the procedure is that it enables us to prove easily the linear homogeneous properties of the integral by a simple application of Theorems 11.2.13 and 11.2.14.

Remark: In contrast to Definition 11.2.5, there is a perhaps more intuitive method for attempting to define the derivative: For each point x select a sequence $\{M_x^1, M_x^2, \ldots\}$ of sets from $\mathcal{R}$ containing x and "shrinking down" upon it. Define f(x) as some one of the limit points of the sequence $\left\{\frac{\mu(M_x^1)}{\nu(M_x^1)}, \frac{\mu(M_x^2)}{\nu(M_x^2)}, \ldots\right\}$ (Because of the absolute continuity, if the denominator of one of these fractions vanishes the numerator will too, and we assign to the fraction any convenient value, say 0.) Under certain assumptions on the character of the system of sets M_x^n -- a fortiori if they have the character of a net, or even if it is possible to prove a suitable analogue of the Vitali covering theorem (Theorem 6.2) -- it can be shown that for the f(x) so obtained we have $f(x) \cong \frac{d\mu(M_x)}{d\nu(M_x)}$ in the sense of Definition 11.2.5. Such a procedure, while it gives a trifle more information about the nature of f(x), has the obvious difficulty that it requires some sort of topology in S to assign a meaning to the phrase "shrinking down" and that not every topology can be used for this purpose. If we permit certain deviations from the usual topological procedure (by using non-overlapping sets M_x^n), the proofs become somewhat simpler and the results more general, but we shall not further discuss this subject here. For our present purposes the derivative of Definition 11.2.5 is ideal.

We shall now give an explicit characterization of the summable functions in terms of their ordinate sets in a certain product space. This is the method of Caratheodory. First we prove:

<u>THEOREM 11.2.15. Let</u> f(x) <u>be a real function defined for</u> $x \in S$. <u>Define</u> $f_+(x) = \mathrm{Max}(f(x), 0)$, $f_-(x) = \mathrm{Max}(-f(x), 0)$, <u>then in order that</u> f(x) <u>be summable</u>

with respect to $\nu(M)$ *it is necessary and sufficient that* $f_+(x)$ *and* $f_-(x)$ *be summable with respect to* $\nu(M)$.

Proof: Suppose $f(x)$ is summable; let $\mu(M) = \int_M f(x)d\,\nu(M_x)$. By Theorem 11.2.13 we have

$$f_+(x) = \text{Max}(f(x), 0) = \frac{d(\text{Max}\ \mu,0)(M_x)}{d\,\nu(M_x)} = \frac{d\mu_+(M_x)}{d\,\nu(M_x)},$$

$$f_-(x) = \text{Max}(-f(x),0) = \frac{d(\text{Max}\ -\mu,0)(M_x)}{d\,\nu(M_x)} = \frac{d\mu_-(M_x)}{d\,\nu(M_x)}.$$

Therefore $f_+(x)$ and $f_-(x)$ are summable.

If conversely $f_+(x)$ and $f_-(x)$ are summable, we have again by Theorem 11.2.13

$$f(x) = f_+(x) - f_-(x) \cong \frac{d(\mu_+(M_x) - \mu_-(M_x))}{d\,\nu(M\ )} = \frac{d\mu(M_x)}{d\nu(M_x)}.$$

Thus $f(x)$ is summable.

This theorem shows that it is sufficient to characterize the non-negative summable functions. We therefore restrict the discussion in the remainder of the section to non-negative everywhere-defined functions $f(x)$.

We shall adhere to the convention that $\nu^*(M)$ is a regular outer measure in S; $\nu(M)$ will be the maximal measure determining $\nu^*(M)$. R_1 will be the Euclidean one-dimensional space. We shall denote the Lebesgue outer measure in R_1 by $m^*(E)$ (for simplicity instead of $\mu^*(E)$).

Definition 11.2.8. Form the direct product space $\mathcal{S} = S \times R_1$. $\mathcal{S}$ *will be called the ordinate space of* S. *The typical point of* S *is an ordered pair* (x, ξ) *where* $x \in S$ *and* $\xi \in R_1$. *Let* $f(x)$ *be a non-negative function defined for* $x \in S$. *The set* $\mathcal{O}_f^+ \subset \mathcal{S}$ *consisting of all pairs* (x, ξ) *for which* $0 \leqq \xi \leqq f(x)$ *is called the maximal ordinate set of* $f(x)$. *The set* $\mathcal{O}_f^-$ *consisting of all pairs*

(x, ξ) _for which_ $0 \leqq \xi < f(x)$ _is called the minimal ordinate set of_ $f(x)$. _Any set_ $\mathcal{O}_f$ _for which_ $\mathcal{O}_f^- \subset \mathcal{O}_f \subset \mathcal{O}_f^+$ _is called an ordinate set of_ $f(x)$.

Definition 11.2.9. Recalling Definition 10.4.3, form in $\mathcal{S} = S \times R_1$ _the half-ring_ $\bar{\mathcal{R}}$ _consisting of all sets_ $\mathcal{M} = M \times E$ _where_ $M \in MS'_{\nu^*}$, $E \in MS'_{m^*}$. _For_ $\mathcal{M} = M \times E$ _define_ $\bar{\nu}(\mathcal{M}) = \nu(M) \cdot m(E)$. _Let_ $\bar{\nu}_1^*(\mathcal{M})$ _be the outer measure determined by_ $\bar{\nu}(\mathcal{M})$. _We shall call_ $\bar{\nu}_1^*(\mathcal{M})$ _the ordinate outer measure corresponding to_ $\nu^*(M)$.

THEOREM 11.2.16. If $f(x)$ _is non-negative and measurable with respect to_ $\nu^*(M)$, _then all ordinate sets_ $\mathcal{O}_f$ _are measurable with respect to_ $\bar{\nu}_I^*(\mathcal{M})$, _and their measures are all equal._

Proof: Denote by $\mathcal{Y}\binom{b}{a}$ the half-open interval $a \leqq \xi < b$ in R_1. Since any $\mathcal{O}_f$ satisfies $\mathcal{O}_f^- \subset \mathcal{O}_f \subset \mathcal{O}_f^+$, and $\mathcal{O}_f \subset S \times \mathcal{Y}\binom{+\infty}{0}$, it is sufficient to prove that the sets $\bar{\mathcal{M}}\mathcal{O}_f^-$ and $\bar{\mathcal{M}}\mathcal{O}_f^+$ are measurable and have the same (finite) measure for all $\bar{\mathcal{M}}$ of the form $\bar{\mathcal{M}} = \bar{M} \times \mathcal{Y}\binom{c}{0}$ where $\bar{M} \in MS'_{\nu^*}$ and $c > 0$. ($\mathcal{O}_f$ is contained in the sum of a countable number of such $\bar{\mathcal{M}}$.)

Take a fixed c and a fixed $\bar{M}$. Let ε be any positive number. Select an integer $\bar{n} > \frac{c \cdot \nu(\bar{M})}{\varepsilon}$. For $0 \leqq n \leqq \bar{n}$, define $c_n = \frac{nc}{\bar{n}}$. Denote by N_n the set $\bar{M} \cdot S_x\{f(x) \geqq c_n\}$ $(n = 0, 1, 2, \ldots, \bar{n})$. Form the sets

$$\mathcal{N}^- = \sum_{n=0}^{\bar{n}-1} N_{n+1} \times \mathcal{Y}\binom{c_{n+1}}{c_n} \text{ and } \mathcal{N}^+ = \sum_{n=0}^{\bar{n}-1} N_n \times \mathcal{Y}\binom{c_{n+1}}{c_n}.$$ We first prove that

$$\mathcal{N}^- \subset \bar{\mathcal{M}}\mathcal{O}_f^- \subset \bar{\mathcal{M}}\mathcal{O}_f^+ \subset \mathcal{N}^+.$$

Suppose $(x, \xi) \in \mathcal{N}^-$, then there is an n such that $\xi \in \mathcal{Y}\binom{c_{n+1}}{c_n}$ and $x \in N_{n+1}$; thus $\xi < c_{n+1} \leqq f(x)$. Therefore $\mathcal{N}^- \subset \bar{\mathcal{M}}\mathcal{O}_f^-$. If $(x, \xi) \in \bar{\mathcal{M}}\mathcal{O}_f^+$, then $0 \leqq \xi < c$, $\xi \leqq f(x)$, and $x \in \bar{M}$. There is thus an n such that $c_n \leqq \xi < c_{n+1}$, and for this n, $f(x) \geqq \xi \geqq c_n$. Therefore $(x, \xi) \in N_n \times \mathcal{Y}\binom{c_{n+1}}{c_n}$ and $\bar{\mathcal{M}}\mathcal{O}_f^+ \subset \mathcal{N}^+$.

The sets $\mathcal{N}^-$, $\mathcal{N}^+$ belong certainly to $R(\mathcal{R})$. Compute their measures:

$$\bar{\nu}_I(\mathcal{N}^-) = \sum_{n=0}^{\bar{n}-1} \nu(N_{n+1}) \cdot m(\mathcal{Y}\left({}^{c_{n+1}}_{c_n}\right)) = \sum_{n=0}^{\bar{n}-1} \nu(N_{n+1})(c_{n+1} - c_n) = \sum_{n=0}^{\bar{n}-1} \nu(N_{n+1}) \frac{c}{\bar{n}},$$

$$\bar{\nu}_I(\mathcal{N}^+) = \sum_{n=0}^{\bar{n}-1} \nu(N_n) \cdot m(\mathcal{Y}\left({}^{c_{n+1}}_{c_n}\right)) = \sum_{n=0}^{\bar{n}-1} \nu(N_n)(c_{n+1} - c_n) = \sum_{n=0}^{\bar{n}-1} \nu(N_n) \frac{c}{\bar{n}}.$$

From these equations it follows that

$$\bar{\nu}_I(\mathcal{N}^+) - \bar{\nu}_I(\mathcal{N}^-) = \frac{c}{\bar{n}}(\nu(N_0) - \nu(N_c)) \leq \frac{c}{\bar{n}} \nu(\bar{M}) < \varepsilon .$$

Now

$$\bar{\nu}_I(\mathcal{N}^-) \leq \bar{\nu}_{I*}(\bar{\mathcal{M}}\mathcal{O}_f^-) \leq \bar{\nu}_I^*(\bar{\mathcal{M}}\mathcal{O}_f^-) \leq \bar{\nu}_I^*(\bar{\mathcal{M}}\mathcal{O}_f^+) \leq \bar{\nu}_I(\mathcal{N}^+),$$

$$\nu_I(\mathcal{N}^-) \leq \bar{\nu}_{I*}(\bar{\mathcal{M}}\mathcal{O}_f^-) \leq \bar{\nu}_{I*}(\bar{\mathcal{M}}\mathcal{O}_f^+) \leq \nu_I^*(\bar{\mathcal{M}}\mathcal{O}_f^+) \leq \bar{\nu}_I(\mathcal{N}^+).$$

From these last three displayed inequalities and the fact that ε is arbitrary, it follows that

$$\bar{\nu}_{I*}(\bar{\mathcal{M}}\mathcal{O}_f^-) = \nu_I^*(\bar{\mathcal{M}}\mathcal{O}_f^-) = \bar{\nu}_{I*}(\bar{\mathcal{M}}\mathcal{O}_f^+) = \bar{\nu}_I^*(\mathcal{M}\mathcal{O}_f^+),$$

and this establishes the result (Theorem 10.2.5).

Corollary: $\bar{\nu}_I(\mathcal{O}_f^+ - \mathcal{O}_f^-) = 0$, if $f(x)$ is measurable.

Proof: If $\mathcal{O}_f^+$ is covered by the sum of a sequence $\{\bar{\mathcal{M}}_n\}$ of sets of the type $\bar{\mathcal{M}}$ above, then $(\mathcal{O}_f^+ - \mathcal{O}_f^-) = \sum_n (\bar{\mathcal{M}}_n \mathcal{O}_f^+ - \bar{\mathcal{M}}_n \mathcal{O}_f^-)$.

The method of Caratheodory for defining the Lebesgue integral suggests that we should hope to find a characterization of summable $f(x)$'s in an expression of the form $\int_M f(x) d\nu(M_x) = \bar{\nu}_I(\mathcal{O}_f^- \cdot (M \times R_1))$. This we now proceed to do.

THEOREM 11.2.17. Let $\mathcal{R}$ be a determining ring for $\nu^*(M)$. If $f(x)$ is a

<u>real measurable function such that for every</u> $M \in \mathcal{R}$ <u>the expression</u> $\bar{\nu}_I(\mathcal{O}_f^- \cdot (M \times R_1))$ <u>is finite, then</u> $f(x)$ <u>is summable with respect to</u> $\nu(M)$ <u>on</u> $\mathcal{R}$, <u>and</u> $\int_M f(x) d\nu(M_x) = \bar{\nu}_I(\mathcal{O}_f^- \cdot (M \times R_1))$.

Proof: The sets M of MS'_{ν^*} for which $\bar{\nu}_I(\mathcal{O}_f^- \cdot (M \times R_1))$ is finite obviously form a restricted Borel-ring $\mathcal{T}$ which contains $\mathcal{R}$ and therefore $BR'(\mathcal{R})$. For $M \in \mathcal{T}$ define $\tilde{\mu}(M) = \bar{\nu}_I(\mathcal{O}_f^- \cdot (M \times R_1))$.

Now if $M_1, M_2, \ldots$ is a finite or infinite sequence of disjunct sets from $\mathcal{T}$ whose sum $M \in \mathcal{T}$, then the total additivity of $\bar{\nu}_I(\mathcal{M})$ gives

$$\sum_i \tilde{\mu}(M_i) = \sum_i \bar{\nu}_I(\mathcal{O}_f^- \cdot (M_i \times R_1)) = \bar{\nu}_I(\mathcal{O}_f^- \cdot (M \times R_1)) = \tilde{\mu}(M).$$

Thus $\tilde{\mu}(M)$ is finite, non-negative and totally additive.

If M is a zero set with respect to $\nu^*(M)$, then by Theorem 10.4.7, $\bar{\nu}_I^*(M \times R_1) = 0$, and thus $\tilde{\mu}(M) = 0$. Therefore $\tilde{\mu}(M)$ is absolutely continuous with respect to $\nu(M)$ on $\mathcal{T}$ (Theorem 11.2.4).

On $\mathcal{R}$ define $\mu(M) = \tilde{\mu}(M)$. Since $\mathcal{R} \subset \mathcal{T}$, $\mu(M)$ is absolutely continuous with respect to $\nu(M)$ on $\mathcal{R}$. Thus if we form the outer measure $\mu_I^*(M)$ determined by $\mu(M)$, Corollary 2 to Theorem 11.2.5 gives $MS_{\nu^*} \subset MS_{\mu_I^*}$. Therefor $\mathcal{R} \subset \mathcal{T} \subset MS'_{\mu_I^*}$ and $\tilde{\mu}(M)$ as a non-negative extension of $\mu(M)$ is equivalent to it.

We now prove that $f(x) \cong \frac{d\mu(M_x)}{d\nu(M_x)}$. Select a fixed $\bar{M} \in \mathcal{R}$. Form the family of sets $M_\alpha = \bar{M} \cdot S_x\{f(x) < \alpha\}$. We have to show that it has the property (1') of Theorem 11.2.11. This is trivial for $\alpha < 0$, for then $M_\alpha = 0$. Assume that $\alpha \geqq 0$. Then if $P \in BR'(\mathcal{R})$ and $P \subset M_\alpha$ we have $\mathcal{O}_f^- \cdot (P \times R_1) \subset (P \times \mathcal{J}\binom{\alpha}{0}) \cdot (P \times R_1) = P \times \mathcal{J}\binom{\alpha}{0}$. Hence

$$\tilde{\mu}(P) = \bar{\nu}_I(\mathcal{O}_f^- \cdot (P \times R_1)) \leqq \bar{\nu}_I(P \times \mathcal{J}\binom{\alpha}{0}) = \alpha \nu(P)$$

Similarly, if $Q \in BR'(\mathcal{R})$ and $Q \subset \bar{M} - M_\alpha$, we get $\mathcal{O}_f^-(Q \times R_1) \supset (Q \times \mathcal{Y}\binom{\alpha}{0})$ and thus

$$\tilde{\mu}(Q) = \bar{\nu}_I(\mathcal{O}_f^- \cdot (Q \times R_1)) \geqq \bar{\nu}_I(Q \times \mathcal{Y}\binom{\alpha}{0}) = \alpha\, \nu(Q).$$

This establishes $f(x) \cong \frac{d\mu(M_x)}{d\nu(M_x)}$. Thus for $M \in \mathcal{R}$,

$$\int_M f(x)d\,\nu(M_x) = \mu(M) = \bar{\nu}_I(\mathcal{O}_f^- \cdot (M \times R_1)).$$

We now establish a converse to this theorem:

THEOREM 11.2.18. *If $f(x)$ is non-negative and summable with respect to $\nu(M)$ on $\mathcal{R}$, then for every $M \in \mathcal{R}$, $\bar{\nu}_I(\mathcal{O}_f^- \circ (M \times R_1))$ is finite.*

Proof: Let $\mu(M) = \int_M f(x)d\,\nu(M_x)$; so that $f(x) \cong \frac{d\mu(M_x)}{d\nu(M_x)}$. Now by Theorem 11.2.13 we have, if c is any real number, $c = \frac{d(c\,\nu(M_x))}{d\,\nu(M_x)}$ and hence $f_c(x) \equiv \mathrm{Min}(f(x),c) \cong \frac{d(\mathrm{Min}\,\mu, c\,\nu)(M_x)}{d\,\nu(M_x)}$, from which $\int_M f_c(x)d\,\nu(M_x) =$ $= (\mathrm{Min}\,\mu,\ c\,\nu)(M) \leqq \mu(M)$. Assume that $c \geqq 0$. By Theorem 11.2.16, $\int_M f_c(x)d\,\nu(M_x) = \bar{\nu}_I(\mathcal{O}_{f_c}^- \cdot (M \times R_1))$. Thus

$$\bar{\nu}_I(\mathcal{O}_{f_c}^- \circ (M \times R_1)) \leqq \mu(M).$$

Now give to c the values 1, 2, ... Clearly $\mathcal{O}_{f_1}^- \subset \mathcal{O}_{f_2}^- \ldots \subset \mathcal{O}_f^- = \lim_{c\to\infty} \mathcal{O}_{f_c}^-$. Therefore (Theorem 10.2.8), $\bar{\nu}_I(\mathcal{O}_f^- \circ (M \times R_1) = \lim_{c\to\infty} \bar{\nu}_I(\mathcal{O}_{f_c}^- \cdot (M \times R_1)) \leqq \mu(M)$. That is, $\bar{\nu}_I(\mathcal{O}_f^- \cdot (M \times R_1))$ is finite for every $M \in \mathcal{R}$.

We now complete the characterization of summable functions in terms of their ordinate sets by proving:

THEOREM 11.2.19. *If any of the ordinate sets $\mathcal{O}_f$ of a non-negative $f(x)$ is measurable with respect to $\bar{\nu}_I^*(\mathcal{M})$, then $f(x)$ is measurable with respect to $\nu^*(M)$.*

Proof: First assume that $f(x)$ is bounded, say $f(x) < c$. Then for any $M \in MS'_{\nu^*}$ define $\mu(M) = \bar{\nu}_I(\mathcal{O}_f \cdot (M \times R_1))$. Since $\mathcal{O}_f \subset S \times \mathcal{Y}\binom{c}{0}$, we have

$\mu(M) \leqq \bar{\nu}_I(M \times \mathfrak{J}\binom{c}{0}) = c\,\nu(M)$. Thus $\mu(M)$ is finite for $M \in MS'_{\nu *}$. It follows from exactly the same argument as used in the proof of Theorem 11.2.17 that $\mu(M)$ is a measure function absolutely continuous with respect to $\nu(M)$ on $MS'_{\nu *}$.

Now define $g(x) \cong \frac{d\mu(M_x)}{d\nu(M_x)}$. Since $g(x)$ is measurable it is sufficient to prove that $f(x) \sim g(x)$. We proceed to do this.

Let α, β be real numbers with $0 \leqq \alpha < \beta$. Let $\bar{M} \in MS'_{\nu *}$. Define $L \cong \bar{M} \cdot S_x\{f(x) > \beta\}$, $N \cong \bar{M} \cdot S_x\{g(x) < \alpha\}$, $K \cong LN$. Since $g(x)$ is measurable, $N \in MS'_{\nu *}$, but we do not know yet that L and K are measurable. Select (Theorem 10.2.11) a K' from $MS'_{\nu *}$ such that $K' \supset K$ and $\nu(K') = \nu^*(K)$, and define $\bar{K} = K'N$. Thus $\bar{K} \in MS'_{\nu *}$, and since $K' \supset \bar{K} \supset K$, $\nu(\bar{K}) = \nu^*(K)$.

Now $\mathcal{O}_f \cdot (\bar{K} \times R_1) \supset \mathcal{O}_f \cdot (K \times R_1) \supset K \times \mathfrak{J}\binom{\beta}{0}$, and therefore $\mu(\bar{K}) = \bar{\nu}_I(\mathcal{O}_f \cdot (\bar{K} \times R_1)) \geqq \bar{\nu}_I^*(\mathcal{O}_f \cdot (K \times R_1)) \geqq \bar{\nu}_I^*(K \times \mathfrak{J}\binom{\beta}{0}) = \nu^*(K) \cdot \beta = \nu(\bar{K}) \cdot \beta$, the next to the last step being a consequence of Theorem 11.4.7. On the other hand, since $\bar{K} \subset N$, we have (Property 1') $\mu(\bar{K}) \leqq \alpha \cdot \nu(\bar{K})$. From these last two inequalities and $\alpha < \beta$ it follows that $\nu(\bar{K}) = 0$, and hence that $\nu^*(K) = 0$.

Now take a countable everywhere-dense set D of real numbers and form the set E of all pairs α, β where $\alpha, \beta \in D$ and $0 \leqq \alpha < \beta$. Let $\{(\alpha_1, \beta_1), (\alpha_2, \beta_2), \dots\}$ be an enumeration of E. Define K_i as the K above with $\alpha = \alpha_i$, $\beta = \beta_i$. Then obviously $\bar{M} \cdot S_x\{g(x) < f(x)\} = \sum K_i$, which is a zero set. Since $\bar{M}$ is any set of $MS'_{\nu *}$ we have $\nu^*(S_x\{g(x) < f(x)\}) = 0$. A similar argument proves that $\nu^*(S_x\{g(x) > f(x)\}) = 0$. Hence $f(x) \sim g(x)$ and is measurable.

If $f(x)$ is not bounded, then $\mathcal{O}_f \cdot (S \times \mathfrak{J}\binom{c}{0})$ $(c = 1, 2, \dots)$, is measurable since $\mathcal{O}_f$ is, and it is an ordinate set for $f_c(x) = \operatorname{Min}(f(x), c)$. Therefore $f_c(x)$ is measurable, and hence so is $f(x) = \lim_{c \to \infty} f_c(x)$.

We conclude this section by remarking that the notions of summability and integral we have thus far defined contain certain unnecessarily stringent restrictions. In the first place, the definition in terms of derivatives requires that the functions f(x) be everywhere defined whereas the method of ordinate sets suggests a perfectly natural definition for the integral of functions which are defined only on a part of the space S. In the second place, summability and the integral were defined in terms of a measure function $\nu(M)$ on a half-ring $\mathcal{R}$ determining $\nu^*(M)$. This would appear something of a disadvantage as one often wishes to treat the integral as a thing depending upon f(x) and $\nu^*(M)$. We shall see in the next section that these difficulties are only apparent.

§3. Elementary properties of the integral

This section will be devoted to deducing mainly from the results of the preceding section a number of properties of the integral. They will for the most part have close analogues in the theorems of Chapter VIII.

We have, however, first to formulate the definition of the integral in what for our purposes will be its final general form. We must remove the restrictions mentioned at the end of the last section.

Unless mention is made to the contrary, $\nu^*(M)$ will denote a regular outer measure in S. $\nu(M)$ will be $\nu^*(M)$ on MS'_{ν^*}. The term "measurable", as well as the equivalence relation $f(x) \sim g(x)$, will be understood as relative to $\nu^*(M)$.

For the first part of this discussion, the functions f(x) will, as in the last section, be everywhere defined in S. Later, we shall consider functions defined on some subset of S. Without exception the domain of any function we consider will be contained in S. We shall denote the domain of f(x) by D_f.

<u>Definition 11.3.1. If</u> M <u>and</u> f(x) <u>are such that</u> $M \subset D_f$, <u>we define</u>

$$f_M(x) = \begin{cases} f(x) & (x \in M) \\ 0 & (x \in S-M). \end{cases}$$

We proceed to the formulation of the general definition of integral by introducing the intermediate

<u>Definition 11.3.2. Let</u> f(x) <u>be summable with respect to</u> $\nu(M)$ <u>in the sense of Definition 11.2.6; let</u> $\bar{M}$ <u>be any set from</u> $M \in MS_{\nu *}$. f(x) <u>is said to be summable with respect to</u> $\nu^*(M)$ <u>over</u> $\bar{M}$ <u>if</u> $\bar{M}$ <u>belongs to the domain of</u> $\tilde{\mu}(M)$, <u>the maximal equivalent extension of</u> $\mu(M) = \int_M f(x)d\,\nu(M_x)$. (Definition 11.2.6. defines $\mu(M)$ only for $M \in MS'_{\nu *}$). <u>If</u> f(x) <u>is summable over</u> $\bar{M}$ <u>we define</u> $\int_{\bar{M}} f(x)d\,\nu(M_x) = \tilde{\mu}(\bar{M})$.

This definition is obviously consistent with Definition 11.2.6, because if $\bar{M}$ happens to be in $MS'_{\nu *}$, it is <u>a fortiori</u> in the domain of $\tilde{\mu}(M)$ and we have $\tilde{\mu}(\bar{M}) = \mu(\bar{M})$.

<u>THEOREM 11.3.1. Let</u> f(x) <u>be summable with respect to</u> $\nu(M)$ <u>in the sense of Definition 11.2.6; let</u> $\bar{M}$ <u>be any set from</u> $MS_{\nu *}$ <u>and</u> M' <u>be any set from</u> $MS_{\nu *}$ <u>over which</u> f(x) <u>is summable in the sense of Definition 11.3.2. Then</u> $f_{\bar{M}}(x)$ <u>is summable with respect to</u> $\nu(M)$, <u>is summable over</u> M', <u>and we have</u>

$$\int_{M'} f_{\bar{M}}(x)d\,\nu(M_x) = \int_{M'\bar{M}} f(x)d\,\nu(M_x). \tag{1}$$

Proof: The summability and hence measurability of f(x) implies the measurability of $f_{\bar{M}}(x)$; for $S_x\{f_{\bar{M}}(x) < \alpha\}$ is equal to $\bar{M} \cdot S_x\{f(x) < \alpha\}$ or to $\bar{M} \cdot S_x\{f(x) < \alpha\} + (S - \bar{M})$, according as $\alpha \leqq 0$ or $\alpha > 0$.

First assume that $M' \in MS'_{\nu *}$; thus that $M'\bar{M} \in MS'_{\nu *}$. Since f(x) is summable with respect to $\nu(M)$, we have by an application of Theorem 11.2.5 the summability with respect to $\nu(M)$ of $f_+(x)$ and $f_-(x)$, and also have the relation

$$\int_{M'\bar{M}} f(x)d\,\nu(M_x) = \int_{M'\bar{M}} f_+(x)d\,\nu(M_x) - \int_{M'\bar{M}} f_-(x)d\,\nu(M_x).$$

An application of Theorems 11.2.18, 11.2.17 gives $\int_{M'\bar{M}} f_+(x)d\nu(M_x) =$ $= \bar{\nu}_I(\mathcal{O}^-_{f^+} \cdot (M'\bar{M} \times R_1))$; $\int_{M'\bar{M}} f_-(x)d\nu(M_x) = \bar{\nu}_I(\mathcal{O}^-_{f_-} \cdot (M'\bar{M} \times R_1))$. Now it is easily seen that $\mathcal{O}^-_{(f_+)_{\bar{M}}} = \mathcal{O}^-_{f_+} \circ (\bar{M} \times R_1)$; $\mathcal{O}^-_{(f_-)_{\bar{M}}} = \mathcal{O}^-_{f_-} \cdot (\bar{M} \times R_1)$ and also that $(f_+)_{\bar{M}}(x) = (f_{\bar{M}})_+(x)$ and $(f_-)_{\bar{M}}(x) = (f_{\bar{M}})_-(x)$. Thus, for every $M \in MS'_{\nu*}$, we have

$$\mathcal{O}^-_{(f_{\bar{M}})_+} \cdot (M \times R_1) = \mathcal{O}^-_{(f_+)_{\bar{M}}} \cdot (M \times R_1) = \mathcal{O}^-_{f_+} \circ (\bar{M} \times R_1) \cdot (M \times R_1) =$$

$= \mathcal{O}^-_{f_+} \cdot (M\bar{M} \times R_1)$, and similarly $\mathcal{O}^-_{(f_{\bar{M}})_-} \cdot (M \times R_1) = \mathcal{O}^-_{f_-} \cdot (M\bar{M} \times R_1)$.

Therefore $\mathcal{O}^-_{(f_{\bar{M}})_+} \cdot (M \times R_1)$ and $\mathcal{O}^-_{(f_{\bar{M}})_-} \cdot (M \times R_1)$ are measurable and of finite measure with respect to $\nu^*_I(\mathcal{M})$ for every $M \in MS'_{\nu*}$. Theorem 11.2.17 with $\mathcal{R} = MS'_{\nu*}$ thus gives the summability of $(f_{\bar{M}})_+(x)$ and $(f_{\bar{M}})_-(x)$ with respect to $\nu(M)$ and also the relations

$$\int_{M'} (f_{\bar{M}})_+(x)d\nu(M_x) = \bar{\nu}_I(\mathcal{O}^-_{(f_{\bar{M}})_+} \circ (M' \times R_1)) = \bar{\nu}_I(\mathcal{O}^-_{f_+} \cdot (M'\bar{M} \times R_1)) =$$

$$= \int_{M'\bar{M}} f_+(x)d\nu(M_x),$$

$$\int_{M'} (f_{\bar{M}})_-(x)d\nu(M_x) = \bar{\nu}_I(\mathcal{O}^-_{(f_M)_-} \cdot (M' \times R_1)) = \bar{\nu}_I(\mathcal{O}^-_{f_-} \cdot (M'\bar{M} \times R_1)) =$$

$$= \int_{M'\bar{M}} f_-(x)d\nu(M_x).$$

Theorem 11.2.15 now gives the summability of $f_{\bar{M}}(x)$ and the relation

$$\int_{M'} f_{\bar{M}}(x)d\nu(M_x) = \int_{M'} (f_{\bar{M}})_+(x)d\nu(M_x) - \int_{M'} (f_{\bar{M}})_-(x)d\nu(M_x) =$$

$$= \int_{M'\bar{M}} f_+(x)d\nu(M_x) - \int_{M'\bar{M}} f_-(x)d\nu(M_x) = \int_{M'\bar{M}} f(x)d\nu(M_x).$$

We now remove the restriction that $M' \in MS'_{\nu*}$ and require only that $f(x)$ be summable over M'. We have to show that M' is in the domain of the maximal equivalent extension of $\mu_{\bar{M}}(M) = \int_M f_{\bar{M}}(x)d\,\nu(M_x)$. If we write $\mu(M) = \int_M f(x)d\,\nu(M_x)$ for $M \in MS'_{\nu*}$, what we have so far proved gives $\mu_{\bar{M}}(M) = \mu(M\bar{M})$ for $M \in MS'_{\nu*}$ and we have to extend this to $M = M'$. Since $\mu_{\bar{M}}(M)$ and $\mu(M)$ are absolutely continuous with respect to $\nu(M)$ and $MS'_{\nu*}$, all sets of $MS_{\nu*}$ are measurable with respect to the outer measures determined by the positive and negative variations respectively of $\mu_{\bar{M}}(M)$ and $\mu(M)$. Now $\mu_{\bar{M}}(M) = \mu_+(M\bar{M}) - \mu_-(M\bar{M})$ is a decomposition of $\mu_{\bar{M}}(M)$ into the difference of two non-negative totally-additive measure functions. Therefore Theorem 11.1.4. gives $(\mu_{\bar{M}})_+(M) \leqq \mu_+(M\bar{M})$, $(\mu_{\bar{M}})_-(M) \leqq \mu_-(M\bar{M})$ for $M \in MS'_{\nu*}$. Now since $M \in MS_{\nu*}$, there is a sequence of disjunct sets $M_i' \in MS'_{\nu*}$ such that $M' = \sum_i M_i'$. Thus we have $\sum_i (\mu_{\bar{M}})_+(M_i') \leqq \sum_i \mu_+(M_i'\bar{M}) \leqq \sum_i \mu_+(M_i')$, and similarly $\sum_i (\mu_{\bar{M}})_-(M_i') \leqq \sum_i \mu_-(M_i')$. Since M' is in the domain of the maximal equivalent extension of $\mu(M)$, the sums on the right hand sides of these inequalities have finite values. This fact, together with the remark on the measurability of M' with respect to the outer measures determined by $(\mu_{\bar{M}})_+(M)$ and $(\mu_{\bar{M}})_-(M)$, shows that M' is in the domain of the maximal equivalent extensions of $(\mu_{\bar{M}})_+(M)$ and $(\mu_{\bar{M}})_-(M)$ and hence in that of the maximal extension of $\mu_{\bar{M}}(M)$. Finally, therefore, we have

$$\tilde{\mu}_{\bar{M}}(M') = \sum_i \mu_{\bar{M}}(M_i') = \sum_i \mu(M_i'\bar{M}) = \tilde{\mu}(M'\bar{M}).$$

This completes the proof.

Definition 11.3.3. An arbitrary real $f(x)$ is called summable with respect to $\nu^*(M)$ over a set $\bar{M} \in MS_{\nu*}$ if

(a) $\bar{M} \subset D_f$

(b) $f_{\bar{M}}(x)$ is summable with respect to $\nu^*(M)$ over $\bar{M}$ in the sense of Definition 11.3.2.

We define the integral of $f(x)$ *over* $\bar{M}$ *by* $\int_{\bar{M}} f(x)d\,\nu(M_x) = \int_{\bar{M}} f_{\bar{M}}(x)d\,\nu(M_x)$.

That this definition is consistent with the previous one is an immediate consequence of Theorem 11.3.1, for if $f(x)$ is summable with respect to $\nu^*(M)$ over $\bar{M}$ in the sense of Definition 11.3.2, we may use Theorem 11.3.1 with $M' = \bar{M}$.

THEOREM 11.3.2. *The system* $\mathcal{T}$ *of sets over which an arbitrary real* $f(x)$ *is summable is a restricted Borel ring.* $\int_M f(x)d\,\nu(M_x)$ *is finite and totally additive for* $M \in \mathcal{T}$. (We can only say that $\int_M f(x)d\,\nu(M_x)$ is a measure function if $\mathcal{T}$ satisfies the condition (b) of Definition 10.1.7. This is the case if and only if $D_f = S$.)

Proof: Suppose that $f(x)$ is summable over $M, N, M_1, M_2, \ldots$. All these sets belong to $MS_{\nu*}$. Therefore so also do $M - MN$, and $\prod M_i$. Apply Theorem 11.3.1 with $\bar{M} = M - MN$, $M' = M$. We get

$$\int_{M'} f_{M-MN}(x)d\,\nu(M_x) = \int_{M-MN} f(x)d\,\nu(M_x),$$

which shows that $f(x)$ is summable over M-MN. A similar application with $\bar{M} = \prod M_i$ and $M' = M_1$ gives the summability of $f(x)$ over $\prod M_i$. Thus conditions (β_2), (γ_2) for restricted Borel rings are satisfied. As for (α_2), we prove that $f(x)$ is summable over $M_1 + M_2$ when it is summable over M_1 and M_2 where $M_1M_2 = 0$. In view of (γ_2), this implies condition (α_2) for any M_1, M_2, because $M_1 + M_2 = M_1 + (M_2 - M_1M_2)$. From Definition 11.3.3 we have that $f_{M_1}(x)$ and $f_{M_2}(x)$ are summable with respect to $\nu(M)$. Write $\mu_1(M) = \int_M f_{M_1}(x)d\,\nu(M_x)$, $\mu_2(M) = \int_M f_{M_2}(x)d\,\nu(M_x)$ for $M \in MS'_{\nu*}$. Now by Theorem 11.3.1 with $\bar{M} = M_2$, $M' = M_1$, we have that $f_{M_2}(x)$ is summable over M_1, and hence that M_1 is in the domain of the maximal equivalent extension of $\mu_2(M)$; M_2 is in this domain by

Definition 11.3.2. By the same argument, M_1 and M_2 are in the corresponding domain for $\mu_1(M)$. Now since $M_1 M_2 = 0$, $f_{M_1}(x) + f_{M_2}(x) = f_{M_1+M_2}(x)$, and from Theorem 11.2.13, 2) and Definition 11.2.7 we have

$$\mu_1(M) + \mu_2(M) = \int_M f_{M_1+M_2}(x) d\,\nu(M_x).$$

It follows from what we proved above about M_1, M_2 that their sum is in the domain of the maximal equivalent extensions of $\mu_1(M)$, $\mu_2(M)$ and hence in that of $\mu_1(M) + \mu_2(M)$. Thus $f_{M_1+M_2}(x)$ is summable over $M_1 + M_2$.

As for the total additivity of $\int_M f(x) d\,\nu(M_x)$, suppose that $\{M_i\}$ is a sequence of disjunct sets from $\mathscr{V}$ whose sum M_o also is in $\mathscr{V}$. Let $\mu(M) = \int_M f_{M_o}(x) d\,\nu(M_x)$ for $M \in MS'_{\nu *}$; let $\tilde{\mu}(M)$ be the maximal equivalent extension of $\mu(M)$. By definition $\int_{M_o} f(x) d\,\nu(M_x) = \tilde{\mu}(M_o)$. Now apply Theorem 11.3.1 with $\bar{M} = M_o$, $M' = M_i$; we get

$$\int_{M_i} f(x) d\,\nu(M_x) = \int_{M_i M_o} f(x) d\,\nu(M_x) = \int_{M_i} f_{M_o}(x) d\,\nu(M_x) = \tilde{\mu}(M_i).$$

The total additivity of $\tilde{\mu}(M)$ gives $\sum_i \int_{M_i} f(x) d\,\nu(M_x) = \sum_i \tilde{\mu}(M_i) = \tilde{\mu}(M_o) = \int_{M_o} f(x) d\,\nu(M_x)$.

It is of use later on to establish an extension of the connection between non-negative summable functions and ordinate sets.

Definition 11.3.4. By an ordinate set of an arbitrary real non-negative $f(x)$, we shall mean an ordinate set of $f_{D_f}(x)$ in the sense of Definition 11.2.8. $\mathcal{O}_f^-$ will be defined as $\mathcal{O}_{f_{D_f}}^-$, $\mathcal{O}_f^+$ as $\mathcal{O}_{f_{D_f}}^+$.

Definition 11.3.5. A function $f(x)$ is called measurable over a set $\bar{M} \subset D_f$ if $f_{\bar{M}}(x)$ is measurable.

THEOREM 11.3.3: Suppose that $f(x)$ is an arbitrary non-negative function, *and suppose that* $\bar{M} \in MS_{\nu *}$ *is such that* $\bar{M} \subset D_f$. *Suppose further that for some ordinate set* $\mathcal{O}_f$, $\mathcal{O}_f \cdot (\bar{M} \times R_1)$ *is measurable and of finite measure with respect to* $\bar{\nu}_I^*(\mathcal{M})$. *Then* $f(x)$ *is summable over* $\bar{M}$, *and* $\int_{\bar{M}} f(x) d\,\nu(M_x) = \bar{\nu}_I(\mathcal{O}_f \cdot (\bar{M} \times R_1))$. *Conversely, if* $f(x)$ *is summable over* $\bar{M}$, *then for every* $\mathcal{O}_f$, $\mathcal{O}_f \cdot (\bar{M} \times R_1)$ *is measurable and of finite measure with respect to* $\nu_I^*(\mathcal{M})$.

Proof: We prove the first statement. Suppose $\mathcal{O}_f \cdot (\bar{M} \times R_1)$ is measurable. $\mathcal{O}_f \cdot (\bar{M} \times R_1)$ is obviously an ordinate set for $f_{\bar{M}}(x)$. Therefore $f_M(x)$ is measurable (Theorem 11.2.19). By Theorem 11.2.16 (Corollary) all ordinate sets differ from each other by zero sets. Hence $\mathcal{O}^-_{f_{\bar{M}}}$ is measurable and $\bar{\nu}_I(\mathcal{O}^-_{f_{\bar{M}}}) = \bar{\nu}_I(\mathcal{O}^-_{f_{\bar{M}}} \cdot (\bar{M} \times R_1)) = \bar{\nu}_I(\mathcal{O}_f \cdot (\bar{M} \times R_1))$. Now if $M \subset S - \bar{M}$, $\mathcal{O}^-_{f_{\bar{M}}} \cdot (M \times R_1) = 0$. Therefore if $M \in MS'_{\nu *}$, $\bar{\nu}_I(\mathcal{O}^-_{f_{\bar{M}}} \cdot (M \times R_1) =$ $= \bar{\nu}_I(\mathcal{O}^-_{f_{\bar{M}}} \cdot (M\bar{M} \times R_1)) + \bar{\nu}_I(\mathcal{O}^-_{f_M} \cdot (M - M\bar{M} \times R_1)) \leqq \bar{\nu}_I(\mathcal{O}^-_{f_{\bar{M}}} \cdot (\bar{M} \times R_1))$. Use of Theorem 11.2.17 with $\mathcal{R} = MS'_{\nu *}$ gives the summability of $f_{\bar{M}}(x)$ with respect to $\nu(M)$ on $MS'_{\nu *}$ and also the relation $\mu(M) = \int_M f_{\bar{M}}(x) d\,\nu(M_x) =$ $= \bar{\nu}_I(\mathcal{O}^-_{f_{\bar{M}}} \cdot (M \times R_1))$ for $M \in MS'_{\nu *}$. Now $\bar{M}$ as a set from $MS_{\nu *}$ is the sum of a sequence of disjunct sets from $MS'_{\nu *}$, say $M_1, M_2, \ldots$. We therefore have $\sum_i \mu(M_i) = \bar{\nu}_I(\mathcal{O}^-_{f_{\bar{M}}} \cdot (\bar{M} \times R_1)) = \nu_I(\mathcal{O}_f \cdot (\bar{M} \times R_1))$. Thus $\tilde{\mu}(\bar{M})$ exists, and we have $\int_{\bar{M}} f(x) d\,\nu(M_x) = \int_{\bar{M}} f_{\bar{M}}(x) d\,\nu(M_x) = \tilde{\mu}(M) = \bar{\nu}_I(\mathcal{O}_f \cdot (\bar{M} \times R_1))$.

For the second part of the theorem, if $f(x)$ is summable over $\bar{M}$, then $f_{\bar{M}}(x)$ is summable with respect to $\nu(M)$ over $MS'_{\nu *}$, and, by Theorem 11.2.18 with $\mathcal{R} = MS'_{\nu *}$, $\mu(M) = \int_M f_{\bar{M}}(x) d\,\nu(M_x) = \bar{\nu}_I(\mathcal{O}^-_{f_M} \circ (M \times R_1))$. We can now write $\bar{M}$ as the sum of a sequence of disjunct sets from $MS'_{\nu *}$ and use the existence of $\tilde{\mu}(\bar{M})$ to prove the finiteness of $\bar{\nu}_I(\mathcal{O}^-_f \cdot (\bar{M} \times R_1))$. Now if $\mathcal{O}_f$ is any ordinate set of $f(x)$, then $\mathcal{O}_f \cdot (\bar{M} \times R_1)$ is an ordinate set of $f_{\bar{M}}(x)$ and thus $\mathcal{O}_f \cdot (\bar{M} \times R_1)$ differs from $\mathcal{O}^-_f \cdot (\bar{M} \times R_1)$ by a zero set. Hence $\mathcal{O}_f \cdot (\bar{M} \times R_1)$

is measurable and of finite measure with respect to $\nu_I^*(\mathcal{M})$.

Corollary: *If $f(x)$ is non-negative and measurable over $\bar{M}$, and if $g(x)$ is summable over $\bar{M}$ and $f(x) \leqq g(x)$ for $x \in \bar{M}$, then $f(x)$ is summable over $\bar{M}$ and* $\int_{\bar{M}} f(x) d\, \nu(M_x) \leqq \int_{\bar{M}} g(x) d\, \nu(M_x)$.

Proof: Under these circumstances $\mathcal{O}^-_{f_{\bar{M}}} \subset \mathcal{O}^-_{g_{\bar{M}}}$ and we can apply the first part of the theorem making use of Theorem 11.2.16 and the fact that $\bar{\nu}_I(\mathcal{O}^-_f \cdot (\bar{M} \times R_1)) \leqq \bar{\nu}_I(\mathcal{O}^-_g \cdot (\bar{M} \times R_1))$.

Definition 11.3.6. *If $f(x)$ is non-negative and measurable over a set $\bar{M} \in MS_{\nu^*}$, and if $\mathcal{O}^+_f \cdot (\bar{M} \times R_1)$ has an infinite measure with respect to $\nu_I^*(\mathcal{M})$, we say that the integral of $f(x)$ over $\bar{M}$ is infinite and write* $\int_{\bar{M}} f(x) d\, \nu(M_x) = +\infty$.

It is to be remarked that with this added convention the formula

$$\int_{\bar{M}} f(x) d\, \nu(M_x) = \nu_I(\mathcal{O}_f \cdot (\bar{M} \times R_1))$$

is valid whenever $f(x)$ is non-negative and is measurable over $\bar{M}$, but it must be emphasized that we allow infinite integrals only in the non-negative case. The word "summable" is reserved strictly for the sense of Definitions 11.3.2, 11.3.3.

THEOREM 11.3.4. *A real $f(x)$ is summable over $\bar{M} \in MS_{\nu^*}$ if and only if $f_+(x)$ and $f_-(x)$ are.*

Proof: Suppose $f(x)$ summable over $\bar{M}$. Then $f_{\bar{M}}(x)$ is summable with respect to $\nu(M)$. Theorem 11.2.15 shows that $(f_M)_+(x)$ and $(f_M)_-(x)$ are also summable, and that if we write for $M \in MS'_{\nu^*}$, $\mu(M) = \int_M f_{\bar{M}}(x) d\, \nu(M_x)$, then $\mu_+(M) = \int_M (f_{\bar{M}})_+(x) d\, \nu(M_x)$ and $\mu_-(M) = \int_M (f_{\bar{M}})_-(x) d\, \nu(M_x)$. Since $\bar{M}$ belongs to the domain of $\tilde{\mu}(M)$, it must also belong to those of $\tilde{\mu}_+(M)$ and $\tilde{\mu}_-(M)$. Since $(f_{\bar{M}})_+(x) = (f_+)_{\bar{M}}(x)$ and $(f_{\bar{M}})_-(x) = (f_-)_{\bar{M}}(x)$, this says that $f_+(x)$ and $f_-(x)$ are summable over $\bar{M}$. The converse is proved by simply retracing these steps, as Theorem 11.2.15 works both ways.

We are now in a position to define summability and the integral for complex valued functions.

Definition 11.3.7. *A complex valued function $f(x)$ is called summable (with respect to $\nu^*(M)$) over a set $\bar{M} \in MS_{\nu*}$ if $\mathcal{R} f(x)$ and $\mathcal{I} f(x)$ are summable over $\bar{M}$. If $f(x)$ is summable over $\bar{M}$ we define the integral over $\bar{M}$ by means of* $\int_{\bar{M}} f(x) d\,\nu(M_x) = \int_{\bar{M}} \mathcal{R} f(x) d\,\nu(M_x) + i \int_{\bar{M}} \mathcal{I} f(x) d\,\nu(M_x)$.

THEOREM 11.3.5. *If $f(x)$, $g(x)$ are real or complex functions whose domains coincide and which are summable over a set $\bar{M} \in MS_{\nu*}$, and if c is any real or complex constant, then $cf(x)$ and $f(x) + g(x)$ are summable over $\bar{M}$ and*

1) $\int_{\bar{M}} cf(x) d\,\nu(M_x) = c \int_{\bar{M}} f(x) d\,\nu(M_x)$

2) $\int_{\bar{M}} (f(x) + g(x)) d\,\nu(M_x) = \int_{\bar{M}} f(x) d\,\nu(M_x) + \int_{\bar{M}} g(x) d\,\nu(M_x)$.

3) *If further, $f(x)$ and $g(x)$ are real, $\mathrm{Max}(f(x), g(x))$, $\mathrm{Min}(f(x), g(x))$ are summable over $\bar{M}$.*

4) *$f(x) \equiv 1$ is summable over every $\bar{M} \in MS'_{\nu*}$ and we have* $\int_{\bar{M}} 1 d\,\nu(M_x) = \nu(\bar{M})$

5) *If $f(x) \sim g(x)$ in $\bar{M}$, then* $\int_{\bar{M}} f(x) d\,\nu(M_x) = \int_{\bar{M}} g(x) d\,\nu(M_x)$.

6) *If $f(x) \geqq 0$ in $\bar{M}$, and if $\int_{\bar{M}} f(x) d\,\nu(M_x) = 0$, then $f(x) \sim 0$ in $\bar{M}$.*

Proof: Assume first that c, $f(x)$, $g(x)$ are real. Form for $M \in MS'_{\nu*}$, the functions $\lambda(M) = \int_M f_{\bar{M}}(x) d\,\nu(M_x)$, $\mu(M) = \int_M g_{\bar{M}}(x) d\,\nu(M_x)$. Theorem 11.2.13, 1), 2) and Definition 11.2.7, then give $c\lambda(M) = \int_M cf_{\bar{M}}(x) d\,\nu(M_x)$ and $\lambda(M) + \mu(M) = \int_M (f_{\bar{M}}(x) + g_{\bar{M}}(x)) d\,\nu(M_x)$. Now $c \cdot f_{\bar{M}}(x) = (cf)_{\bar{M}}(x)$ and $(f_{\bar{M}}(x) + g_{\bar{M}}(x)) = (f + g)_{\bar{M}}(x)$. If $\bar{M}$ is in the domain of $\tilde{\lambda}(M)$ and $\tilde{\mu}(M)$, it is certainly in that of $\widetilde{\lambda + \mu}(M)$ and $\widetilde{c\lambda}(M)$. This is an easy consequence of Theorem 11.1.12. Therefore we have the summability of $cf(x)$ and $f(x) + g(x)$ over $\bar{M}$, and furthermore we have

$$\int_{\bar{M}} c \cdot f(x) d\nu(M_x) = \int_{\bar{M}} (c \cdot f)_{\bar{M}}(x) d\nu(M_x) = \int_{\bar{M}} c \cdot f_{\bar{M}}(x) d\nu(M_x) =$$

$$= c\,\tilde{\lambda}(\bar{M}) = c \int_{\bar{M}} f(x) d\nu(M_x).$$

Similarly

$$\int_{\bar{M}} (f(x) + g(x)) d\nu(M_x) = \int_{\bar{M}} f(x) d\nu(M_x) + \int_{\bar{M}} g(x) d\nu(M_x).$$

For the complex case we use what we have just proved on the real and imaginary parts separately.

As for the summability of Max $(f(x),g(x))$ and Min $(f(x),g(x))$, we simply apply what we have just proved together with Theorem 11.3.4 to the identities

$$\text{Max } (f(x),g(x)) = f(x) + h_+(x)$$

$$\text{Min } (f(x),g(x)) = f(x) - h_-(x)$$

where $h(x) = g(x) - f(x)$.

The statement 4) is a direct consequence of 5), Theorem 11.2.13.

If $f(x) \sim g(x)$ in $\bar{M}$, then $\lambda(M)$, $\mu(M)$ above are identical in $MS'_{\nu*}$ and thus have identical maximal equivalent extensions. This proves 5).

To prove statement 6), we observe that since $f(x) \geqq 0$, we must have $\tilde{\lambda}(\bar{M}) \geqq 0$ (Corollary to Theorem 11.3.3). Therefore, for $M \subset \bar{M}$, $\tilde{\lambda}(M) = 0$. Obviously $\tilde{\lambda}(M)$ vanishes for $M \subset S - \bar{M}$. Thus $\tilde{\lambda}(M) = \lambda(M) \equiv 0$, and by Theorem 11.2.12, $f_{\bar{M}}(x) \sim 0$; i.e., $f(x) \sim 0$ in $\bar{M}$.

<u>THEOREM 11.3.6. A real or complex $f(x)$ is summable over $\bar{M} \in MS_{\nu*}$ if and only if it is measurable over $\bar{M}$ and $|f(x)|$ is summable over $\bar{M}$.</u>

Proof: First suppose $f(x)$ real. If $f(x)$ is summable over $\bar{M}$, it is measurable over $\bar{M}$, $f_+(x)$ and $f_-(x)$ are summable over $\bar{M}$ and, by Theorem 11.3.5, $|f(x)| = f_+(x) + f_-(x)$ is summable over $\bar{M}$. If, conversely, $f(x)$ is measurable over $\bar{M}$ and $|f(x)|$ is summable over $\bar{M}$, then $f_+(x)$, $f_-(x)$ are measurable over $\bar{M}$

and, since $f(x) \leqq f_+(x)$, $f_-(x) \leqq f(x)$, the Corollary to Theorem 11.3.3 proves that $f_+(x)$, $f_-(x)$ are both summable over $\bar{M}$. Hence by Theorem 11.3.4, $f(x)$ is summable over $\bar{M}$.

If $f(x)$ is complex and summable over $\bar{M}$, then $\mathcal{R}f(x)$ and $\mathcal{J}f(x)$ are summable over $\bar{M}$. Thus $\mathcal{R}f_{\bar{M}}(x) = (\mathcal{R}f)_{\bar{M}}(x)$ and $\mathcal{J}f_{\bar{M}}(x) = (\mathcal{J}f)_M(x)$ are measurable. Now $|f|_{\bar{M}}(x) = |f_{\bar{M}}(x)| = \sqrt{|\mathcal{R}f_{\bar{M}}(x)|^2 + |\mathcal{J}f_{\bar{M}}(x)|^2}$ which by Theorem 11.2.7, (ii),(iii), is measurable. Therefore $|f(x)|$ is measurable over $\bar{M}$. Since $|f(x)| \leqq |\mathcal{R}f(x)| + |\mathcal{J}f(x)|$, and since by the present theorem for real $f(x)$'s, $|\mathcal{R}f(x)|$ and $|\mathcal{J}f(x)|$ are both summable over $\bar{M}$, it follows from Theorem 11.3.5 that $|\mathcal{R}f(x)| + |\mathcal{J}f(x)|$ is summable over $\bar{M}$. Therefore, from the Corollary to Theorem 11.3.3, $|f(x)|$ is summable over $\bar{M}$. If conversely $f(x)$ is measurable over $\bar{M}$ and $|f(x)|$ is summable over $\bar{M}$, then by observing that $|\mathcal{R}f(x)| \leqq |f(x)|$ and $|\mathcal{J}f(x)| \leqq |f(x)|$ we are able to deduce the summability of $|\mathcal{R}f(x)|$, $|\mathcal{J}f(x)|$, hence that of $\mathcal{R}f(x)$, $\mathcal{J}f(x)$, hence that of $f(x)$.

We now turn our attention to the behavior of the integral under limiting processes on $f(x)$.

THEOREM 11.3.7. *Let $\bar{M}$ be a fixed set from $\mathbf{MS}_{\nu*}$. Let $f_1(x), f_2(x), \ldots$ be a sequence of real or complex functions each summable over $\bar{M}$. Assume that there exists a function $f(x)$ such that $D_f \supset \bar{M}$ and such that $\lim_{n\to\infty} f_n(x) = f(x)$ for all $x \in \bar{M}$ except possibly a zero set. Assume that there exists a non-negative function $g(x)$ summable over $\bar{M}$ and such that $|f_n(x)| \leqq g(x)$. Under these circumstances, $f(x)$ is summable over $\bar{M}$ and we have*

$$\lim_{n\to\infty} \int_{\bar{M}} f_n(x)d\,\nu(M_x) = \int_{\bar{M}} f(x)d\,\nu(M_x).$$

Proof: We may always suppose that for every $x \in \bar{M}$, $\lim_{n\to\infty} f_n(x) = f(x)$, for, by redefining all the functions as zero on the exceptional set, we obtain an $f'_n(x)$, $f'(x)$ satisfying the conditions of the theorem and such that in $\bar{M}$, $f'_n(x) \sim f_n(x)$, $f'(x) \sim f(x)$. If the theorem holds for $f'_n(x)$, $f'(x)$, then it holds for $f_n(x)$, $f(x)$. We may further assume that all the $f_n(x)$'s have the same domain, for we could replace them all by $f_{n\bar{M}}(x)$. We proceed in several steps.

(i) Suppose that $f_n(x)$, $f(x)$ are real and non-negative and that $f_1(x) \leqq f_2(x) \leqq \dots \leqq f(x) = \lim_{n\to\infty} f_n(x)$. Form the sets $\mathcal{M}_n = \mathcal{O}^-_{f_n} \cdot (\bar{M} \times R_1)$ $(n = 1, 2, \dots)$, $\mathcal{M} = \mathcal{O}^-_f \cdot (\bar{M} \times R_1)$, $\mathcal{N} = \mathcal{O}^+_g \cdot (\bar{M} \times R_1)$. We then have $\mathcal{M}_1 \subset \mathcal{M}_2 \subset \dots \subset \mathcal{M} = \lim_{n\to\infty} \mathcal{M}_n$. Now by Theorem 11.3.3, $\mathcal{M}_n$ are measurable with respect to $\bar{\nu}^*_I(\mathcal{M})$ and we have $\int_{\bar{M}} f_n(x) d\,\nu(M_x) = \bar{\nu}_I(\mathcal{M}_n)$. Hence by Theorem 10.2.8 $\lim_{n\to\infty} \int_{\bar{M}} f_n(x) d\,\nu(M_x) = \lim_{n\to\infty} \bar{\nu}_I(\mathcal{M}_n) = \bar{\nu}_I(\mathcal{M})$. Since $\mathcal{M} \subset \mathcal{N}$ and $g(x)$ is summable over $\bar{M}$, $\bar{\nu}_I(\mathcal{M})$ is finite; and therefore, again by Theorem 11.3.3, $f(x)$ is summable over $\bar{M}$, and $\int_{\bar{M}} f_n(x) d\,\nu(M_x) = \bar{\nu}_I(\mathcal{M})$. This settles this case.

(ii) Suppose $f_n(x)$, $f(x)$ real and non-negative, and suppose that $f_1(x) \geqq f_2(x) \geqq \dots \geqq f(x) = \lim_{n\to\infty} f_n(x)$. Form the sets $\mathcal{M}_n = \mathcal{O}^+_{f_n} \cdot (\bar{M} \times R_1)$ $(n = 1, 2, \dots)$, $\mathcal{M} = \mathcal{O}^+_f \cdot (\bar{M} \times R_1)$. Then $\mathcal{M}_1 \supset \mathcal{M}_2 \supset \dots \supset \mathcal{M} = \lim_{n\to\infty} \mathcal{M}_n$. We now repeat a similar argument to that used in (i) above, using the finiteness of $\bar{\nu}_I(\mathcal{M}_1)$ to apply Theorem 10.2.9 instead of 10.2.8.

(iii) Suppose merely that $f_n(x)$ are real and non-negative. Define $g_{n,m}(x) = \mathrm{Min}(f_n(x), f_{n+1}(x), \dots, f_{n+m}(x))$, $h_{n,m}(x) =$ $= \mathrm{Max}\,(f_n(x), f_{n+1}(x), \dots, f_{n+m}(x))$, $g_n(x) = \lim_{m\to\infty} g_{n,m}(x)$, $h_n(x) = \lim_{m\to\infty} h_{n,m}(x)$. Since $g_{n,m}(x) \leqq f_n(x) \leqq h_{n,m}(x) \leqq g(x)$ and since the $g_{n,m}(x)$'s fall under case (ii) above, and the $h_{n,m}(x)$'s under case (i) above, we have (cf. Corollary to Theorem 11.3.3)

$$\int_{\bar{M}} g_n(x)d\,\nu(M_x) = \lim_{m\to\infty} \int_{\bar{M}} g_{n,m}(x)d\,\nu(M_x) \leqq \int_{\bar{M}} f_n(x)d\,\nu(M_x) \leqq$$

$$\leqq \lim_{m\to\infty} \int_{\bar{M}} h_{n,m}(x)d\,\nu(M_x) = \int_{\bar{M}} h_n(x)d\,\nu(M_x) \leqq \int_{\bar{M}} g(x)d\,\nu(M_x).$$

Now the $g_n(x)$'s fall under (i) above and the $h_n(x)$'s under (ii). Furthermore, $\lim_{n\to\infty} g_n(x) = \lim_{n\to\infty} h_n(x) = f(x)$. Therefore

$$\int_{\bar{M}} f(x)d\,\nu(M_x) = \lim_{n\to\infty} \int_{\bar{M}} g_n(x)d\,\nu(M_x) \leqq \underline{\lim}_{n\to\infty} \int_{\bar{M}} f_n(x)d\,\nu(M_x) \leqq$$

$$\leqq \overline{\lim}_{n\to\infty} \int_{\bar{M}} f_n(x)d\,\nu(M_x) \leqq \lim_{n\to\infty} \int_{\bar{M}} h_n(x)d\,\nu(M_x) = \int_{\bar{M}} f(x)d\,\nu(M_x).$$

This proves that $\int_{\bar{M}} f(x)d\,\nu(M_x) = \lim_{n\to\infty} \int_{\bar{M}} f_n(x)d\,\nu(M_x)$.

(iv) Suppose that the $f_n(x)$'s are real. Then we have $f_+(x) = \lim_{n\to\infty} f_{n+}(x)$ and $f_-(x) = \lim_{n\to\infty} f_{n-}(x)$ together with $|f_{n+}(x)| \leqq g(x)$, $|f_{n-}(x)| \leqq g(x)$. Applying (iii) we get

$$\int_{\bar{M}} f(x)d\,\nu(M_x) = \int_{\bar{M}} f_+(x)d\,\nu(M_x) - \int_{\bar{M}} f_-(x)d\,\nu(M_x) = \lim_{n\to\infty} \int_{\bar{M}} f_{n+}(x)d\,\nu(M_x) -$$

$$- \lim_{n\to\infty} \int_{\bar{M}} f_{n-}(x)d\,\nu(M_x) = \lim_{n\to\infty} \left(\int_{\bar{M}} f_{n+}(x)d\,\nu(M_x) - \int_{\bar{M}} f_{n-}(x)d\,\nu(M_x)\right) =$$

$$= \lim_{n\to\infty} \int_{\bar{M}} f_n(x)d\,\nu(M_x).$$

(v) Suppose finally that the $f_n(x)$'s are complex. Then $\lim_{n\to\infty} \mathcal{R} f_n(x) = \mathcal{R} f(x)$, $\lim_{n\to\infty} \mathcal{J} f_n(x) = \mathcal{J} f(x)$, and $|\mathcal{R} f_n(x)| \leqq g(x)$, $|\mathcal{J} f_n(x)| \leqq g(x)$. We can now apply the argument of (iv) to the real and imaginary parts separately.

Corollary 1: If $f_n(x)$ are real and non-negative, if $f_1(x) \leqq f_2(x) \leqq \dots \leqq f(x) = \lim_{n\to\infty} f_n(x)$, and if the $f_n(x)$ are measurable over $\bar{M}$, then we have

$$\lim_{n\to\infty} \int_M f_n(x)d\,\nu(M_x) = \int_{\bar{M}} f(x)d\,\nu(M_x)$$

irrespective of the intervention of a summable majorant $g(x)$.

Proof: By the remark following Definition 11.3.6, the argument in (i) of the proof of the theorem extends to the proof of this corollary by the omission of reference to $g(x)$.

Corollary 2: If $f_n(x)$ are real, non-negative and measurable over $\bar{M}$, then

$$\int_{\bar{M}} (\underline{\lim}_{n\to\infty} f_n(x))d\,\nu(M_x) \leqq \underline{\lim}_{n\to\infty} \int_{\bar{M}} f_n(x)d\,\nu(M_x).$$

Proof: Let $f(x) = \underline{\lim}_{n\to\infty} f_n(x)$. Define $g_n(x) = \text{g.l.b.}_{k>n} (f_k(x))$. Then we have $g_1(x) \leqq g_2(x) \leqq \dots \leqq f(x) = \lim_{n\to\infty} g_n(x)$. Since $g_n(x) \leqq f_n(x)$, we have by an application of Corollary 1

$$\int_{\bar{M}} (\underline{\lim}_{n\to\infty} f_n(x)d\,\nu(M_x) = \int_{\bar{M}} f(x)d\,\nu(M_x) = \lim_{n\to\infty} \int_{\bar{M}} g_n(x)d\,\nu(M_x) =$$

$$= \underline{\lim}_{n\to\infty} \int_{\bar{M}} g_n(x)d\,\nu(M_x) \leqq \underline{\lim}_{n\to\infty} \int_{\bar{M}} f_n(x)d\,\nu(M_x).$$

This completes the discussion of the integral with respect to a fixed outer measure $\nu^*(M)$. We now investigate the behavior of the integral as we allow $\nu^*(M)$ to vary.

Lemma 1. Let $\mathcal{R}$ be a fixed half-ring in $\mathcal{P}(S)$ determining each

$\nu_n^*(M)$ of a sequence of outer measures. Let $\{a_n\}$ be a sequence of positive numbers. Suppose that for every $M \in \mathcal{R}$, $\sum_n a_n \nu_n^*(M)$ converges. Let $\nu^*(M) = \sum_n a_n \nu_n^*(M)$. (By Lemma 1 of §1. $\nu^*(M)$ is an outer measure determined by $\mathcal{R}$.) Now form in $\mathcal{S} = S \times R_1$ the outer measures $\bar{\nu}_{nI}^*(\mathcal{M})$ $(n = 1, 2, \ldots)$, $\bar{\nu}_I^*(\mathcal{M})$ corresponding respectively to $\nu_n^*(M)$, $\nu^*(M)$ in the sense of Definition 11.2.9. Form in $\mathcal{S}$ the half-ring $\bar{\mathcal{R}}$ consisting of all sets $M \times E$ where $M \in \mathcal{R}$, $E \in MS'_{m*}$. Then $\bar{\nu}_I^*(\mathcal{M}) = \sum_n a_n \bar{\nu}_{nI}^*(\mathcal{M})$ and a set is measurable with respect to $\bar{\nu}_I^*(\mathcal{M})$ if and only if it is measurable with respect to each $\bar{\nu}_{nI}^*(\mathcal{M})$.

Proof: Let $\mathcal{M} \in \bar{\mathcal{R}}$; that is, $\mathcal{M} = M \times E$ where $M \in \mathcal{R}$, $E \in MS'_{m*}$. Then $\sum_n a_n \nu_{nI}^*(\mathcal{M}) = \sum_n a_n \nu_n^*(M) \cdot m^*(E) = \nu^*(M) \cdot m^*(E) = \nu_I^*(\mathcal{M})$. Thus $\sum_n a_n \nu_{nI}^*(\mathcal{M})$ converges for every $\mathcal{M} \in \bar{\mathcal{R}}$. Define for any $\mathcal{M}$, $\bar{\mu}_I^*(\mathcal{M}) = \sum_n a_n \nu_{nI}^*(\mathcal{M})$. By Lemma 1 of §1, $\bar{\mu}_I^*(\mathcal{M})$ is determined by $\bar{\mathcal{R}}$. But $\bar{\nu}_I^*(\mathcal{M})$ is also determined by $\bar{\mathcal{R}}$, and on $\bar{\mathcal{R}}$ we have seen that $\bar{\mu}_I^*(\mathcal{M})$ and $\nu_I^*(\mathcal{M})$ agree. Eence $\bar{\mu}_I^*(\mathcal{M}) = \bar{\nu}_I^*(\mathcal{M})$. Therefore Lemma 1 of §1 applies to prove the remaining statement about measurability..

THEOREM 11.3.8. Let $\{\nu_n^*(M)\}$ be a sequence of outer measures, $\{a_n\}$ a sequence of positive numbers such that $\nu_n^*(M)$ and $\nu^*(M) = \sum_n a_n \nu_n^*(M)$ have a common determining ring. Let $\bar{M}$ be a set measurable with respect to $\nu^*(M)$ and let $f(x)$ be summable with respect to $\nu^*(M)$ over $\bar{M}$. Then $f(x)$ is summable over $\bar{M}$ with respect to every $\nu_n^*(M)$ and we have

$$\int_{\bar{M}} f(x)d\,\nu(M_x) = \sum_n a_n \int_{\bar{M}} f(x)d\,\nu_n(M_x).$$

Proof: Suppose first that $f(x)$ is non-negative. By Lemma 1 of §1, $\bar{M}$ is measurable with respect to every $\nu_n^*(M)$. By the same lemma, $f_{\bar{M}}(x)$ is easily seen to be measurable with respect to every $\nu_n^*(M)$. Now form the $\bar{\nu}_I^*(\mathcal{M})$,

$\nu^*_{nI}(\mathcal{M})$ as above. By Theorem 11.3.3, $\mathcal{O}^-_f \circ (\bar{M} \times R_1)$ is measurable and of finite measure with respect to $\bar{\nu}^*_I(\mathcal{M})$. By Lemma 1 of the present section the same is true with respect to $\bar{\nu}^*_{nI}(\mathcal{M})$. Therefore by Theorem 11.3.3, $f(x)$ is summable with respect to each $\nu^*_n(M)$ and we have

$$(2) \quad \int_{\bar{M}} f(x) d\,\nu(M_x) = \bar{\nu}_I(\mathcal{O}^-_f \cdot (\bar{M} \times R_1)) = \sum_n a_n \bar{\nu}_{nI}(\mathcal{O}^-_f \circ (\bar{M} \times R_1)) =$$

$$= \sum_n a_n \int_{\bar{M}} f(x) d\,\nu_n(M_x) \;.$$

For real $f(x)$ summable over $\bar{M}$ with respect to $\nu^*(M)$, we have the summability of $f_+(x)$ and $f_-(x)$ over $\bar{M}$ with respect to $\nu^*(M)$ (Theorem 11.3.2). Thus we get from the above argument the summability over $\bar{M}$ of $f_+(x)$ and $f_-(x)$ with respect to each $\nu^*_n(M)$ and also get the equation (2) for $f_+(x)$ and $f_-(x)$. Hence

$$\int_{\bar{M}} f(x) d\,\nu(M_x) = \int_{\bar{M}} f_+(x) d\,\nu(M_x) - \int_{\bar{M}} f_-(x) d\,\nu(M_x) = \sum_n a_n \int_{\bar{M}} f_+(x) d\,\nu_n(M_x) -$$

$$- \sum_n a_n \int_{\bar{M}} f_-(x) d\,\nu_n(M_x) = \sum_n a_n \int_{\bar{M}} f(x) d\,\nu_n(M_x).$$

For the complex case we do the usual trick of splitting into real and imaginary parts.

<u>THEOREM 11.3.9.</u> <u>Let</u> a_n, $\nu^*_n(M)$, $\nu^*(M)$ <u>be as in Theorem 11.3.8. Let</u> $\bar{M}$ <u>be measurable with respect to every</u> $\nu^*_n(M)$, <u>and let</u> $f(x)$ <u>be a function summable over</u> $\bar{M}$ <u>with respect to every</u> $\nu^*_n(M)$. <u>If now the series</u> $\sum_n a_n \int_{\bar{M}} |f(x)| d\,\nu_n(M_x)$ <u>is convergent, then</u> $f(x)$ <u>is summable over</u> $\bar{M}$ <u>with respect to</u> $\nu^*(M)$.

Proof: By Lemma 1 above, $\bar{M}$ is measurable with respect to $\nu^*(M)$; also $f(x)$ is measurable over $\bar{M}$ with respect to $\nu^*(M)$. We have $\int_{\bar{M}} |f(x)| d\,\nu_n(M_x) =$

$= \bar{\nu}_{nI}(\mathcal{O}^{-}_{|f|} \cdot (\bar{M} \times R_1))$. Therefore $\bar{\nu}_I(\mathcal{O}^{-}_{|f|} \cdot (\bar{M} \times R_1)) =$

$= \sum_n a_n \bar{\nu}_{nI}(\mathcal{O}^{-}_{|f|} \cdot (\bar{M} \times R_1))$ is finite. By Theorem 11.3.3, $|f(x)|$ is summable over $\bar{M}$ with respect to $\nu^*(M)$. By Theorem 11.3.6.then, $f(x)$ is summable over $\bar{M}$ with respect to $\nu^*(M)$.

Corollary: If the $\nu_n^*(M)$ are finite in number, then $f(x)$ is summable over $\bar{M}$ with respect to $\nu^*(M)$ if and only if it is summable over $\bar{M}$ with respect to each $\nu_n^*(M)$.

Proof: By Theorem 11.3.8, 11.3.9 and the fact that the series mentioned in the hypotheses of Theorem 11.3.9 is a finite sum.

We now extend the definition of integral to the case where $\nu(M)$ is not necessarily negative. $\nu(M)$ will be required merely to be totally additive, finite, and of bounded variation over some half-ring $\mathcal{R}$.

Since the case where $\nu(M)$ is real may be regarded as a special instance of that in which $\nu(M)$ is complex, we shall state the theorems for the complex case. We shall preserve the notation $\nu(M) = \sum_{\rho=0}^{3} i^\rho \nu_\rho^{(o)}(M)$ for the special decomposition of Theorem 11.1.6. The decomposition $\nu(M) = \nu_+(M) - \nu_-(M)$ is a special case of this with $\nu_1^{(o)}(M) = \nu_3^{(o)}(M) = 0$. Recalling further that $\nu_0^{(o)}(M) = (\mathcal{R}\nu)_+(M)$, $\nu_2^{(o)}(M) = (\mathcal{R}\nu)_-(M)$, $\nu_1^{(o)}(M) = (\mathcal{J}\nu)_+(M)$, $\nu_3^{(0)}(M) = (\mathcal{J}\nu)_-(M)$, we shall sometimes give the proofs for the real case and merely indicate the extension to the complex case.

Definition 11.3.8. A function $f(x)$ is called summable with respect to $\nu(M)$ over a set $\bar{M}$ if it is summable over $\bar{M}$ with respect to each of the four outer measures $\nu_{\rho\, I}^{(o)*}(M)$ $(\rho = 0, 1, 2, 3)$, determined by $\nu_\rho^{(o)}(M)$ respectively.

Note: This definition does not conflict with any of the preceding definitions. It is true that we defined summability with respect to a non-negative totally-additive measure function $\nu(M)$ in Definition 11.2.6, and summability over a set $\bar{M}$ with respect to an outer measure $\nu^*(M)$ in Definition 11.3.3, and that since Definition 11.3.8 above is more nearly an extension of the second of these definitions, we should perhaps invent some denotation that would partake more nearly of the form of the second.

Since summability over a set $\bar{M}$ does not enter into Definition 11.2.6, no ambiguity need arise. It is obvious that in case the $\nu(M)$ in Definition 11.3.8 happens to be non-negative, then summability of $f(x)$ over $\bar{M}$ with respect to $\nu(M)$ in the sense of Definition 11.3.8, means summability of $f(x)$ over $\bar{M}$ with respect to $\nu_I^*(M)$ in the sense of Definition 11.3.3.

Note further that in the non-negative case summability over $\bar{M}$ was not defined for $\bar{M}$'s non-measurable with respect to $\nu^*(M)$. Hence Definition 11.3.8 implicitly requires that $\bar{M}$ be measurable with respect to $\nu_{\rho}^{(o)}{}_I^*(M)$ $(\rho = 0, 1, 2, 3)$.

Definition 11.3.9. If $f(x)$ is summable over $\bar{M}$ with respect to $\nu(M)$ we define the integral $\int_{\bar{M}} f(x)d\,\nu(M_x)$ as $\sum_{\rho=0}^{3} i^{\rho} \int_{\bar{M}} f(x)d\,\nu_{\rho}^{(o)}(M_x)$.

Definition 11.3.10. $f(x)$ is said to be measurable over $\bar{M}$ with respect to a $\nu(M)$ if it is measurable over $\bar{M}$ with respect to $\nu_{\rho}^{(o)}{}_I^*(M)$ $(\rho = 0, 1, 2, 3)$.

Definition 11.3.11. We say that $f(x) \sim g(x)$ in $\bar{M}$ with respect to $\nu(M)$ if $f(x) \sim g(x)$ with respect to $\nu_{\rho}^{(o)}{}_I^*(M)$ $(\rho = 0,1,2,3)$.

THEOREM 11.3.10. If $\nu(M) = \sum_{\rho=0}^{3} i^{\rho}\, \nu_{\rho}(M)$ is any decomposition of $\nu(M)$ into finite non-negative and totally-additive measure functions on $\mathcal{R}$, and if $f(x)$ is summable over $\bar{M}$ with respect to $\nu_{\rho I}^*(M)$ $(\rho = 0, 1, 2, 3)$, the outer measures determined by $\nu_{\rho}(M)$ respectively, then $f(x)$ is summable with respect to $\nu(M)$ over $\bar{M}$ and we have

$$\int_{\bar{M}} f(x)d\,\nu(M_x) = \sum_{\rho=0}^{3} i^{\rho} \int_{\bar{M}} f(x)d\,\nu_{\rho}(M_x).$$

Proof: By Theorem 11.1.6 we have $\nu_\rho(M) = \begin{cases} \nu_\rho^{(o)}(M) + \alpha(M) & (\rho \text{ even}) \\ \nu_\rho^{(o)}(M) + \beta(M) & (\rho \text{ odd}) \end{cases}$

By Theorem 11.1.7, therefore, $\nu^*_{\rho I}(M) = \begin{cases} \nu_\rho^{(o)*}{}_I(M) + \alpha^*_I(M) & (\rho \text{ even}) \\ \nu_\rho^{(o)*}{}_I(M) + \beta^*_I(M) & (\rho \text{ odd}) \end{cases}$ Thus by the corollary to Theorems 11.3.9, f(x) is summable over $\bar{M}$ with respect to $\nu_\rho^{(o)*}{}_I(M)$ and $\alpha^*_I(M)$ (resp. $\beta^*_I(M)$). This proves that f(x) is summable over $\bar{M}$ with respect to $\nu(M)$. Further by Theorem 11.3.8 we have $\int_{\bar{M}} f(x)d\,\nu_\rho(M) = \int_{\bar{M}} f(x)d\,\nu_\rho^{(o)}(M_x) + \int_{\bar{M}} f(x)d\alpha(M_x)$ (resp. $\beta(M_x)$). If we form the expression $\sum_{\rho=0}^{3} i^\rho \int_{\bar{M}} f(x)d\,\nu_\rho(M_x)$ and use these equations we obviously get $\int_{\bar{M}} f(x)d\,\nu(M_x)$ as its value.

THEOREM 11.3.11. *If f(x) is summable over $\bar{M}$ with respect to $\nu(M)$ and $\nu'(M)$, then it is also summable with respect to $\nu(M) + \nu'(M)$, and we have $\int_{\bar{M}} f(x)d(\nu(M_x) + \nu'(M_x)) = \int_{\bar{M}} f(x)d\,\nu(M_x) + \int_{\bar{M}} f(x)d\,\nu'(M_x)$.*

Proof: $\nu(M) + \nu'(M) = \sum_{\rho=0}^{3} i^\rho(\nu_\rho^{(o)}(M) + \nu_\rho'^{(o)}(M))$ is obviously a decomposition of $\nu(M) + \nu'(M)$ of the type described in the preceding theorem. By Theorems 11.3.8, 11.3.9, we have the summability of f(x) over $\bar{M}$ with respect to $\nu_\rho^{(o)}(M) + \nu_\rho'^{(o)}(M)$ (ρ = 0, 1, 2, 3), and the relations

$\int_{\bar{M}} f(x)d(\nu_\rho^{(o)}(M_x) + \nu_\rho'^{(o)}(M_x)) = \int_{\bar{M}} f(x)d\,\nu_\rho^{(o)}(M_x) + \int_{\bar{M}} f(x)d\,\nu_\rho'^{(o)}(M_x)$. If we multiply the ρ'th of these equations by i^ρ, add over ρ = 0, 1, 2, 3, and apply Theorem 11.3.10, we get the relation $\int_{\bar{M}} f(x)d(\nu(M_x) + \nu'(M_x)) = \int_{\bar{M}} f(x)d\nu(M_x) + \int_{\bar{M}} f(x)d\,\nu'(M_x)$.

THEOREM 11.3.12. *If c is any constant, and if f(x) is summable over*

$\bar{M}$ _with respect to_ $\nu(M)$, _then it is summable with respect to_ $c\,\nu(M)$ _and we have_

$$\int_{\bar{M}} f(x)d(c\,\nu(M_x)) = c\int_{\bar{M}} f(x)d\,\nu(M_x).$$

Proof: Let $\lambda(M) = c\,\nu(M)$. If $c \geqq 0$, then $\lambda_\rho^{(o)}(M) = c\,\nu_\rho^{(o)}(M)$, and by Theorem 11.3.8 we have $\int_{\bar{M}} f(x)d\,\lambda_\rho^{(o)}(M) = c\int_{\bar{M}} f(x)d\,\nu_\rho^{(o)}(M)$. Multiplying i^ρ and summing $\rho = 0, 1, 2, 3$, we get the theorem for this case. If $c = i$, the $\lambda_\rho^{(o)}(M)$ are permuted cyclically, and it turns out easily that $\int_{\bar{M}} f(x)d(i\nu(M_x)) = i\int_{\bar{M}} f(x)d\nu(M_x)$.

The result for any complex c follows from these cases by an appeal to Theorem 11.3.11.

THEOREM 11.3.13. $f(x)$ is summable over $\bar{M}$ with respect to $\nu(M)$ if and only if $f(x)$ is measurable with respect to $\nu(M)$ over $\bar{M}$, and $|f(x)|$ is summable with respect to $\bar{\nu}(M)$ over $\bar{M}$, where $\bar{\nu}(M)$ is the total variation function of $\nu(M)$.

Proof: Suppose that $f(x)$ is measurable with respect to $\bar{\nu}(M)$ and $|f(x)|$ is summable with respect to $\bar{\nu}(M)$ over $\bar{M}$. Then since $\bar{\nu}(M) = \sum_{\rho=0}^{3}\nu_\rho^{(o)}(M)$ and thus $\bar{\nu}_I^*(M) = \sum_{\rho=0}^{3}\nu_{\rho\ I}^{(o)*}(M)$, we have by Theorem 11.3.8 the measurability of $f(x)$ over $\bar{M}$ with respect to each $\nu_{\rho\ I}^{(o)*}(M)$ and the summability of $|f(x)|$ over $\bar{M}$ with respect to each $\nu_{\rho\ I}^{(o)*}(M)$. By Theorem 11.3.6 $f(x)$ is summable with respect to each $\nu_{\rho\ I}^{(o)*}(M)$. But this means summability with respect to $\nu(M)$.

Conversely, if $f(x)$ is summable over $\bar{M}$ with respect to $\nu(M)$, it is summable over $\bar{M}$ with respect to each $\nu_{\rho\ I}^{(o)*}(M)$; hence by Theorem 11.3.8 with respect to $\bar{\nu}_I^*(M)$. But by Theorem 11.3.6 this implies that $|f(x)|$ is summable with respect to $\bar{\nu}_I^*(M)$, i.e., with respect to $\nu(M)$. Also since $f(x)$ is measurable over $\bar{M}$ with respect to each $\nu_{\rho\ I}^{(o)*}(M)$, it is by Theorem 11.3.8 measurable with respect to $\bar{\nu}^*_I(M)$ over $\bar{M}$.

THEOREM 11.3.14. *If $f(x)$, $g(x)$ are summable over $\bar{M}$ with respect to $\nu(M)$, and if c is any constant, then $cf(x)$ and $f(x) + g(x)$ are summable with respect to $\nu(M)$ over $\bar{M}$, and we have*

1) $\int_{\bar{M}} cf(x)d\,\nu(M_x) = c\int_{\bar{M}} f(x)d\,\nu(M_x)$

2) $\int_{\bar{M}}(f(x) + g(x))d\,\nu(M_x) = \int_{\bar{M}} f(x)d\,\nu(M_x) + \int_{\bar{M}} g(x)d\,\nu(M_x)$.

Proof: The summability of $cf(x)$ and $f(x) + g(x)$ over $\bar{M}$ with respect to $\nu^{(o)}_\rho(M)$ imply by Theorem 11.3.5 the relations 1), 2), above, with $\nu(M)$ replaced by $\nu^{(o)}_\rho(M)$. We then multiply by the ρ'th of these with i^ρ and add.

THEOREM 11.3.15. *If $\{f_n(x)\}$ is a sequence of functions each summable with respect to $\nu(M)$ over $\bar{M}$ and if $g(x)$ is summable over $\bar{M}$ and $|f_n(x)| \leqq g(x)$, and if $f(x) = \lim_{n\to\infty} f_n(x)$ except on a zero set with respect to $\bar{\nu}^*_I(M)$, then $f(x)$ is summable over $\bar{M}$ and*

$$\lim_{n\to\infty}\int_{\bar{M}} f_n(x)d\,\nu(M_x) = \int_{\bar{M}} f(x)d\,\nu(M_x).$$

Proof: Observing that a zero set with respect to $\bar{\nu}^*_I(M)$ is a zero set with respect to each $\nu^{(o)*}_{\rho\ I}(M)$, we employ Theorem 11.3.7 to prove

$$\lim_{n\to\infty}\int_{\bar{M}} f_n(x)d\,\nu^{(o)}_\rho(M_x) = \int_{\bar{M}} f(x)d\,\nu^{(o)}_\rho(M_x).$$

We can then multiply the ρ'th relation by i^ρ and add.

THEOREM 11.3.16. *If $\{\nu_n(M)\}$ is a sequence of totally-additive measure functions of bounded variation over $\mathcal{R}$, and is further such that for $M \in \mathcal{R}$, $\sum_n \bar{\nu}_n(M)$ is convergent, and if $f(x)$ is summable over $\bar{M}$ with respect to each $\nu_n(M)$, and the series $\sum_n \int_{\bar{M}} |f(x)|\, d\,\bar{\nu}_n(M_x)$ is convergent, then*

$$\sum_n \int_{\bar{M}} f(x)d\, \nu_n(M_x) = \int_{\bar{M}} f(x)d\, \nu(M_x)$$

where $\nu(M) = \sum_n \nu_n(M)$.

Proof: On the assumption of the convergence of $\sum_n \bar{\nu}_n(M)$ for every $M \in \mathcal{R}$, we have that of $\sum_n \nu_{n\rho}^{(o)}(M)$ for $M \in \mathcal{R}$; on the assumption of the convergence of $\sum_n \int_{\bar{M}} |f(x)|\, d\, \bar{\nu}_n(M_x)$ we have that of $\sum_n \int_{\bar{M}} |f(x)|\, d\, \nu_{n\rho}^{(o)}(M_x)$. Thus an application of Theorem 11.3.9 gives

$$\sum_n \int_{\bar{M}} f(x)d\, \nu_n(M_x) = \sum_{\rho=o}^{3} i^{\rho} \sum_n \int_{\bar{M}} f(x)d\, \nu_{n\rho}^{(o)}(M_x) =$$

$$= \sum_{\rho=o}^{3} i^{\rho} \int_{\bar{M}} f(x)d(\sum_n \nu_{n\rho}^{(o)}(M_x)).$$

Since $\nu(M) = \sum_{\rho=o}^{3} i^{\rho}(\sum_n \nu_{n\rho}^{(o)}(M_x))$ is a decomposition of $\nu(M)$, Theorem 11.3.10 shows that the last expression is equal to $\int_{\bar{M}} f(x)d\, \nu(M_x)$.

We conclude this section by listing the elementary properties of the integral:

1) $\int_{\bar{M}} c \cdot f(x)d\, \nu(M_x) = c \int_{\bar{M}} f(x)d\, \nu(M_x)$

2) $\int_{\bar{M}} (f(x) + g(x))d\, \nu(M_x) = \int_{\bar{M}} f(x)d\, \nu(M_x) + \int_{\bar{M}} g(x)d\, \nu(M_x)$

3) $\int_{\bar{M}} f(x)d(c\, \nu(M_x)) = c \int_{\bar{M}} f(x)d\, \nu(M_x)$

4) $\int_{\bar{M}} f(x)d(\mu(M_x) + \nu(M_x)) = \int_{\bar{M}} f(x)d\mu(M_x) + \int_{\bar{M}} f(x)d\, \nu(M_x)$

5) $\int_{\bar{M}} 1\, d\, \nu(M_x) = \nu(\bar{M})$

6) $\int_{\bar{M}} f_M(x)d\, \nu(M_x) = \int_{\bar{M}M} f(x)d\, \nu(M_x)$

7) If $\nu(M)$ is non-negative and $f(x) \geqq 0$, then $\int_{\bar{M}} f(x)d\,\nu(M_x) \geqq 0$. If equality holds here, then $f(x) \sim 0$ in $\bar{M}$.

8) $\sum_i \int_{\bar{M}_i} f(x)d\,\nu(M_x) = \int_{\sum \bar{M}_i} f(x)d\,\nu(M_x) \quad (M_iM_j = 0,\ i \neq j)$

9) $\int_{\bar{M}} f(x)d\,\nu(M_x)$ exists in the strict sense (i.e,, infinite integrals disallowed) if and only if $f(x)$ is measurable with respect to $\bar{\nu}(M)$ over $\bar{M}$ and $\int_{\bar{M}} |f(x)|\,d\,\bar{\nu}(M_x)$ is finite, where $\bar{\nu}(M)$ is the total variation function of $\nu(M)$.

10) If $\int_{\bar{M}} f_n(x)d\,\nu(M_x)$ exists for every n, and if in $\bar{M}$ except possibly on a zero set $\lim_{n\to\infty} f_n(x) = f(x)$, and if $|f_n(x)| < g(x)$, where $\int_{\bar{M}} g(x)d\,\bar{\nu}(M_x)$ is finite, then $\int_{\bar{M}} f(x)d\,\nu(M_x)$ exists and $\lim_{n\to\infty} \int_{\bar{M}} f_n(x)d\,\nu(M_x) =$
$= \int_{\bar{M}} f(x)d\,\nu(M_x)$.

11) If $\nu(M)$ is non-negative and $f_n(x)$ are non-negative and measurable with respect to $\nu(M)$ over $\bar{M}$, then

$$\int_{\bar{M}} (\underline{\lim}_{n\to\infty} f_n(x))d\,\nu(M_x) \leqq \underline{\lim}_{n\to\infty} \int_{\bar{M}} f_n(x)d\,\nu(M_x).$$

12) If $\{\nu_n(M)\}$ is a sequence of non-negative functions on a half-ring $\mathcal{R}$ and if $\nu(M) = \sum_n \nu_n(M)$ is finite for every $M \in \mathcal{R}$, then

$$\sum_n \int_{\bar{M}} f(x)d\,\nu_n(M_x) = \int_{\bar{M}} f(x)d\,\nu(M_x)$$

provided either that the right-hand side makes sense, or that the left-hand side makes sense and $\sum_n \int_{\bar{M}} |f(x)|\,d\,\nu_n(M_x)$ is convergent. If the $\nu_n(M)$ are not non-negative but if the $\bar{\nu}_n(M)$ satisfy the conditions imposed above and $\sum_n \int_{\bar{M}} |f(x)|\,d\,\bar{\nu}_n(M_x)$, then the above equality holds.

If infinite integrals are not admitted, then equations 1), 2), 3), 4), 5), 8) are to be interpreted from right to left, i.e., the left-hand sides make sense if the right-hand sides do. If infinite integrals are admitted and all functions occurring are non-negative, then the same thing is true. Equation 6) may be interpreted from both sides.

§4. The Fubini Theorem and related theorems

In this section we derive a generalization of the well known Fubini Theorem on the relation between iterated and multiple Lebesgue integrals. In view of the close connection between measure and the integral we may as well formulate the problem for measures. The Fubini theorem for Lebesgue measure in the plane is as follows: In the x, y plane let M be a superficially measurable set. On each line x = const. lies a certain portion of M. Denote the set of the y-coordinates of the points of this portion by M_x; i.e., M_x is the set of all y for which $(x,y) \in M$. Let $m^*(M_x)$ be the (linear) outer measure of M_x. Under these circumstances M_x is measurable with respect to m^* except possibly for an x-set of linear measure zero and $\int m^*(M_x)dx$ is equal to the superficial measure of M.

We now formulate the general problem: Consider two spaces S and T. x will denote a typical point of S, y a typical point of T. Form the direct product space $\mathcal{S} = S \times T$; a point of $\mathcal{S}$ is a pair (x,y) $(x \in S, y \in T)$. Now let let $\mu^*(M)$ be a regular outer measure in S, $\nu^*(N)$ a regular outer measure in T. Form in $\mathcal{S}$ the half-ring $\mathcal{R}$ consisting of all sets $\mathcal{M} = M \times N$ where $M \in MS'_{\mu^*}$, $N \in MS'_{\nu^*}$. On $\mathcal{R}$ form the measure function $\lambda(\mathcal{M}) = \mu(M) \cdot \nu(N)$ where $\mathcal{M} = M \times N$. Extend $\lambda(\mathcal{M})$ to its outer measure $\lambda^*(\mathcal{M}) = \lambda_I^*(\mathcal{M})$. So far everything we have done is a special instance of the procedure in Theorem 10.4.7.

Definition 11.4.1. Let $\mathcal{M}$ be any set of $\mathcal{S}$, x any point of S. By $\mathcal{M}_x$ we shall mean $S_y\{y \in T;\ (x,y) \in \mathcal{M}\}$. Thus $\mathcal{M}_x \subset T$. Similarly we define $\mathcal{M}_y = S_x\{x \in S;\ (x,y) \in \mathcal{M}\}$.

THEOREM 11.4.1. Let $\mathcal{M}$ be measurable with respect to $\lambda^*(\mathcal{M})$. Then

1) The set of all x's for which $\mathcal{M}_x$ is non-measurable with respect to $\nu^*(N)$ is a zero set with respect to $\mu^*(M)$.

2) The function $f(x) = \nu^*(\mathcal{M}_x)$ is measurable with respect to $\mu^*(M)$ and $\lambda(\mathcal{M}) = \int_S f(x)d\mu(M_x)$.

Proof: Call any measurable $\mathcal{M}$ for which the conclusions of the theorem hold a Fubini set ($\equiv \mathcal{F}$-set). What we have thus to prove is that every measurable $\mathcal{M}$ is an $\mathcal{F}$-set. We shall carry out the proof in several steps.

(i) Every $\mathcal{M} \in \mathcal{R}$ is an $\mathcal{F}$-set. Let $\mathcal{M} = M \times N$ where $M \in MS'_{\mu^*}$, $N \in MS'_{\nu^*}$. Now $\mathcal{M}_x = \begin{cases} N & (x \in M) \\ 0 & (x \in S-M) \end{cases}$. This shows that 1) is satisfied. As for 2) we have $f(x) = \nu^*(\mathcal{M}_x) = \begin{cases} \nu(N) & (x \in M) \\ 0 & (x \in S-M) \end{cases}$; that is, $f(x) = \nu(N) \cdot 1_M(x)$ which is measurable. Furthermore $\int_S f(x)d\mu(M_x) = \nu(N) \cdot \int_S 1_M d\mu(M_x) = \nu(N) \int_M 1\ d\mu(M_x) = \nu(N) \cdot \mu(M) = \lambda(\mathcal{M})$. Thus 2) is satisfied.

(ii) The sum of a finite or infinite sequence of disjunct $\mathcal{F}$-sets is again an $\mathcal{F}$-set. Let $\mathcal{M}^{(1)}, \mathcal{M}^{(2)}, \ldots$ be disjunct $\mathcal{F}$-sets, $\mathcal{M}$ their sum. For each x the sets $\mathcal{M}_x^{(n)}$ are disjunct and $\mathcal{M}_x = \sum_n \mathcal{M}_x^{(n)}$. Let $M_o^{(n)}$ be the set of x's where $\mathcal{M}_x^{(n)}$ is non-measurable, M_o the set of x's where $\mathcal{M}_x$ is non-measurable. If $x \in S - \sum_n M_o^{(n)}$ then each $\mathcal{M}_x^n$ is measurable and hence $\mathcal{M}_x$ is. Therefore $M_o \subset \sum_n M_o^{(n)}$. Since each $M_o^{(n)}$ is a zero set with respect to $\mu^*(M)$ it follows that M_o is also. This proves that $\mathcal{M}$ satisfies condition 1). We now prove that $\mathcal{M}$ satisfies condition 2). Suppose first that the number of $\mathcal{M}^{(n)}$'s is finite, say $\mathcal{M} = \sum_{n=1}^{\bar{n}} \mathcal{M}^{(n)}$. Form the functions $f(x) = \nu^*(\mathcal{M}_x)$,

$f_n(x) = \nu^*(\mathcal{M}_x^{(n)})$. For each $x \in S - \sum_k M_o^{(k)}$ we have $f(x) = \nu^*(\mathcal{M}_x) = \nu^*(\sum_1^{\bar{n}} \mathcal{M}_x^{(n)}) = \sum_1^{\bar{n}} \nu^*(\mathcal{M}_x^{(n)}) = \sum_1^{\bar{n}} f_n(x)$. Therefore we get, using condition 2) on each $\mathcal{M}^{(n)}$:

$$\lambda(\mathcal{M}) = \sum_1^{\bar{n}} \lambda(\mathcal{M}^{(n)}) = \sum_1^{\bar{n}} \int_S f_n(x)d\mu(M_x) = \sum_1^{\bar{n}} \int_{S-\sum_k M_o^{(k)}} f_n(x)d\mu(M_x) =$$

$$= \int_{S-\sum_k M_o^{(k)}} f(x)d\mu(M_x) = \int_S f(x)d\mu(M_x).$$

This proves condition 2) for $\mathcal{M}$ in this case. If the number of $\mathcal{M}^{(n)}$'s is infinite define $\bar{\mathcal{M}}^{(n)} = \sum_{m=1}^{n} \mathcal{M}^{(m)}$, $\bar{f}_n(x) = \nu^*(\bar{\mathcal{M}}_x^{(n)})$. Now since the $\bar{\mathcal{M}}^{(n)}$'s form an ascendant sequence, the same is true of the $\bar{\mathcal{M}}_x^{(n)}$'s for each x. Hence $\{\bar{f}_n(x)\}$ is a non-decreasing sequence. Furthermore, for $x \in S - \sum_k M_o^{(k)}$, $f(x) = \nu^*(\mathcal{M}_x) = \nu^*(\sum_n \mathcal{M}_x^{(n)}) = \sum_n \nu^*(\mathcal{M}_x^{(n)}) = \lim_{n\to\infty} \sum_{m=1}^{n} \nu^*(\mathcal{M}_x^{(m)}) = \lim_{n\to\infty} \nu^*(\bar{\mathcal{M}}_x^{(n)}) = \lim_{n\to\infty} \bar{f}_n(x)$. By what we have proved above, each $\bar{\mathcal{M}}^{(n)}$ is an $\mathcal{F}$-set and so

$$\lambda(\mathcal{M}) = \lim_{n\to\infty} \lambda(\bar{\mathcal{M}}^{(n)}) = \lim_{n\to\infty} \int_S \bar{f}_n(x)d\mu(M_x) = \int_S f(x)d\mu(M_x).$$

We have thus established the statement at the outset of this paragraph.

(iii) <u>If $\mathcal{M}$, $\mathcal{M}'$ are $\mathcal{F}$-sets with $\mathcal{M} \subset \mathcal{M}'$ and if $\lambda(\mathcal{M}')$ is finite, then $\mathcal{M}' - \mathcal{M}$ is an $\mathcal{F}$-set.</u> Let $\mathcal{M}'' = \mathcal{M}' - \mathcal{M}$. Then $\mathcal{M}''_x = \mathcal{M}'_x - \mathcal{M}_x$. Let M_o, M'_o, M''_o be the sets of all x where $\mathcal{M}_x$, $\mathcal{M}'_x$, $\mathcal{M}''_x$ are non-measurable. Obviously $M''_o \subset M'_o + M_o$ so that M''_o is a zero set with respect to $\mu^*(M)$. Thus $\mathcal{M}''$ satisfies condition 1). Now let $f(x) = \nu^*(\mathcal{M}_x)$, $f'(x) = \nu^*(\mathcal{M}'_x)$, $f''(x) = \nu^*(\mathcal{M}''_x)$. Since $\mu(\mathcal{M}')$ is finite, condition 2) on $\mathcal{M}'$ gives the finiteness of $\int_S f'(x)d\mu(M_x)$; i.e., $f'(x)$ is summable over

S with respect to $\mu^*(M)$. Hence except possibly for a zero set with respect to $\mu^*(M)$ in S, $f'(x)$ is finite. If $\bar{M}'_0$ denotes the set of x's for which $f'(x)$ is infinite, we have for $x \in S - (M'_0 + M_0 + \bar{M}'_0)$, $f''(x) = \nu^*(\mathcal{M}''_x) = \nu^*(\mathcal{M}'_x - \mathcal{M}_x) = \nu^*(\mathcal{M}'_x) - \nu^*(\mathcal{M}_x) = f'(x) - f(x)$. Thus $f''(x) \sim f'(x) - f(x)$. Since $f'(x) \geqq f(x)$ and $f'(x) \geqq f''(x)$, $f''(x)$ and $f(x)$ are summable over S with respect to $\mu^*(M)$. We then have

$$\lambda(\mathcal{M}'') = \lambda(\mathcal{M}') - \lambda(\mathcal{M}) = \int_S f'(x)d\mu(M_x) - \int_S f(x)d\mu(M_x) =$$

$$= \int_S (f'(x) - f(x))d\mu(M_x) = \int_S f''(x)d\mu(M_x)$$

and thus condition 2) is satisfied.

(iv) <u>If $\mathcal{M}$ is measurable and of finite measure with respect to $\lambda^*(\mathcal{M})$ and if there exist $\mathcal{F}$-sets $\mathcal{M}'$, $\mathcal{M}''$ such that $\mathcal{M}' \subset \mathcal{M} \subset \mathcal{M}''$ and $\lambda(\mathcal{M}'' - \mathcal{M}') = 0$, then $\mathcal{M}$ is an $\mathcal{F}$-set.</u> Let $f(x) = \nu^*(\mathcal{M}_x)$, $f'(x) = \nu^*(\mathcal{M}'_x)$, $f''(x) = \nu^*(\mathcal{M}''_x)$. Since $\mathcal{M}'_x \subset \mathcal{M}_x \subset \mathcal{M}''_x$, we have $f'(x) \leqq f(x) \leqq f''(x)$. Now using condition 2) on $\mathcal{M}'$, $\mathcal{M}''$ we have

$$0 = \lambda(\mathcal{M}'' - \mathcal{M}') = \lambda(\mathcal{M}'') - \lambda(\mathcal{M}') = \int_S f''(x)d\mu(M_x) - \int_S f'(x)d\mu(M_x) =$$

$$= \int_S (f''(x) - f'(x))d\mu(M_x).$$

Since $(f''(x) - f'(x)) \geqq 0$, we have $f''(x) - f'(x) \sim 0$, or $f''(x) \sim f'(x)$ in S. From this and $f'(x) \leqq f(x) \leqq f''(x)$ it follows that $f'(x) \sim f(x) \sim f''(x)$ in S. Denote by $\bar{M}''_0$ the set of x's where $f''(x)$ is infinite, by M'_0, M''_0 respectively the x's where $\mathcal{M}'_x$, $\mathcal{M}''_x$ are non-measurable with respect to $\nu^*(N)$. Condition 1) on $\mathcal{M}'$, $\mathcal{M}''$ shows that M'_0, M''_0 are zero sets with respect to $\mu^*(M)$. Since $\lambda(\mathcal{M}'') = \lambda(\mathcal{M})$, $\lambda(\mathcal{M}'')$ is finite; this implies that $f''(x)$ is summable over S and hence that $\bar{M}''_0$ is a zero set with respect to $\mu^*(M)$. Finally, let

$\bar{\bar{M}}_o$ be the set of x's for which $f''(x) > f'(x)$. Since $f''(x) \sim f'(x)$, $\bar{\bar{M}}_o$ is also a zero set with respect to $\mu^*(M)$. Now for $x \in S - (M'_o + M''_o + \bar{M}''_o + \bar{\bar{M}}_o)$ we have $\mathcal{M}'_x \subset \mathcal{M}_x \subset \mathcal{M}''_x$, $\mathcal{M}'_x$, $\mathcal{M}''_x$ measurable with respect to $\nu^*(N)$, $\nu^*(\mathcal{M}''_x)$ finite, $\nu^*(\mathcal{M}'_x) = \nu^*(\mathcal{M}''_x)$. From this it follows that for these x, $\nu^*(\mathcal{M}''_x - \mathcal{M}'_x) = 0$. Thus by the remark following Theorem 10.2.12, $\mathcal{M}_x$ is measurable with respect to $\nu^*(N)$ for $x \in S - (M'_o + M''_o + \bar{M}''_o + \bar{\bar{M}}_o)$. Thus condition 1) is satisfied by $\mathcal{M}$. As for condition 2), we use the fact that $f(x) \sim f''(x)$ and get

$$\lambda(\mathcal{M}) = \lambda(\mathcal{M}'') = \int_S f''(x)d\mu(M_x) = \int_S f(x)d\mu(M_x).$$

(v) <u>Every</u> $\mathcal{M} \in BR(\mathcal{R})$ <u>is an</u> $\mathcal{F}$<u>-set.</u> Denote by $\mathcal{Y}$ the system of sets $\mathcal{M}$ which are measurable with respect to $\lambda^*(\mathcal{M})$ and have the property that if $\mathcal{P} \in \mathcal{R}$, then $\mathcal{M}\mathcal{P}$ is an $\mathcal{F}$-set. By (i) above, $\mathcal{Y} \supset \mathcal{R}$; has furthermore the properties:

(α) The sum of a sequence of mutually disjunct sets from $\mathcal{Y}$ belongs again to $\mathcal{Y}$.

(γ) If $\mathcal{P} \in \mathcal{R}$ and $\mathcal{M} \in \mathcal{Y}$ then $\mathcal{P} - \mathcal{M}\mathcal{P} \in \mathcal{Y}$.

(α) is a consequence of (ii) above, for if $\mathcal{M} = \sum_n \mathcal{M}^{(n)}$ where the $\mathcal{M}^{(n)}$ are mutually disjunct sets from $\mathcal{Y}$, then for any $\mathcal{P} \in \mathcal{R}$, $\mathcal{M}\mathcal{P} = \sum_n \mathcal{M}^{(n)}\mathcal{P}$. Thus the sets $\mathcal{M}^{(n)}\mathcal{P}$ are mutually disjunct $\mathcal{F}$-sets and by (ii) $\mathcal{M}\mathcal{P}$ is an $\mathcal{F}$-set. As for (γ), let $\mathcal{P}, \mathcal{P}' \in \mathcal{R}$, $\mathcal{M} \in \mathcal{Y}$; we have to prove that $\mathcal{P}'(\mathcal{P} - \mathcal{M}\mathcal{P})$ is an $\mathcal{F}$-set. Put $\mathcal{P}'' = \mathcal{P}\cdot\mathcal{P}'$, then $\mathcal{P}'(\mathcal{P} - \mathcal{M}\mathcal{P}) = \mathcal{P}'' - \mathcal{M}\mathcal{P}''$, $\mathcal{P}'' \in \mathcal{R}$. $\mathcal{P}''$ and $\mathcal{M}\mathcal{P}''$ are $\mathcal{F}$-sets, therefore by (iii) $\mathcal{P}'' - \mathcal{M}\mathcal{P}''$ is also.

Now Theorem 10.1.3 shows that $\mathcal{Y} \supset BR(\mathcal{R})$. Let $\mathcal{M}$ be a set from $BR(\mathcal{R})$. Then (Theorem 10.1.5, Corollary 2) $\mathcal{M}$ is contained in the sum of a sequence $\{\mathcal{P}^{(n)}\}$ of mutually disjunct sets from $\mathcal{R}$. Thus $\mathcal{M} = \sum_n \mathcal{M}\mathcal{P}^{(n)}$. As $\mathcal{M} \in \mathcal{Y}$, every $\mathcal{M}\mathcal{P}^{(n)}$ is an $\mathcal{F}$-set. By (ii) above, $\mathcal{M}$ is an $\mathcal{F}$-set.

(vi) Every $\mathcal{M}$ measurable and of finite measure with respect to $\lambda^*(\mathcal{M})$ is an $\mathcal{F}$-set. By Theorem 10.2.12 there are sets $\mathcal{M}'$, $\mathcal{M}'' \in BR(\mathcal{R})$ such that $\mathcal{M}' \subset \mathcal{M} \subset \mathcal{M}''$ and $\lambda(\mathcal{M}'' - \mathcal{M}') = 0$. By (v), $\mathcal{M}'$, $\mathcal{M}''$ are $\mathcal{F}$-sets; by (iv) therefore, $\mathcal{M}$ is an $\mathcal{F}$-set.

The theorem itself now follows from (ii) and (vi) by observing that every measurable $\mathcal{M}$ is the sum of a sequence of mutually disjunct measurable $\mathcal{M}$'s with finite measures. This completes the proof.

Corollary: If $\mathcal{M}$ is measurable with respect to $\lambda^*(\mathcal{M})$, then

1') The set of y's for which $\mathcal{M}_y$ is non-measurable with respect to $\mu^*(M)$ is a zero set with respect to $\nu^*(N)$.

2') The function $g(y) \equiv \mu^*(\mathcal{M}_y)$ is measurable with respect to $\nu^*(N)$ and $\lambda(\mathcal{M}) = \int_T g(y)d\,\nu(N_y)$. Thus $\int_T g(y)d\,\nu(N_y) = \int_S f(x)d\mu(M_x) = \lambda(\mathcal{M})$.

Proof: Reverse the roles of S, T and apply the theorem.

We now apply this theorem to the relation between iterated and multiple integrals.

THEOREM 11.4.2. Preserve the notations introduced above, the spaces S, T, $\mathcal{S} = S \times T$, and the outer measures $\mu^*(M)$, $\nu^*(N)$, $\lambda^*(\mathcal{M})$. Let $f(x,y) = f([x,y])$ be a non-negative function measurable with respect to $\lambda^*(\mathcal{M})$ over $\mathcal{S}$. Then

1) The set M_o of x's for which the function $f(x,y)$ quâ function of y is non-measurable with respect to $\nu^*(N)$ over T, is a zero set with respect to $\mu^*(M)$

2) The function $g(x)$ defined as $\int_T f(x,y)d\,\nu(N_y)$ in case $x \in S - M_o$ and defined arbitrarily if $x \in M_o$ is measurable with respect to $\mu^*(M)$ over S and we have

$$\int_S g(x)d\mu(M_x) = \int_{\mathcal{S}} f(x,y)d\,\lambda(\mathcal{M}_{(x,y)}).$$

Proof: Let $\bar{\mathcal{S}}$ be the direct product space $S \times T \times R_1$, $\mathcal{T}$ the ordinate space of T that is $T \times R_1$. A point of $\bar{\mathcal{S}}$ is a triad (x, y, ξ) where $x \in S$, $y \in T$, $\xi \in R_1$; a point of $\mathcal{T}$ is a pair (y, ξ) with $y \in T$, $\xi \in R_1$.

Consider the direct product space $\mathcal{S} \times R_1$ and $S \times \mathcal{T}$. Each of these is isomorphic with $\bar{\mathcal{S}}$ under the correspondence $((x, y), \xi) \leftrightarrow (x, y, \xi) \leftrightarrow (x, (y, \xi))$. We shall regard the three as identical.

Now $\bar{\mathcal{S}}$ as $\mathcal{S} \times R_1$ is the ordinate space of $\mathcal{S}$ (cf.Definition 11.2.8). In $\bar{\mathcal{S}}$ we form the ordinate outer measure $\bar{\lambda}^*_I(\bar{\mathcal{M}})$ (cf.Definition 11.2.9) associated with $\lambda^*(\mathcal{M})$ in $\mathcal{S}$. $\bar{\lambda}^*_I(\bar{\mathcal{M}})$ is the outer measure determined by the measure function $\bar{\lambda}(\bar{\mathcal{M}}) = \lambda(\mathcal{M}) \cdot m(E)$, for $\bar{\mathcal{M}} = \mathcal{M} \times E$ with $\mathcal{M} \in MS'_{\lambda*}$, $E \in MS'_{m*}$. Now $\lambda(\mathcal{M})$ on the half-ring $MS'_{\lambda*}$ is by the definition of $\lambda^*(\mathcal{M})$ an equivalent extension of $\lambda(\mathcal{M})$ on the half-ring of all sets $\mathcal{M} = M \times N$ where $M \in MS'_{\mu*}$, $N \in MS'_{\nu*}$. It follows from Theorem 10.4.8 that the half-ring $\bar{\mathcal{R}}$ consisting of all sets $(M \times N) \times E$, $M \in MS'_{\mu*}$, $N \in MS'_{\nu*}$, $E \in MS'_{m*}$ and the half-ring $\bar{\mathcal{R}}'$ consisting of all sets $\mathcal{M} \times E$, $\mathcal{M} \in MS'_{\lambda*}$, $E \in MS'_{m*}$ both determine the outer measure $\bar{\lambda}^*_I(\bar{\mathcal{M}})$.

In $\mathcal{T}$ we form the ordinate outer measure $\bar{\nu}^*_I(\mathcal{N})$ associated with the outer measure $\nu^*(N)$ in T. Consider in $\bar{\mathcal{S}} = S \times \mathcal{T}$ the outer measure $\rho^*_I(\bar{\mathcal{M}})$ determined by the measure function $\rho(\bar{\mathcal{M}}) = \mu(M) \cdot \bar{\nu}_I(\mathcal{N})$ for $\bar{\mathcal{M}} = M \times \mathcal{N}$ with $M \in MS'_{\mu*}$, $\mathcal{N} \in MS'_{\bar{\nu}_I*}$. By a similar argument to that used above $\rho^*_I(\bar{\mathcal{M}})$ is determined also by the half-ring consisting of all sets $M \times (N \times E)$ where $M \in MS'_{\mu*}$, $N \in MS'_{\nu*}$, $E \in MS'_{m*}$.

In $\bar{\mathcal{S}}$ quâ $S \times T \times R_1$, $\bar{\lambda}^*_I(\bar{\mathcal{M}})$ and $\rho^*_I(\bar{\mathcal{M}})$ are both determined by the half-ring $\bar{\mathcal{R}}$ of all $M \times N \times E$, $M \in MS'_{\mu*}$, $N \in MS'_{\nu*}$, $E \in MS'_{m*}$. In $\bar{\mathcal{R}}$ they obviously agree. Hence we can forget about $\rho^*_I(\bar{\mathcal{M}})$.

Now let $\mathcal{O}$ be $\mathcal{O}_f^-$, that is $\mathcal{O} \equiv S_{(x,y,\xi)}(f(x,y) < \xi)$. Since $f(x,y)$ is measurable with respect to $\lambda^*(\mathcal{M})$, $\mathcal{O}$ is measurable with respect to $\bar{\lambda}_I^*(\bar{\mathcal{M}})$ (Theorem 11.2.16). Apply Theorem 11.4.1 to $\mathcal{S} = S \times \mathcal{T}$. We get:

1') The set M_o of all x's for which $\mathcal{O}_x$ is non-measurable with respect to $\bar{\nu}_I^*(\mathcal{N})$ is a zero set with respect to $\mu^*(M)$.

2') The function $\bar{g}(x) \equiv \bar{\nu}_I^*(\mathcal{O}_x)$ is measurable with respect to $\mu^*(M)$ and we have

$$\bar{\lambda}_I^*(\mathcal{O}) = \int_S \bar{g}(x)d\mu(M_x).$$

If, however, we examine $\mathcal{O}_x$ we see that it is nothing but an ordinate set in $\mathcal{T}$ of the function $f(x,y)$ quâ function of y. Thus as a consequence of Theorem 11.2.19 we know that for $x \in S - M_o$, $f(x,y)$ is measurable in y with respect to $\nu^*(N)$. Thus statement 1) of the theorem is proved. As for the second statement, it follows directly from 2') above, the fact that if $x \in S - M_o$, $\bar{\nu}_I^*(\mathcal{O}_x) = \int_T f(x,y)d\nu(N_y)$, that $\bar{\lambda}_I^*(\mathcal{O}) = \int_{\mathcal{S}} f(x,y)d\lambda(\mathcal{M}(x,y))$, and that the $g(x)$ of the Theorem is equivalent to the $\bar{g}(x)$ above with respect to $\mu^*(M)$.

<u>Corollary 1. With the hypotheses of the theorem we have</u>

$$\int_{S\times T} f(x,y)d\lambda(\mathcal{M}_{(x,y)}) = \int_S [\int_T f(x,y)d\nu(N_y)]d\mu(M_x) =$$

$$= \int_T [\int_S f(x,y)d\mu(M_x)]d\nu(N_y),$$

<u>where the expressions in square brackets are defined arbitrarily when they fail to make sense.</u>

Proof: The first equality is the theorem. The second equality results from interchanging S and T.

Theorem 11.4.2 can be amended as follows:

Theorem 11.4.3. Alter the hypothesis of Theorem 11.4.2 by assuming that $f(x)$ *is real or complex and is summable over* $\mathcal{S}$ *with respect to* $\lambda^*(\mathcal{M})$*. Then*

1) *The set* M_o *of* x*'s for which the function* $f(x,y)$ *quâ function of* y *is non-summable with respect to* $\nu^*(N)$ *over* T *is a zero set with respect to* $\mu^*(M)$.

2) *The function* $g(x)$ *defined as* $\int_T f(x,y)d\nu(N_y)$ *if* $x \in S - M_o$ *and defined arbitrarily if* $x \in M_o$ *is summable with respect to* $\mu^*(M)$ *over* S *and we have*

$$\int_S g(x)d\mu(M_x) = \int_{\mathcal{S}} f(x,y)d\lambda(\mathcal{M}_{(x,y)}).$$

Proof: First take a non-negative summable $f(x,y)$. We have the equation deduced in the above proof $\bar{\lambda}_I^*(\mathcal{O}) = \int_S \nu_I^*(\mathcal{O}_x)d\mu(M_x)$. The summability of $f(x,y)$ with respect to $\lambda^*(\mathcal{M})$ implies the finiteness of $\bar{\lambda}_I^*(\mathcal{O})$, which in turn implies that $\bar{\nu}_I^*(\mathcal{O}_x)$ is finite except on a zero set with respect to $\mu^*(M)$. The finiteness of $\bar{\nu}_I^*(\mathcal{O}_x)$ and the measurability of $\mathcal{O}_x$ together imply the summability of $f(x,y)$ as a function of y.

If $f(x,y)$ is real but not non-negative, consider $f_+(x,y)$ and $f_-(x,y)$; both are summable. The set of x's where $f(x,y)$ is not summable in y with respect to $\nu^*(N)$ is the sum of the corresponding sets for $f_+(x,y)$ and $f_-(x,y)$. Call this M_o. For $x \in S - M_o$ we have

$$\int_T f(x,y)d\nu(N_y) = \int_T f_+(x,y)d\nu(N_y) - \int_T f_-(x,y)d\nu(N_y).$$

The latter two expressions as functions of x are summable over S and hence their difference is. We get

$$\int_S [\int_T f(x,y)d\nu(N_y)]d\mu(M_x) =$$

$$= \int_S [\int_T f_+(x,y)d\nu(N_y) - \int_T f_-(x,y)d\nu(N_y)]d\mu(M_x) =$$

$$= \int_{\mathcal{S}} f_+(x,y)d\lambda(\mathcal{M}_{(x,y)}) - \int_{\mathcal{S}} f_-(x,y)d\lambda(\mathcal{M}_{(x,y)}) = \int_{\mathcal{S}} f(x,y)d\lambda(\mathcal{M}_{(x,y)}).$$

A similar argument extends this to complex $f(x,y)$.

By interchanging S and T we get:

Corollary: If $f(x,y)$ is summable with respect to $\lambda(\mathcal{M}_{x,y})$ over $\mathcal{S}$

$$\int_{\mathcal{S}} f(x,y)d\lambda(\mathcal{M}_{x,y}) = \int_S [\int_T f(x,y)d\nu(N_y)]d\mu(M_x) =$$

$$= \int_T [\int_S f(x,y)d\mu(M_x)]d\nu(N_y),$$

where the bracketed expressions are defined arbitrarily when they make no sense.

We now consider some results of a different character. They are generalizations in integral form of the formula from the elementary calculus $\frac{d}{dx} f(g(x)) = \frac{df}{dg} \cdot \frac{dg}{dx}$.

THEOREM 11.4.4. Let $\mathcal{R}$ be a half-ring in $\mathcal{P}(S)$ and let $\mu(M)$, $\rho(M)$, $\nu(M)$ be real measure functions on $\mathcal{R}$ such that $\rho(M)$ and $\nu(M)$ are both non-negative, $\rho(M)$ is absolutely continuous with respect to $\nu(M)$, and $\mu(M)$ is absolutely continuous with respect to $\rho(M)$. Then $\mu(M)$ is absolutely continuous with respect to $\nu(M)$ and we have $\frac{d\mu(M_x)}{d\rho(M_x)} \cdot \frac{d\rho(M_x)}{d\nu(M_x)} \cong \frac{d\mu(M_x)}{d\nu(M_x)}$.

Proof: The absolute continuity of $\mu(M)$ with respect to $\nu(M)$ can easily be proved from the definition much in the same manner that one proves that a continuous function of a continuous function is continuous. It can be seen more quickly by an appeal to Theorem 11.2.4: Extend $\mu(M)$, $\rho(M)$, $\nu(M)$ to $BR'(\mathcal{R})$; let $\tilde{\mu}(M)$, $\tilde{\rho}(M)$, $\tilde{\nu}(M)$ be the extensions. Then $\tilde{\nu}(M) = 0$ implies $\tilde{\rho}(M) = 0$ and $\tilde{\rho}(M) = 0$ implies $\tilde{\mu}(M) = 0$; thus $\tilde{\nu}(M) = 0$ implies $\tilde{\mu}(M) = 0$. Hence $\tilde{\mu}(M)$ is absolutely continuous with respect to $\tilde{\nu}(M)$ and thus $\mu(M)$ with respect to $\nu(M)$.

Now assume temporarily that $\mu(M)$ is non-negative. Define $f(x) \cong \frac{d\mu(M_x)}{d\rho(M_x)}$, $g(x) \cong \frac{d\rho(M_x)}{d\nu(M_x)}$. By Theorem 11.2.9 we can always choose $f(x)$, $g(x)$ so that they are finite and that $S_x\{f(x) < \alpha\}$ and $S_x\{g(x) < \alpha\}$ belong to $BR(\mathcal{R})$; furthermore, by Corollary to Theorem 11.2.12, we can choose them non-negative. Assume this done. **Form** the equivalent extensions $\tilde{\mu}(M)$, $\tilde{\rho}(M)$, $\tilde{\nu}(M)$ of $\mu(M)$, $\rho(M)$, $\nu(M)$ from $\mathcal{R}$ to $BR(\mathcal{R})$. Select an $\bar{M} \in \mathcal{R}$ and for each real α define the systems $L_\alpha = \bar{M} \cdot S_x\{f(x)\, g(x) < \alpha\}$, $M_\alpha = \bar{M} \cdot S_x\{f(x) < \alpha\}$, $N_\alpha = \bar{M} \cdot S_x\{g(x) < \alpha\}$. We want to prove that the systems of sets has the property (1') of Theorem 11.2.11 with respect to the functions $\tilde{\mu}(M)$ $\tilde{\nu}(M)$.

Let D be a countable everywhere dense set of α's. For a given α let E be the set of all pairs $\beta, \gamma \in D$ for which $\beta \geqq 0$, $\gamma \geqq 0$ and $\beta \cdot \gamma < \alpha$. E is countable: $\{(\beta_1, \gamma_1), (\beta_2, \gamma_2), \ldots\}$.

Now $L_\alpha = \sum_n M_{\beta_n} N_{\gamma_n}$. For if $x_0 \in L_\alpha$, then $f(x_0)g(x_0) < \alpha$ and there is an n such that $f(x_0) < \beta_n$ and $g(x_0) < \gamma_n$. Thus $L_\alpha \subset \sum_n M_{\beta_n} N_{\gamma_n}$. Conversely, if $x_0 \in M_{\beta_n} N_{\gamma_n}$ then $f(x_0) < \beta_n$, $g(x_0) < \gamma_n$, and thus $f(x_0)g(x_0) < \beta_n\gamma_n < \alpha$. Now let $J_n = M_{\beta_n} N_{\gamma_n} - M_{\beta_n} N_{\gamma_n} (\sum_1^{n-1} M_{\beta_k} N_{\gamma_k})$. The J_n are disjunct, their sum is L_α, and $J_n \subset M_{\beta_n} N_{\gamma_n}$; furthermore $J_n \in BR(\mathcal{R})$. If now $P \in BR'(\mathcal{R})$ and $P \in L_\alpha$, we have

$$\tilde{\mu}(P) = \tilde{\mu}(PL_\alpha) = \tilde{\mu}(P \sum_n J_n) = \sum_n \tilde{\mu}(PJ_n).$$

Recalling property (1') of Theorem 11.2.11, we have, since $PJ_n \subset M_{\beta_n}$, $\tilde{\mu}(PJ_n) \leqq \beta_n \tilde{\rho}(PJ_n)$, and since $PJ_n \subset N_{\gamma_n}$, $\tilde{\rho}(PJ_n) \leqq \gamma_n \tilde{\nu}(PJ_n)$. Since $\beta_n, \gamma_n \geqq 0$, these together give $\tilde{\mu}(PJ_n) \leqq \beta_n \gamma_n \tilde{\nu}(PJ_n) \leqq \alpha \tilde{\nu}(PJ_n)$. Therefore

$$\tilde{\mu}(P) \leqq \sum_n \alpha \tilde{\nu}(PJ_n) = \alpha \tilde{\nu}(P)$$

and the first condition of property (1') is satisfied.

To show the second we define $L'_\alpha = \bar{M} \cdot S_x\{f(x)g(x) > \alpha\}$. For a fixed α denote by E' the set of all pairs $\beta', \gamma' \in D$ with $\beta, \gamma' \geqq 0$, $\beta' \gamma' > \alpha$. Enumerate E': $\{(\beta'_1, \gamma'_1), (\beta'_2, \gamma'_2), \ldots\}$. By an analogous argument to that used above we get $L'_\alpha = \sum_n (\bar{M} - M_{\beta'_n}) \cdot (\bar{M} - N_{\gamma'_n})$. We can now define J'_n analogously to the J_n above in such a way that they are disjunct, $J'_n \subset (\bar{M} - M_{\beta'_n}) \cdot (\bar{M} - N_{\gamma'_n})$ and $L'_\alpha = \sum_n J'_n$. Now if $Q' \in BR'(\mathcal{R})$ and $Q' \subset L'_\alpha$ we get by a similar argument to that above $\tilde{\mu}(Q') \geqq \alpha \tilde{\nu}(Q')$. Then since clearly $\bar{M} - L_\alpha = \prod_n L'_{\alpha - \frac{1}{n}}$ we have, if $Q \subset \bar{M} - L_\alpha$, $\tilde{\mu}(Q) \geqq (\alpha - \frac{1}{n})\tilde{\nu}(Q)$. Since this holds for all n, we have $\tilde{\mu}(Q) \geqq \alpha \tilde{\nu}(Q)$. This is the second of the conditions (1').

We have proved now for non-negative $\mu(M)$ that $f(x) \cdot g(x) \cong \frac{d\mu(M_x)}{d\nu(M_x)}$. If $\mu(M)$ is real but not non-negative, then $\mu_+(M)$ and $\mu_-(M)$ are absolutely continuous with respect to $\rho(M)$, and the theorem applies to them. Since $g(x) \geqq 0$, $(f \cdot g)_+(x) = f_+(x) \circ g(x)$ and $(f \circ g)_- = f_-(x)g(x)$. Hence, using Theorem 11.2.13,

$$\frac{d\mu(M_x)}{d\nu(M_x)} \cong (f_+(x) - f_-(x))g(x) = \left(\frac{d\mu_+(M_x)}{d\,\rho(M_x)} - \frac{d\mu_-(M_x)}{d\,\rho(M_x)}\right) \circ \frac{d\,\rho(M_x)}{d\,\nu(M_x)} =$$

$$= \frac{d\mu(M_x)}{d\,\rho(M_x)} \circ \frac{d\,\rho(M_x)}{d\,\nu(M_x)} .$$

This completes the proof.

THEOREM 11.4.5. *Let $\nu(M)$ be a finite real or complex totally additive measure of bounded variation over a half-ring $\mathcal{R}$. Let $g(x)$ be summable with respect to $\nu(M)$ for every $M \in \mathcal{R}$. For $M \in \mathcal{R}$ define $\rho(M) = \int_M g(x) d\,\nu(M_x)$. Let $\bar{M}$ be any set measurable with respect to the outer measure determined by $\nu(M)$. Then a function $f(x)$ is summable with respect to $\rho(M)$ over $\bar{M}$ if and only if $f(x) \cdot g(x)$ is summable with respect to $\nu(M)$ over $\bar{M}$. If $f(x)$ satisfies either of these conditions, then*

$$(1) \qquad \int_{\bar{M}} f(x) d\rho(M_x) = \int_{\bar{M}} f(x) \cdot g(x) d\nu(M_x).$$

(in forming the product $f(x) \cdot g(x)$ we employ the convention $+\infty \cdot 0 = 0$.)

Proof: Since the relation $\rho(M) = \int_M g(x) d\nu(M_x)$ is invariant under equivalent extensions of $\rho(M)$, $\nu(M)$ from $\mathcal{R}$ to $BR'(\mathcal{R})$, we may assume that $\mathcal{R}$ is a restricted Borel-ring.

We proceed to establish the theorem in several steps:

(i) We show first that the theorem is valid when $\nu(M)$ and $g(x)$ are non-negative and $f(x)$ is real. Assume that $f(x)$ is summable over $\bar{M}$ with respect to $\rho(M)$. For $M \in \mathcal{R}$ define $\mu(M) = \int_M f_{\bar{M}}(x) d\rho(M_x)$. (Cf. Definition 11.3.1.) Then $\mu(M)$, $\rho(M)$, $\nu(M)$ satisfy the hypotheses of Theorem 11.4.4 and $g(x) \cong \frac{d\rho(M_x)}{d\nu(M_x)}$, $f_{\bar{M}}(x) \cong \frac{d\mu(M_x)}{d\rho(M_x)}$. By that theorem therefore we have $f_{\bar{M}}(x) \cdot g(x) \cong \frac{d\mu(M_x)}{d\nu(M_x)}$. Now for $M \in \mathcal{R}$

$$(2) \quad \mu(M) = \int_M f_{\bar{M}}(x) d\rho(M_x) = \int_M f_{\bar{M}}(x) g(x) d\nu(M_x) = \int_M (f \cdot g)_{\bar{M}}(x) d\,\nu(M_x).$$

Since $\bar{M}$ is in the domain of the maximal equivalent extension of $\mu(M)$, we may write $M = \bar{M}$ in the last three expressions in (2); (2) then reduces to (1) and shows incidentally that $f(x) \cdot g(x)$ is summable with respect to $\nu(M)$ over $\bar{M}$.

Conversely, suppose $f(x) \cdot g(x)$ is summable with respect to $\nu(M)$ over $\bar{M}$. Then $(f \cdot g)_{\bar{M}}(x) = f_{\bar{M}}(x) \cdot g(x)$ is summable with respect to $\nu(M)$. Define $\lambda(M) = \int_M f_{\bar{M}}(x)g(x)d\,\nu(M_x)$ for $M \in \mathcal{R}$. Now, by Theorem 11.2.4, $\lambda(M)$ is absolutely continuous with respect to $\nu(M)$, for if $M_o \in \mathcal{R}$ and $\rho(M_o) = 0$, then since $\nu(M) \geqq 0$, $g(x) \geqq 0$, we must have $g(x) \sim 0$ with respect to $\nu(M)$ in M_o, hence $f_{\bar{M}}(x) \circ g(x) \sim 0$ in M_o, hence $\lambda(M_o) = 0$. Let $h(x) \cong \frac{d\,\lambda(M_x)}{d\,\rho(M_x)}$. Now as $h(x)$ is summable with respect to $\rho(M)$ over every $M \in \mathcal{R}$ we may apply to $h(x)$, $g(x)$ what we have just proved for $f(x)$, $g(x)$ and obtain for all $M \in \mathcal{R}$

$$(2') \qquad \lambda(M) = \int_M h(x)d\rho(M_x) = \int_M h(x)g(x)d\nu(M_x).$$

From the definition of $\lambda(M)$,

$$\int_M h(x)g(x)d\nu(M_x) = \int_M f_{\bar{M}}(x)g(x)d\nu(M_x).$$

Therefore $h(x)y(x) \sim f_{\bar{M}}(x)g(x)$ with respect to $\nu(M)$. Now

$$S_x\{h(x) \neq f_{\bar{M}}(x)\} \subset S_x\{h(x)g(x) \neq f_{\bar{M}}(x)g(x)\} + S_x\{g(x) = 0\}.$$

By what we have just proved, the first of the latter two sets is a zero set with respect to $\nu_I^*(M)$ and hence with respect to $\rho_I^*(M)$; the second is obviously a zero set with respect to $\rho_I^*(M)$. Thus $S_x\{h(x) \neq f_{\bar{M}}(x)\}$ is a zero set with respect to $\rho_I^*(M)$; i.e., $h(x) \sim f_M(x)$ with respect to $\rho(M)$. It is clear from the definition of $\lambda(M)$ that $h(x)$ is summable with respect to $\rho(M)$ over $\bar{M}$; hence so is $f_{\bar{M}}(x)$, and therefore $f(x)$ also.

(ii) We observe that (i) extends immediately to the situation in which $f(x)$ is complex, $\nu(M) \geqq 0$, $g(x) \geqq 0$. This is obtained simply by applying (i) to $\mathcal{R}f(x)$ and $\mathcal{J}f(x)$ separately and observing that

$\mathcal{R}(f(x)g(x)) = \mathcal{R}f(x) \cdot g(x)$, $\mathcal{J}(f(x) \cdot g(x)) = \mathcal{J}f(x) \cdot g(x)$.

(iii) We now establish the result with the mere restriction that $\nu(M)$ be non-negative. Let $g_0(x) = (\mathcal{R}g)_+(x)$, $g_1(x) = (\mathcal{J}g)_+(x)$, $g_2(x) = (\mathcal{R}g)_-(x)$, $g_3(x) = (\mathcal{J}g)_-(x)$, so that $g(x) = \sum_{\tau=0}^{3} i^\tau g_\tau(x)$. It follows from the results of Theorem 11.2.13 and the definition of the special decomposition $\rho(M) = \sum_{\tau=0}^{3} i^\tau \rho_\tau^{(o)}(M)$ that $\rho_\tau^{(o)}(M) = \int_M g_\tau(x)d\nu(M_x)$. If furthermore we define $\bar{g}(x) = \sum_{\tau=0}^{3} g_\tau(x)$, then $\bar{\rho}(M) = \int_M \bar{g}(x)d\nu(M_x)$ is the total variation function of $\rho(M)$.

Now assume that $f(x)$ is summable over $\bar{M}$ with respect to $\rho(M)$; it is then also summable over $\bar{M}$ with respect to each $\rho_\tau^{(o)}(M)$. Applying (i), (ii) to each pair $f(x)$, $g_\tau(x)$, we get the summability of $f(x) \cdot g_\tau(x)$ with respect to $\nu(M)$ over $\bar{M}$ and the relation

$$\int_{\bar{M}} f(x)d\rho_\tau^{(o)}(M_x) = \int_{\bar{M}} f(x)g_\tau(x)d\nu(M_x).$$

From these, the relation (1) follows by the homogeneous linear properties of the integral.

Conversely, assume that $f(x) \cdot g(x)$ is summable over $\bar{M}$ with respect to $\nu(M)$. We wish to prove that $f(x)$ is summable over $\bar{M}$ with respect to $\rho(M)$. For this, it is by Theorem 11.3.13. sufficient that $f(x)$ be measurable over $\bar{M}$ with respect to $\rho(M)$ and $|f(x)|$ be summable with respect to $\rho(M)$ over $\bar{M}$. $(f \cdot g)_{\bar{M}}(x) = f_{\bar{M}}(x) \cdot g(x)$ and $g(x)$ are measurable with respect to $\nu(M)$. Consequently, from the identity

$$\mathcal{R}(f_{\bar{M}}(x)) = \frac{\mathcal{R}(f_{\bar{M}}(x) \cdot g(x)) \cdot \mathcal{R}g(x) + \mathcal{J}(f_{\bar{M}}(x) \cdot g(x)) \cdot \mathcal{J}g(x)}{(\mathcal{R}g(x))^2 + (\mathcal{J}g(x))^2},$$

a similar expression for $\mathcal{J}(f_{\bar{M}}(x))$, and the results of Theorem 11.2.7, we obtain the measurability of $f_{\bar{M}}(x)$, hence of $f(x)$ over $\bar{M}$, with respect to $\nu(M)$.

From the absolute continuity of the $\rho_\tau^{(o)}(M)$ with respect to $\nu(M)$, it follows at once that f(x) is measurable with respect to $\rho(M)$ over $\bar{M}$. On the other hand the summability of $f(x) \cdot g(x)$ over $\bar{M}$ implies that of $|f_{\bar{M}}(x)| \cdot |g(x)|$ and, since $g_\tau(x) \leqq |g(x)|$, that of $|f_{\bar{M}}(x)| \cdot g_\tau(x)$ ($\tau = 0, 1, 2, 3$). Hence $|f_{\bar{M}}(x)|\, \bar{g}(x)$ is summable over $\bar{M}$ with respect to $\nu(M)$. If we now apply the converse statement in (i) to the functions $f_{\bar{M}}(x)$, $\bar{g}(x)$ we obtain the summability of $|f_{\bar{M}}(x)|$ with respect to $\bar{\rho}(M)$ over $\bar{M}$.

(iv) We now prove the theorem without restrictions. Let $\nu(M) = \sum_{\tau=0}^{3} i^\tau \nu_\tau^{(o)}(M)$ be the special decomposition of $\nu(M)$. Since $\bar{\nu}(M) = \sum_{\tau=0}^{3} \nu_\tau^{(o)}(M)$, each $\nu_\tau^{(o)}(M)$ is absolutely continuous with respect to $\bar{\nu}(M)$. Define $h_\tau(x) \cong \dfrac{d\,\nu_\tau^{(o)}(M_x)}{d\,\bar{\nu}(M_x)}$, $h(x) = \sum_{\tau=0}^{3} i^\tau h_\tau(x)$. Obviously $\nu(M) = \int_M h(x) d\,\bar{\nu}(M_x)$. We apply (iii) above to g(x), h(x), and obtain

$$\rho(M) = \int_M g(x)\,\nu(M_x) = \int_M g(x)h(x)d\,\bar{\nu}(M_x).$$

If now f(x) is summable with respect to $\rho(M)$ over $\bar{M}$, it follows from (iii) that $f(x) \circ (g(x) \cdot h(x))$ is summable over $\bar{M}$ with respect to $\bar{\nu}(M)$. The converse statement in (iii) applied to $(f(x) \cdot g(x))$, h(x) gives the summability over $\bar{M}$ of $f(x) \cdot g(x)$ with respect to $\nu(M)$; also we get

$$\int_{\bar{M}} f(x) d\,\rho(M_x) = \int_{\bar{M}} f(x) \cdot (g(x) \circ h(x)) d\,\bar{\nu}(M_x) =$$

$$= \int_{\bar{M}} (f(x) \cdot g(x)) \circ h(x) d\bar{\nu}(M_x) = \int_{\bar{M}} f(x) \circ g(x) d\nu(M_x).$$

The converse statement in the present situation is obtained by simply retracing these last few steps.

Remark: The function $h(x)$ defined in (iv) of the proof of the last theoremis of some interest. The theorem itself applied to $f(x)$, $h(x)$ shows that if $f(x)$ is summable over $\bar{M}$ with respect to $\nu(M)$, then

$$\int_{\bar{M}} f(x) d\,\nu(M_x) = \int_{\bar{M}} f(x)h(x) d\,\bar{\nu}(M_x).$$

Thus we have a representation of the integrals with respect to not necessarily non-negative measure functions in terms of the integrals with respect to non-negative measure functions. If $\nu(M)$ is real, then, $|h(x)| = h_+(x) + h_-(x)$.

Consequently $\int_M |h(x)| d\,\bar{\nu}(M_x) = \int_M h_+(x) d\,\bar{\nu}(M_x) + \int_M h_-(x) d\bar{\nu}(M_x) =$ $= \nu_+(M) + \nu_-(M) = \bar{\nu}(M)$, and thus $|h(x)| \sim 1$. If $\nu(M)$ is complex, we get $\int_M |h(x)| d\,\bar{\nu}(M_x) \leqq \int_M \sum_{\tau=0}^{3} h_\tau(x) d\,\bar{\nu}(M_x) = \bar{\nu}(M)$; hence for complex $\nu(M)$, $|h(x)| \leqq 1$ except possibly on a zero set with respect to $\nu(M)$. As a matter of fact, if we had employed the alternative definition for the total variation of a complex measure function (cf. Note after Definition 11.1.5), we should have obtained $|h(x)| \sim 1$ in the complex case as well.

GPSR Authorized Representative: Easy Access System Europe - Mustamäe tee 50, 10621 Tallinn, Estonia, gpsr.requests@easproject.com

www.ingramcontent.com/pod-product-compliance
Lightning Source LLC
Chambersburg PA
CBHW061357210326
41598CB00035B/6016

* 9 7 8 0 6 9 1 0 7 9 6 6 0 *